TD 405 .W6624 2003

W odson, R. Dodge 1955-

Wa r wells and septic
 ms handbook

# Water Wells and Septic Systems Handbook

# Water Wells and Septic Systems Handbook

R. Dodge Woodson

NEW ENGLAND INSTITUTE OF TECHNOLOGY
LIBRARY

McGraw-Hill

New York  Chicago  San Francisco  Lisbon  London  Madrid
Mexico City  Milan  New Delhi  San Juan  Seoul
Singapore  Sydney  Toronto

*The McGraw·Hill Companies*

**Cataloging-in-Publication Data is on file with the Library of Congress**

Copyright 2003 by The McGraw-Hill Companies, Inc. All rights reserved. Printed in the United States of America. Except as permitted under the United States Copyright Act of 1976, no part of this publication may be reproduced or distributed in any form or by any means, or stored in a database or retrieval system, without the prior written permission of the publisher.

1 2 3 4 5 6 7 8 9 0   DOC/DOC   0 9 8 7 6 5 4 3

ISBN 0-07-140200-4

*The sponsoring editor for this book was Larry S. Hager and the production supervisor was Pamela A. Pelton. It was set in Century Schoolbook by Lone Wolf Enterprises, Ltd.*

*Printed and bound by RR Donnelley.*

*This book is printed on recycled, acid-free paper containing a minimum of 50% recycled, de-inked fiber.*

McGraw-Hill books are available at special quantity discounts to use as premiums and sales promotions, or for use in corporate training programs. For more information, please write to the Director of Special Sales, McGraw-Hill Professional, Two Penn Plaza, New York, NY 10121-2298. Or contact your local bookstore.

---

Information contained in this work has been obtained by The McGraw-Hill Companies, Inc. ("McGraw-Hill") from sources believed to be reliable. However, neither McGraw-Hill nor its authors guarantee the accuracy or completeness of any information published herein, and neither McGraw-Hill nor its authors shall be responsible for any errors, ommissions, or damages arising out of use of this information. This work is published with the understanding that McGraw-Hill and its authors are supplying information but are not attempting to render engineering or other professional services. If such services are required, the assistance of an appropriate professional should be sought.

---

*This book is dedicated to the memory of my mother.*

# Contents

Introduction ... xix

**Chapter 1. Administration** ... 1
  General Regulations ... 1
    Public Sewer Connections ... 2
    Discontinued Disposal Systems ... 2
    Faulty Systems ... 2
  Applicability ... 2
    Maintenance ... 3
    Modifications ... 3
    Change in Occupancy ... 3
    Historic Buildings ... 3
    Relocated Buildings ... 4
  Code Officials ... 4
  Approval ... 5
  Permits ... 5
    Plans and Specifications ... 5
    Soil Reports ... 6
    Site Plans ... 6
    Permit Expiration ... 6
    Suspension ... 7
  Inspections ... 7
  Violations ... 7
    Stop Work Order ... 7
  Unsafe Conditions ... 8
  Appeal ... 8

**Chapter 2. Definitions** ... 11

**Chapter 3. General Regulations** ... 15
  Specific Limitations ... 15
  Flood Plain ... 16

**Chapter 4. Site Evaluation and Requirements** ... 17
  Slope ... 18
  Test Boring ... 18
    Soil Mottles ... 19
    Ground Water ... 19

|  |  |
|---|---|
| Color Patterns | 20 |
| Bedrock | 21 |
| Alluvial and Colluvial Deposits | 21 |
| Percolation Tests | 21 |
| Sandy Soils | 21 |
| Other Soils | 22 |
| Verification | 22 |
| Design of Monitoring Wells | 23 |
| Observation Requirements | 23 |
| Site Requirements | 24 |
| Bedrock | 25 |
| Perk Rate | 26 |
| Filled Areas | 26 |
| Design Requirements | 27 |

## Chapter 5. Materials — 29

|  |  |
|---|---|
| Septic Tanks | 29 |
| Pipe | 31 |
| Joints and Connections | 31 |
| Prohibited Joints and Connections | 34 |

## Chapter 6. Soil Absorption Systems — 35

|  |  |
|---|---|
| Excavating Seepage Trenches | 36 |
| Seepage Beds | 38 |
| Seepage Pits | 38 |
| Rules of Excavation and Construction | 38 |
| Aggregate and Backfill | 39 |
| Piping Requirements | 39 |
| Observation Pipes | 40 |
| Weather | 40 |

## Chapter 7. Pressure Distribution Systems — 43

|  |  |
|---|---|
| System Design | 43 |
| Pumps | 44 |
| Dosing | 45 |

## Chapter 8. Tanks — 47

|  |  |
|---|---|
| Inlets and Outlets | 47 |
| Manholes | 48 |
| Manhole Covers | 49 |
| Inspection Openings | 49 |
| Sizing Tanks | 49 |
| Installation | 49 |
| Backfill Material | 51 |
| Riser Joints | 51 |
| Dosing or Pumping Chambers | 52 |
| Treatment Tanks | 52 |

|  |  |
|---|---|
| Maintenance | 53 |
| Chemical Restoration | 53 |
| Holding Tanks | 53 |

## Chapter 9. Mound Systems — 55
- System Design — 56
  - Trenches — 69
  - Beds — 69
- Construction Methods — 70
  - Plowing — 70
  - Fill — 71
  - Absorption Area — 71
  - Distribution System — 71
  - Cover — 71
  - Maintenance — 72

## Chapter 10. Cesspools — 73

## Chapter 11. Inspections — 75

## Chapter 12. Referenced Standards — 77

## Chapter 13. Site Evaluation for Water Wells — 79
- Location — 81
- Reading Signs To Find Water — 83
  - Maps — 83
  - Plants — 85
- Dowsing — 85
  - High-Tech Stuff — 86
- It Doesn't Take Very Long — 86

## Chapter 14. Testing and Protecting Potable Water — 89
- Confusion — 90
- Testing Recommendations — 96
- New Wells — 99
- Who Tests The Water? — 100
- Pesticides in Potable Water — 103
  - Health Effects — 104
- Protecting Private Wells — 104

## Chapter 15. Drilled Wells — 107
- Depth — 108
- Well-Drilling Equipment — 108
- The Basics — 110
  - Access — 113
  - Working On Your Site — 114

| | |
|---|---:|
| Pump Installers | 114 |
| Quantity And Quality | 115 |
|     Quality | 116 |
| Supervising the Drilling | 117 |
| Trenching | 118 |
| Installing A Pump System | 119 |
|     At The Well Casing | 120 |
|     The Water Service | 123 |
|     Inside the Foundation | 124 |
| The Finer Points | 126 |
|     Problems | 126 |
| **Chapter 16. Shallow Wells** | **129** |
| Shallow Wells | 131 |
| When Should You Use A Shallow Well? | 131 |
|     True Dug Wells | 131 |
|     Bored Wells | 133 |
|     Drilled Wells | 133 |
|     Jetted Wells | 134 |
| Characteristics | 135 |
| Look Around | 137 |
| The Preliminary Work | 138 |
| After A Well Is In | 138 |
| Pump Selection | 139 |
|     Single-Pipe Jet Pumps | 139 |
|     Two-Pipe Jet Pumps | 141 |
|     Submersible Pumps | 142 |
| Installing a Single-Pipe Jet Pump | 143 |
| Two-Pipe Pumps | 147 |
| Pressure Tanks | 147 |
|     Small Tanks | 147 |
|     Large Tanks | 148 |
|     In-Line Tanks | 148 |
|     Stand Models | 149 |
|     Pump-Stand Models | 150 |
|     Underground Tanks | 151 |
| Diaphragm Tanks | 151 |
|     Sizing | 152 |
| Installation Procedures | 152 |
|     A Relief Valve | 153 |
|     Keep It Dry | 153 |
| **Chapter 17. Alternative Water Sources** | **155** |
| Driven Wells | 156 |
|     The Point | 157 |
|     The Well Pipe | 157 |
|     Drive Caps | 158 |

|  |  |
|---|---|
| Power Driving | 158 |
| Getting It Out | 158 |
| Driving the Well Point | 159 |
| Towards the End | 159 |
| Suitable Soils | 160 |
| Quantity | 160 |
| Quality | 160 |
| Cisterns | 160 |
| Ponds and Lakes | 162 |
| Springs | 162 |
| Many Needs | 163 |

## Chapter 18. Troubleshooting Well Systems — 165

|  |  |
|---|---|
| Basic Well Problems | 166 |
| Driven Wells | 166 |
| Sand | 167 |
| Other Contaminants | 167 |
| Shallow Wells | 167 |
| Drilled Wells | 168 |
| Troubleshooting Jet Pumps | 169 |
| Will Not Run | 169 |
| Runs But Gives No Water | 171 |
| Cycles Too Often | 172 |
| Won't Develop Pressure | 172 |
| Switch Fails | 173 |
| Troubleshooting Submersible Pumps | 173 |
| Won't Start | 174 |
| Will Not Run | 177 |
| Doesn't Produce Water | 178 |
| Tank Pressure | 178 |
| Frequent Cycling | 178 |
| Water Quality | 179 |
| Physical Characteristics | 179 |
| Chemical Characteristics | 180 |
| Chlorides | 181 |
| Copper | 181 |
| Fluorides | 181 |
| Iron | 181 |
| Lead | 181 |
| Manganese | 181 |
| Nitrates | 182 |
| Pesticides | 182 |
| Sodium | 182 |
| Sulfates | 182 |
| Zinc | 182 |
| Hard Water | 183 |
| Acidic Water | 183 |

| | |
|---|---:|
| Biological Factors | 183 |
| Radiological Factors | 184 |
| Solving Water-Quality Problems | 184 |
|     Bacteria | 184 |
|     Acid Neutralizers | 184 |
|     Iron | 184 |
|     Water Softeners | 185 |
|     Activated Carbon Filters | 185 |
|     Turbidity | 185 |

## Chapter 19. Site Limitations for Septic Systems     187

| | |
|---|---:|
| As They Appear | 189 |
| Grade | 190 |
| A Safe Way | 190 |
| Standard Systems | 191 |
| Chamber Systems | 191 |
| Pump Stations | 192 |
| Trees | 192 |
| Burying A Septic Tank | 192 |
| Underground Water | 193 |
| Driveways and Parking Areas | 194 |
| Erosion | 194 |
| Setbacks | 194 |
| Read The Fine Print | 195 |
| Site Visits | 196 |

## Chapter 20. Septic Designs and Soil Studies     199

| | |
|---|---:|
| Septic Designs | 199 |
| Design Criteria | 204 |
|     Trench and Bed Systems | 204 |
|     Mound and Chamber Systems | 208 |
| Soil Types | 208 |
|     The Best | 208 |
|     Pretty Good | 209 |
|     Not So Good | 209 |
|     Just Won't Do | 209 |

## Chapter 21. Gravel-and-Pipe Septic Systems     211

| | |
|---|---:|
| The Components | 213 |
| Types Of Septic Tanks | 217 |
|     Precast Concrete | 217 |
|     Metal | 217 |
|     Fiberglass | 218 |
|     Wood | 218 |
|     Brick and Block | 218 |
| Installing a Simple Septic System | 218 |
| Do It Yourself | 224 |

|  |  |
|---|---|
| Money | 225 |
| Work Is Work | 226 |
| Control and Quality | 226 |
| The Technical Side | 227 |
| Bad Ground | 227 |
| **Chapter 22. Chamber-Type Septic Systems** | **229** |
| Chamber Systems | 230 |
| Trench Systems | 231 |
| More Land Area | 232 |
| Mound Systems | 233 |
| Pump Systems | 234 |
| Holding Tanks | 234 |
| Other Types of Systems | 235 |
| **Chapter 23. Wastewater Pump Systems** | **237** |
| Single-Fixture Pumps | 238 |
| Gray-Water Sumps | 240 |
| Black-Water Sumps | 241 |
| Whole-House Pump Stations | 244 |
| **Chapter 24. Landscaping Septic-System Areas** | **247** |
| Appearance | 247 |
| General Information | 248 |
| Plant Selection | 248 |
| Grass | 249 |
| Flowers | 249 |
| Other Options | 249 |
| **Chapter 25. Septic-System Maintenance** | **251** |
| Garbage Disposals | 252 |
| Unwanted Objects | 252 |
| Temperature | 253 |
| Pathogens | 253 |
| Failure Symptoms | 253 |
| Beneficial Bacteria | 253 |
| Additives | 254 |
| Excess Water | 254 |
| Cleaning | 254 |
| **Chapter 26. Troubleshooting Septic Systems** | **255** |
| An Overflowing Toilet | 256 |
| Whole-House Backups | 257 |
| The Problem is in the Tank | 258 |
| Problems With a Leach Field | 260 |
| Older Fields | 261 |
| Clogged With Solids | 261 |

xiv  Contents

|  |  |
|---|---|
| Too Much Pitch | 261 |
| Garbage Disposers | 262 |
| Other Causes | 262 |
| House Stinks | 262 |
| Outside Odors | 263 |
| Puddles and Odors | 264 |
| Supervision | 264 |
| When a Problem Occurs | 265 |
| Don't Be Afraid to Ask For Help | 265 |

**Chapter 27. Low-Cost Septic Systems** — 271

|  |  |
|---|---|
| Knowledge | 272 |
| Try Installing a System Yourself | 272 |
| Choose Sites Selectively | 273 |
| A Spec Builder | 273 |
| Cost Overruns | 275 |
| Shop Around | 275 |

**Chapter 28. Low-Cost Well Systems** — 277

|  |  |
|---|---|
| Cutting Too Many Corners | 277 |
| Using a Spring | 278 |
| Cheap Pumps | 280 |
| Smart Ways | 280 |
| Well Selection | 281 |
| Shallow Wells | 281 |
| Driven Wells | 282 |
| Pumps | 282 |
| Disadvantages of Submersible Pumps | 282 |
| Two-Pipe Pumps | 283 |
| Single-Pipe Pumps | 284 |
| Pressure Tanks | 284 |
| Shopping Prices | 284 |
| Read the Specs | 286 |
| Do Your Own Installations | 288 |
| Negotiations | 288 |
| Volume Deals | 289 |

**Chapter 29. Suggestions for Bidding Jobs** — 291

|  |  |
|---|---|
| Disclosure | 293 |
| Danger Spots | 294 |
| Septic Systems | 294 |
| Septic Permits | 295 |
| A Specific Design | 295 |
| Clearing | 295 |
| Access | 296 |
| Rock | 296 |
| Materials | 296 |

|   |   |
|---|---|
| The Sewer | 297 |
| Landscaping | 297 |
| Perimeter Trees | 297 |
| **Wells** | 297 |
|   By the Job | 298 |
|   Drilled or Dug | 299 |
|   Water Quantity | 300 |
|   Flow Rate | 300 |
|   Water Quality | 300 |
|   Location | 301 |
|   Access | 301 |
|   Permission | 302 |
|   The Trench | 302 |
|   The Pump System | 302 |
| **A Spec Sheet** | 302 |
| **What You Don't Know** | 303 |
| **Chapter 30. Cost Overruns** | **305** |
| **Setting Up a Job Budget** | 305 |
|   Septic Systems | 306 |
|   Wells | 307 |
| **Take-Offs** | 307 |
| **Accuracy** | 308 |
| **Sorting Through Bids** | 308 |
|   Well Prices | 308 |
|   Guaranteed Prices | 310 |
|   Reading Between the Lines | 310 |
|   Septic Prices | 311 |
|   Suppliers | 312 |
| **Once a Job Starts** | 314 |
|   Extras | 315 |
|   Written Records | 315 |
| **As a Job is Winding Down** | 315 |
| **Pulling It All Together** | 316 |
| **A Money Diet** | 316 |
| **Here We Are** | 317 |
| **Chapter 31. Worksite Safety** | **319** |
| **Very Dangerous** | 319 |
| **General Safety** | 322 |
|   Vehicles | 322 |
| **Clothing** | 323 |
| **Jewelry** | 324 |
| **Eye And Ear Protection** | 324 |
| **Pads** | 324 |
| **Tool Safety** | 325 |
|   Torches | 326 |

xvi Contents

|  |  |
|---|---|
| Lead Pots And Ladles | 327 |
| Drills And Bits | 327 |
| Power Saws | 328 |
| Hand-Held Drain Cleaners | 328 |
| Large Sewer Machines | 329 |
| Power Pipe Threaders | 329 |
| Air-Powered Tools | 329 |
| Powder-Actuated Tools | 330 |
| Ladders | 330 |
| Screwdrivers And Chisels | 331 |
| Co-Worker Safety | 331 |
| **Chapter 32. First Aid** | **333** |
| Open Wounds | 334 |
| Bleeding | 334 |
| Super Serious | 335 |
| Tourniquets | 336 |
| Infection | 336 |
| Splinters And Such | 337 |
| Eye Injuries | 338 |
| Scalp Injuries | 339 |
| Facial Injuries | 339 |
| Nose Bleeds | 340 |
| Back Injuries | 340 |
| Legs And Feet | 340 |
| Blisters | 340 |
| Hand Injuries | 341 |
| Shock | 341 |
| Burns | 343 |
| Treatment | 343 |
| Heat-Related Problems | 344 |
| Cramps | 344 |
| Exhaustion | 344 |
| **Chapter 33. Mathematics for the Trade** | **347** |
| **Appendix. Reference and Conversion Tables and Data** | **407** |
| Index. | 445 |

# Acknowledgements

I would like to thank the United States Government, the International Conference of Building Officials, A.Y. McDonald Mfg. Co., American Tank Company, Inc., and Goulds Pumps, Inc. for their generous contribution of illustrations found in this book.

# Introduction

According to the U.S. Census Bureau's 1990 census, there are 15.9 million individual household wells in the United States. Since 1970 there has been an average of 239,000 wells installed annually. Nearly 15 percent of American homes are served by private wells. This same census says that there are 24.7 million household septic systems in use. The average number of private septic systems installed annually since 1970 is 400,000. Close to 25 percent of all American homes depend on private septic systems for their sewage disposal. These are big numbers, and they are only based on individual households. There are some 282,828 public supply wells in the United States. The private water and sewer system market is clearly a large one.

How much do you know about the installation of wells and septic systems? Could you expand your business income if you were more knowledgeable about this growing field? Wouldn't you like to cash in on the profits of jobs that you are presently uncomfortable working with? This book is your guide into the professional world of working with wells and septic systems.

The author, R. Dodge Woodson, is a master plumber and a builder. He has built as many as 60 homes a year during his long career of more than 25 years. His experience in building homes that depend on private water and sewage systems is vast. As a plumber, he has installed countless systems himself. This is your chance to learn from an expert.

Whether you are a developer, a builder, a general contractor, a plumber, a well installer, or a septic-system installer, you will learn a lot from this valuable tool that you are now holding in your hands. The wealth of experience that Woodson brings to this topic is invaluable on the job site. Whether you are doing the hands-on work, planning the job, or supervising the installation, what you need to know is contained within these pages.

As an author, Woodson has written dozens of books, many of which have been noted as bestsellers. He brings a style of writing to his readers that is easy to understand. His words are to the point, and his instructions are easy to follow. Speaking from experience, he clarifies issues that are often very confusing. The illustrations you will find here are a tremendous aid in solving on-the-job problems. This is the total package for anyone working with well and septic systems.

Take a moment to review the table of contents. It won't take long for you to see the depth of coverage that you will gain with this indispensable guide. From site selection to soil testing to installation to business practices, it's all here.

Flip through the pages and take notice of the illustrations, the bulleted lists, and the easy access to information. What more could you ask for? If you want to know more about wells and septic systems, you want this book. If there is a definitive guide to the practical application of wells, this is it.

# Chapter 1

# Administration

The administration of code requirements for septic systems has changed over the last few years. Under the guidance of the International Code Council, there is now an International Private Sewage Disposal Code®. Unlike the plumbing code, the septic code is small, but don't let the size fool you into thinking that it is not important. Rules and regualtions for septic systems are needed to safegard public safety.

## General Regulations

Understanding the general regulations of the septic code is not difficult. However, as with any code, there are exceptions to many of the rules and regulations. The septic code is not nearly as complex as the plumbing code, but it can still be a bit daunting for those who are not familiar with its provisions.

The scope of the septic code encompasses septic tanks and effluent absorption systems. Treatment tanks and effluent disposal systems are also covered. These systems are generally not permitted, unless a public sewer system is not available.

Private waste-disposal systems are required to be entirely separate from and independent of any other building. It is possible to gain an official variance to this rule, but there must be compelling reasons for it. When the job uses a common system or a system on a parcel other than the one where the structure is being served, the installation must be in full compliance with code requirements.

## Public Sewer Connections

Public sewer connections must be used when they are available. If a property is operating on a private waste-disposal system and a public sewer is installed and accessible to the property, the private system must be abandoned. The time allowed for such a conversion is subject to local law, but must not exceed 1 year.

## Discontinued Disposal Systems

Discontinued disposal systems must be plugged or capped at the time of abandonment. Any abandoned treatment tank or seepage pit must be pumped free of its contents. All waste must be disposed of properly. Once all waste is removed, the tank must either be removed or filled with a permanent filler, such as concrete.

## Faulty Systems

Faulty systems must be repaired or abandoned within the time allowed by local code requirements. In no case may a faulty system remain in operation for more than 1 year. It is essential to avoid environmental contamination.

What constitutes a failing system? If a system is unable to accept sewage discharge and backup into the structure served by the system, you must consider the system to be faulty. When sewage is discharged to the surface of the ground or drain tile, a system is also deemed faulty. Any system that discharges sewage to any surface or ground water must be repaired or abandoned. Another form of failure is the introduction of sewage into saturation zones, adversely affecting the operation of a private disposal system. The septic code exists to protect public health, safety, and welfare to any extent that they may be at risk due to the installation or maintenance of a private sewage-disposal system.

## Applicability

Existing systems may be grandfathered. Private sewage systems that were in lawful existence when the septic code was put into effect are allowed to be used in their existing condition. Maintenance and repair are allowed, so long as the systems are in accordance with the original design and do not pose a risk to public safety, life, health or property.

**Maintenance**

Routine maintenance is required to assure the proper and legal functioning of private sewage systems. Property owners, or their agents, are responsible for the maintenance of systems that serve their property. It is the privilege of the local code-enforcement officer to require an inspection of any private waste-disposal system.

**Modifications**

Modifications to existing sewage systems must be made in compliance with code requirements. Additions, alterations, repairs, and renovations must be made in compliance with this code just as if the work was being performed on a new system. No modification will be allowed if the change may cause an existing system to become unsafe, unsanitary, or overloaded. An exception to this rule pertains to minor modifications.

When minor modifications are made, they may be done in a manner consistent with the existing system. Special permission may be required to deviate from the code requirements. Check with your local code-enforcement office to obtain approval of your plans. Generally speaking, minor alterations are allowed, even if not code-compliant, when the changes are minimal and do not present a hazard.

**Change in Occupancy**

If a change in occupancy is planned that will subject a structure to any special provision of the septic code, approval for the change must be obtained from a code officer. A code officer will have to determine that the change of use will not result in any hazard to public health, safety, or welfare.

**Historic Buildings**

Historic buildings are generally exempt from the septic code. However, if there is a risk to health, safety, or public welfare, special requirements may be dictated by the local code official. To qualify for special treatment, the building must be identified and classified by the state or local jurisdiction as a historic building.

### Relocated Buildings

Relocated buildings generally must be treated as new construction in terms of the septic systems used. There can be exceptions to this rule, so check with your local code office if you will be working with relocated buildings.

### Code Officials

Code officials have significant power. These officials enforce the local code requirements. When questions arise about the installation, alteration, repair, maintenance, or operation of a private sewage-disposal system, a code official must respond. If circumstances pertaining to public health, safety and general welfare indicate a risk, the local code officer is empowered to interpret and implement the provisions of the code to eliminate the threat.

Code officials receive permit applications and decide their disposition. Additionally, the local code officer is responsible for site inspections of new installations and alterations.

It is not always mandatory for a code officer to perform site inspections. Approved agencies and professionals can perform required inspections and submit their reports to the code officers. The reports must be in writing and certified by a responsible officer, an approved agency, or the responsible individual.

If code officials have reasonable cause to believe that a building or property contains unsafe, unsanitary, dangerous, or hazardous conditions, they may enter the building or property to inspect the installation. Such access to the building or property is required to be exercised at reasonable times. If a building is occupied, the code officer must present proper credentials to gain entry to the building. When a building is vacant, a code officer must make a reasonable effort to locate the owner of the building or the person responsible for the building. In the event that access to a building or property is denied, the code officer may exercise every remedy provided by law to secure entry.

All code officials are required to maintain full records of permit applications, permits, certificates issued, fees collected, reports of inspections, notices, and orders. These documents must be maintained for as long as the building remains in existence, unless otherwise provided for by other regulations.

## Approval

Variances are sometimes allowed to meet approval standards under the code. A code officer has the power to authorize a variance when strict compliance with the code creates an unusual hardship. Variances are considered on a case-by-case basis. As long as the intent of the code is maintained and no health, safety, or welfare problems are created, a variance may be approved for extenuating circumstances.

## Permits

When a permit is required for the work to be done, work must not be started until it has been issued. Permit applications are filed with the local code-enforcement office. The application will require basic information, such as the following:

- A description of the type of system to be installed

- The system location

- The occupancy of all parts of the structure and all portions of the site or lot not covered by the structure

- The maximum number of bedrooms for residential occupancies

- Additional information that may be required by the local code-enforcement office

## Plans and Specifications

Two sets of plans and specifications must be submitted with a permit application. The drawings must be drawn to scale and show adequate detail for the work proposed. Any equipment, such as pumps or controls, must be identified and described completely. When a job involves only minor work, a code officer may waive the requirement for plans and specs.

Once a permit is issued, the plans and specifications will be marked with an official approval. Once the plans and specs are approved, they must not be altered without further permission from a code official. Most code officials require one set of approved plans and specifications to remain on the job site.

## Soil Reports

Soil reports must be provided with applications for septic permits. The reports must detail boring procedures and percolation rates. Surface elevations are required for all boring details. Reference points, final grade elevation, and other pertinent data may also be required prior to the approval and issuance of a permit.

## Site Plans

Site plans are another requirement when applying for a septic permit. The plans must be drawn to scale and show the proposed locations of septic tanks, holding tanks, treatment tanks, building sewers, wells, water mains, water services, streams, lakes, dosing or pumping chambers, distribution boxes, effluent systems, dual disposal systems, replacement system areas, and the locations of all buildings and structures.

Site plans are required to show distances and dimensions between items on the plans. Vertical and horizontal reference points must be noted. Grade slopes with contours are required on site plans for systems serving structures other than single-family homes.

## Permit Expiration

A permit will expire if the work is not started within 180 days of the issuance of the permit. Expiration will also occur if work begun is abandoned or suspended for a period of 180 days or more. If a permit expires, a new permit must be issued. When a time extension is needed, it can be requested. A code officer can approve the extension for a period of time up to 180 days, assuming that there is reasonable cause for the extension. There can only be one extension per permit.

## Suspension

Code officers have the authority to suspend or revoke active permits. If false information was given in the application process, a code officer may revoke an approved permit.

## Inspections

Inspections are usually conducted by code officers at various points during an installation. It is possible for an independent agency or inspector to conduct site inspections and provide written reports to the local code officer. Independent inspectors must meet requirements set forth by the local code-enforcement office. Once all required inspections are complete with satisfactory results, a certificate of approval will be given to the permit holder.

## Violations

It is unlawful for any person, firm, or corporation to erect, construct, alter, repair, remove, demolish, or utilize any private sewage-disposal system, or cause the same to be done, in conflict with or in violation of any code requirements. Anyone creating a violation will receive notification of the violation from a code official. If a violation is ignored and not corrected, prosecution for the violation will be considered. Punishment for violations can include the levying of fines, revocation of permits, and even imprisonment.

## Stop Work Order

A stop work order can be given when a code officer feels that such action is warranted. The notice will be in writing and given to the property owner or the owner's agent. Reasons for the issuance of a stop work order will be explained in the notification. If an emergency situation exists, a code officer can post a stop work order without giving written notice. Once a stop work order is issued, all work must stop immediately, with the exception of work required to remove a violation or safety hazard. Failure to comply with a stop work order can result in fines and other authoritative action.

## Unsafe Conditions

When unsafe conditions are found, they must be dealt with. Such conditions may result in the need to repair, rehabilitate, demolish, or remove an entire septic system or the defective elements of a system. Potentially unsafe conditions include the following:

- Inadequate maintenance

- Dilapidation

- Obsolescence

- Disaster

- Damage

- Abandonment

A code officer has the power to condemn a septic system that is deemed to be unsafe. If this type of action is required, the code officer will provide written notice to the owner or the owner's agent. The notification may require repair or removal of a defective system. A time period for required action will be identified in the official notification. If a notice of this type is issued, the septic system that is found to be unsafe must cease to be used. In an emergency situation, a code officer may disconnect a system immediately and without prior written notice.

## Appeal

Anyone can file an appeal to a decision made by code officials. Such appeals are made to the board of appeals. An application for an appeal must be filed on a form obtained from a code officer. A person must file an appeal request within 20 days of receiving the notice that is being appealed.

Reason for appeal must be based on a claim that the true intent of the code has been interpreted incorrectly by the code official. Another reason for an appeal could be that the code does not apply fully to the situation being appealed. If equally good, or better, construction is used and denied, there is reason for an appeal.

This concludes our view of administration matters. We will now move to the next chapter and gain a full understanding of definitions provided within the septic code.

# Chapter 2

# Definitions

Aggregate – Hard rock that is graded with a value of 3 or more and washed with pressurized water over a screen to remove the fine material. A hardness value of 3 or more on the Mohs' Scale of Hardness means that it will scratch a copper penny without leaving a residue.

Alluvium – Soil deposits due to floodwaters.

Bedrock – Rock underneath or at the earth's surface. It is composed of weathered, in-place consolidated material, larger than 0.08 (2 mm) inches in size and greater than 50 percent by volume.

Cesspool – A covered excavation designed to retain organic matter and solids while receiving sewage or other organic wastes from a drainage system.. The liquids then seep into soil cavities.

Clear-water wastes – Liquid having no impurities or impurities reduced below a harmful level. This liquid can come from cooling water and condensate drainage, water used for equipment chilling purposes, or cooled condensate from steam heating systems.

Code official – The designated authority charged with administration and enforcement of a code. This can also include a duly authorized representative.

Colluvium – Soil moved by gravity.

Color – Based on the Munsell soil color charts, the color of soil when moist.

Construction documents – All the documents necessary for obtaining a building permit.

Conventional soil absorption system – A system in which effluent is elevated or distributed to a seepage trench or bed.

Effluent – Liquid discharged from a septic or treatment tank.

Flood fringe – The part of the flood plain located outside of the floodway and usually covered with standing water during regional floods.

Floodway – The channel that carries and discharges flood water in a river or stream. Can also include portions of the flood plain.

High ground water – Soil saturation zones, aquifers, or periodically saturated zones.

High-water level – The highest known floodwater elevation of a water body as established by a state or federal agency.

Holding tank – An approved, watertight container for holding sewage.

Horizontal reference point – A stationary point to which horizontal dimensions are related.

Legal description – A lot and block number, a recorded assessor's plat, a public land survey description, or an accurate metes and bounds description.

Manhole – An opening that allows a person to gain access to a sewer or sewage-disposal system.

Mobile unit – A portable structure with or without a permanent foundation designed to be moved from one site to another.

Mobile unit park – Plots of ground upon which two or more mobile units are located and which owned by a person, state or local government.

Nuisance – In common law or equity jurisprudence, whatever is dangerous to human life or health; a structure which is not well ventilated, sewered, cleaned or lighted; or whatever renders air, food, drink or the water supply unwholesome.

Pan – A soil horizon cemented with an agent such as iron, organic matter, silica, or a combination of chemicals. It resists penetration with a knife blade and is impermeable or slowly permeable.

Percolation test – A way of testing soil absorption qualities.

Permeability – The ease with which liquids move through soil; listed in soil survey reports.

Pressure distribution system – The introduction of effluent into the soil by a pump or automatic siphon, using piping with small-diameter perforations.

Private sewage-disposal system – A sewage treatment system consisting of a septic tank and soil absorption field which serves a single structure; also, an alternative sewage-disposal system, a system serving more than one structure, or a system located on a different parcel of land than the structure. Such a system can be owned by the property owner or a special-purpose district.

Privy – A structure used by persons for disposing of body waste but not connected to a plumbing system.

Registered design professional – A person licensed to practice a design profession as defined in statutory requirements of the jurisdiction in which the project is to be constructed.

Seepage bed – With a bedding of aggregate and more than one distribution line, an excavated area of larger than 5 feet.

Seepage pit – A pit that allows disposal of effluent by means of an underground receptacle with soil absorption through its floor and walls.

Seepage trench – A trench containing a bedding of aggregate and a single distribution line over an excavated area of 1 to 5 feet in width.

Septage – Material removed from a private sewage treatment and disposal system.

Septic tank – A tank that separates sewage into solids and liquids by the processes of sedimentation, floatation and bacterial action. The liquid is then discharged into a soil absorption system.

Soil – Porous material over bedrock of 0.08 inch and smaller.

Soil boring – An observation pit, an augured hole, or a soil core taken intact and undisturbed with a probe.

Soil mottles – Contrasting soil colors caused by saturation over a normal year with a color value of 4 or more and chroma of 2 or less.

Soil saturation – A condition in which all pores in a soil are filled with water and the water flows into a bore hole.

Vent cap – An approved cover for the vent terminal of an effluent disposal system to prevent accidental closure, and still permit circulation of air.

Vertical elevation reference point – A visible stationary point for establishing the relative elevation of percolation tests and soil borings.

Watercourse – A stream flowing in a definite channel discharging into another body of water. A watercourse is more than mere surface drainage in legal terms. It has a bed and banks and may sometimes dry up.

# Chapter 3

# General Regulations

The general regulations of the septic code are not extensive. They are, however, important to know and observe. Let's take a look at them now.

## Specific Limitations

Domestic waste, which includes all wastes and sewage derived from ordinary living uses, must terminate in an approved septic or treatment tank if a public sewer connection is not available. It is possible for a code officer to create a variance to this requirement. Privies are not allowed. Except where specifically approved by a code officer, cesspools are prohibited.

The disposal of industrial wastes must be approved by a code officer. Certain objects must be prevented from entering a private sewage system. Some of the prohibited materials include:

- Ashes
- Cinders
- Rags

- Flammable substances

- Poisonous substances

- Explosive liquids and gases

- Oil

- Grease

- Insoluble materials

Clear water, such as surface water, rain or other clear water, shall not be allowed to enter a private sewage-disposal system. Water from a water softener or iron filter backwash must discharge into a septic system, unless it can be disposed of on the ground surface without creating a nuisance.

**Flood Plain**

A flood plain can create a serious problem for the installation of a septic system. Soil absorption systems must not be installed in a floodway. This type of system is also prohibited in a flood fringe, unless specific approval is given by a code official. No new sewage-disposal system may be installed in a floodway.

Replacement systems utilized to abate health hazards in floodways are considered and evaluated on a case-by-case basis. When a holding tank is installed in a floodway, the tank must be made floodproof. A soil absorption system can be replaced outside a flood plain boundary connected to the development by a force main or an approved site location outside the floodway but in the flood fringe area. Any septic tank in a floodway must be floodproofed. A malfunctioning soil absorption system must be replaced in the flood fringe, so long as the soil conditions and other site factors are suitable.

# Chapter 4

# Site Evaluation and Requirements

Site evaluation is a very important step in the installation of a septic system. Most people know that a perk test is needed before a suitable location can be established, but there is more to the process than just the test. When a site is under consideration for a septic system, the following items must be evaluated:

- Soil conditions

- Properties

- Permeability

- Depth to zones of soil saturation

- Depth to bedrock

- Slope

- Landscape position

- Setback requirements

- Flooding potential

Soil test data must be taken from undisturbed elevations. A vertical-elevation reference point must be established. Reports on site evaluation must be prepared using approved forms and filed within 30 days of the site testing.

When a site is being evaluated for a replacement system, a percolation test is generally not required. Care must be taken when replacing a system, however. It is not acceptable to disturb the existing site in such a way as to compromise the performance of the location for its intended use. A replacement system cannot be used for the following purposes:

- Building construction

- Parking lots

- Parking areas

- Below-ground swimming pools

- Other uses that may adversely affect the replacement area

### Slope

If a potential site for a soil absorption system has a slope of more than 20 percent, the system cannot be installed in the location. Conventional soil absorption systems must be located a minimum of 20 feet from the crown of a site with a slope greater than 20 percent, except where the top of the aggregate of a system is at or below the bottom of an adjacent roadside ditch.

### Test Boring

Test boring is done to retrieve soil samples for evaluation and testing. This type of testing is required on any site where a septic system is to be installed. A boring must extend at least 3 feet below the bottom of the proposed septic system. It is common for the boring samples to be taken prior to the perk test. Power augers must not be used to collect boring samples; boring must be done with a backhoe or dug by hand.

A minimum of three borings per intended site is required. Accurate details of the borehole locations must be logged and maintained, and reports on the locations must be drawn to scale.

Soil profile descriptions must be written for all borings. All differences in the thickness of different soil horizons observed shall be indicated. Horizons are differentiated on the basis of color, texture, soil mottles, or bedrock. Depths are to be measured from the ground surface.

## Soil Mottles

The highest level of soil mottles must be used when estimating the seasonal or periodic soil saturation zone. On occasion, a code officer may require a detailed description of soil mottling. This is normally required only on marginal sites. Abundance, size, contrast, and color of soil mottles must be described consistently.

When mottled color occupies less than 2 percent of an exposed surface, a sample is said to have an abundance of "few." When the percentage ranges from 2 to 20 percent, the rating is listed as "common." An abundance rating of "many" is given to a sample whose percentage is in excess of 20 percent.

The length of the mottle must be measured. This determines the "size" of the mottle. When the mottle is less than 5 millimeters in length, the sample is said to be "fine." A length of 5 to 15 millimeters is noted as "medium." Samples that are longer than 15 millimeters are rated as "coarse."

Contrast is another factor in soil evaluation. This refers to the difference in color between the soil mottle and the background color of the soil. When the mottle is evident only on close examination, the rating is "faint." Mottle that can be seen readily but is not striking is given a rating of "distinct." A rating of "prominent" is given when the mottle is obvious and one of the outstanding features of the horizon. Reports must include the color or colors of the mottle or mottles.

## Ground Water

When ground water is present, it must be reported. The depth to the ground water must be measured and recorded. Measurements are made from ground level. Ground water is measured from its highest location in a bore hole or at the highest level of seepage through the side of a bore hole. Soil above the level of existing ground water must be tested for the presence of soil mottles.

### Color Patterns

Color patterns may not be indicative of soil saturation. Soil profiles that have an abrupt textural change with finer textured soils overlying more than 4 feet of unmottled, loamy sand or coarser soils can have a mottled zone for the finer textured material. If the mottled zone is less than 12 inches thick and located immediately above the textural change, a soil absorption system is permitted in the loamy sand or coarser material below the mottled layer. When soil mottle occurs in sandy material, a site is considered unsuitable for a conventional system. Certain coarse, sandy loam soils may be considered as coarse material.

Not all soil mottles occur due to seasonal or periodic soil-saturation zones. Many natural occurrences can result in soil mottles. Some conditions that result in soil mottles but are not related to seasonal or periodic soil saturation zones include the following:

- Residual sandstone deposits

- Uneven weathering of glacially deposited material

- Glacially deposited material that is naturally gray in color

- Concretionary material in various stages of decomposition

- Deposits of lime derived from highly calcareous parent material

- Light-colored silt coats deposited on soil bed faces

- Soil mottles that are usually vertically oriented along old or decayed root channels with a dark organic stain usually present in the center of the mottled area

Any mottled soil conditions must be reported. A code official will evaluate the report and determine if a site is suitable for a proposed use.

## Bedrock

With the exception of sandstone, the depth of bedrock must be established at the depth in a soil boring where more than 50 percent of the weathered-in-place material is consolidated. When the bedrock is sandstone, the depth is determined by the point where increased resistance to the penetration of a knife blade is established.

## Alluvial and Colluvial Deposits

If you encounter alluvial or colluvial deposits, you may have a problem. Subsurface soil absorption systems are not allowed in alluvial or colluvial deposits when any of the following conditions exist:

- Alluvial or colluvial deposits have shallow depths.

- Extended periods of saturation exist or are expected.

- Flooding is possible.

## Percolation Tests

Percolation tests are required when evaluating a potential site for a private sewage-disposal system. A minimum of three perk tests are required. Test holes must be labeled and recorded. The depth of the test holes must extend at least to the bottom of the proposed absorption system.

Test holes can be bored or dug. The holes must have a horizontal dimension of 4 to 8 inches. A sharp instrument must be used to scrape the bottom and sides of a test hole. This is done to expose natural soil. Any loose material collected in a test hole must be removed. The bottom of a test hole must be covered with 2 inches of gravel or coarse sand.

## Sandy Soils

When testing sandy soils, the test hole must be filled with clear water to a minimum of 12 inches above the bottom of the hole. Once the hole is filled with water, the seepage of the water into the soil must be timed and recorded. Clear water is added to the hole until it is not more than 6 inches above the gravel or coarse sand. A timetable of 10 minutes is used to record the absorption rate. This testing goes on for 60 minutes. If 6 inches of water seep away in less than 10 minutes, a shorter interval between measurements will be used. It is important that the water depth is never more than 6 inches. If 6 inches of water seep away in less than 2 minutes, the testing can be stopped, and a rate of less than 3 minutes per inch can be reported.

### Other Soils

Test holes for soils other than sandy soils should be filled with 12 inches of clear water. The water level must be maintained at this depth for 4 hours. Once the 4-hour period passes, any water remaining should be left in the hole. The soil should be allowed to swell not less than 16 hours and not more than 30 hours. Once the soil-swelling process is complete, the measurements for determining the perk rate can be made.

To conduct a test, all loose material filling a test hole must be removed. The water level in the hole has to be adjusted to a depth of 6 inches above the gravel or coarse sand. From a fixed reference point, the water level must be monitored in 30-minute intervals for a period of 4 hours. An exception to this is if two successive water-level drops do not vary by more than 0.62 inch.

A minimum of three water level drops must be observed and recorded. A test hole must be filled with clear water to a point not more than 6 inches above the gravel or coarse sand whenever a hole becomes nearly empty. It is not acceptable to make adjustments to the water level during a test, except to the limits of the last measured water-level drop. Once the first 6 inches of water seep away in less than 30 minutes, the time interval between measurements shall be 10 minutes, and the test must last for an hour. Water depth is not to exceed 5 inches at any time during the measurement period. When the final drop occurs during the final measurement period, it should be used in calculating the percolation rate.

Mechanical test equipment must be of an approved type. Soil evaluations must establish the estimated percolation based on structure and texture in accordance with accepted soil evaluation practices.

### Verification

Code officials can require the verification of soil evaluations. They may ask that soil testing be done under the supervision of a code official.

When monitoring levels of ground water, a property owner or developer has an option to provide documentation that soil mottling or other color patterns at a particular site are not an indication of seasonally saturated soil conditions of high ground water levels.

The monitoring of ground-water levels must be done at a time of the year when maximum ground-water elevation occurs. The area must be investigated for any artificial drainage, such as drainage tile or open ditches. If artificial drainage is present, full documentation on the drainage system is required.

When an owner or an owner's agent plans to monitor ground water, the individual must notify a code official of the monitoring. The code official may perform a field check. A minimum of three monitoring sites is required for an official monitoring test.

### Design of Monitoring Wells

A minimum of two monitoring wells is required to extend to a depth of at least 6 feet below the ground surface and a minimum of 3 feet below the designed system design. When layered, mottled soil exists, at least one monitoring well must terminate within the mottled layer. Depending on site conditions, a deeper depth for monitoring holes may be required.

Monitoring wells are to be made with solid pipe that is installed in a bored hole. A minimum pipe size is 1 inch, and a maximum size is 4 inches. The bore hole shall be a minimum of 4 inches and a maximum of 8 inches larger than the pipe to be inserted in the hole.

### Observation Requirements

Observations must be done every seven days or until a site is found to be unacceptable. If water is observed above the critical depth, another observation is required one week later. If water is found at that time to be above the critical depth, the site is determined unacceptable. When rainfall of 0.5 inch or more occurs in a 24-hour period during monitoring, observations must be made at more frequent intervals.

If monitoring reveals saturated soil conditions, the following data must be submitted in writing:

- Test locations

- Ground elevations at wells

- Soil profile descriptions

- Soil series

- Dates observations were made

- Depths of water observed

- Local precipitation data

If a site is monitored and found to be acceptable, the following data is required to be supplied in writing:

- Location of test holes

- Depth of test holes

- Ground elevations at wells

- Soil profile descriptions

- Soil series

- Dates observations were made

- Results of observations

- Information on artificial drainage, if any

- Local precipitation data

**Site Requirements**

The surface grade of all soil absorption systems must be located at a point lower than the surface grade of any nearby water well or reservoir on the same or adjoining property. In cases where this is not possible, the system must be located in a way to prevent surface water from draining towards a well or reservoir. There are minimum horizontal distances for the location of soil absorption systems and other elements. The following list reveals many of the horizontal distance requirements:

| | |
|---|---|
| Cistern | Minimum of 50 feet |
| Habitable building with a below-grade foundation | Minimum of 25 feet |
| Habitable building, slab-on-grade | Minimum of 15 feet |
| Lake, high-water mark | Minimum 50 feet |
| Lot line | Minimum 5 feet |
| Reservoir | Minimum 50 feet |
| Roadway ditch | Minimum 10 feet |
| Spring | Minimum 100 feet |
| Stream | Minimum 50 feet |
| Swimming pool | Minimum 15 feet |
| Uninhabited building | Minimum 10 feet |
| Water main | Minimum 50 feet |
| Water service | Minimum 10 feet |
| Water well | Minimum 50 feet |

It is not acceptable to install a private sewage-disposal system in a compacted area, such as a parking lot. At no time shall surface water be allowed to be diverted to a private sewage system on the same or a neighboring property.

## Bedrock

Bedrock, ground water, and slowly permeable soils require special consideration when creating a private sewage system. A minimum of 3 feet of soil is required between the bottom of the soil absorption system and high ground water or bedrock. Any soil that has a perk rate of 60 minutes per inch, or faster, of the proposed soil absorption system and at least 3 feet below the proposed bottom of the system. There must be a minimum of 56 inches of suitable soil from original grade for a conventional soil absorption system.

## Perk Rate

Trench and bed-type septic systems cannot be installed if the perk rate for any one of the three perk tests is slower than 60 minutes for water to fall 1 inch. The slowest perk rate on any test hole is the rate that must be used when establishing the perk rate of soil.

Seepage pits require perk tests in each horizon penetrated below the inlet pipe. Soil strata where perk rates are slower than 30 minutes per inch shall not be included in computing the absorption area.

Some parcels of land suffer from severe limitations for on-site liquid-waste disposal systems. These limitations might be determined through the use of soil maps. If permission to install a system is denied based on data from a soil map, a property owner has a right to present any evidence that will prove a suitable site is available on the property.

## Filled Areas

Filled areas are not acceptable sites for sewage systems. However, it is possible to obtain written approval to install a system in a filled area under certain conditions. If there is evidence that proves a filled site can conform to code requirements, perk rates, and elevation, the site might be approved by the code officials.

There are times when filled sites are created for a septic system. This seems like a contradiction to the code, which typically indicates that filled sites are not suitable. When are you allowed to fill a site? When there is less than 56 inches, but at least 30 inches, of soil over bedrock, you can fill the site. The existing soil must be sand or loamy sand to fit the protocol. Any fill that is used must be of the same soil texture as the existing soil. This same basic procedure can be used when a site with high ground water is encountered. When fill is installed, the installation must be inspected by a code official.

## Design Requirements

Any filled area for a septic system must be large enough to accommodate a shallow trench system and a replacement system. The installation area must be approved based on the perk rate of the soil and the use of the building that will be served by the system. If any portion of the trench system or its replacement is placed in the filled area, the fill must extend 20 feet beyond all sides of both systems before the slope begins. Topsoil and vegetation must be removed from any area that will be filled. Maximum slope requirements must have one vertical unit to three horizontal units, provided that the 20-foot spacing distance can be maintained. A variance to slope rates may be available under special circumstances.

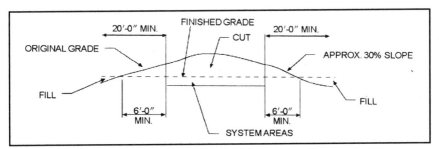

**Figure 4.1** Excavation of complete hilltop: *(Courtesy of 2000 International Private Sewage Disposal Code).*

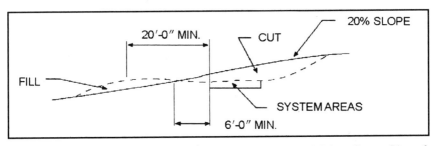

**Figure 4.2** Excavation into hillside: *(Courtesy of 2000 International Private Sewage Disposal Code).*

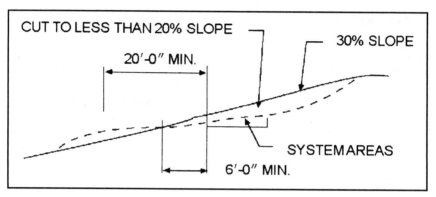

**Figure 4.3** Regrade of hillside: (*Courtesy of 2000 International Private Sewage Disposal Code*).

# Chapter 5

# Materials

Materials for private sewage systems must meet minimum standards as set forth by the code used by local jurisdictions. A manufacturer's mark or name and the quality of the product or identification must be cast, embossed, stamped, or indelibly marked on each material used in an installation. Septic tanks must be marked to identify the tank's capacity.

Any materials used in an installation must be of an approved type. Professional and protective handling of materials is required during the installation of materials. Defective or damaged materials may not be installed.

## Septic Tanks

All septic tanks installed must be in compliance with the local code requirements. Steel septic tanks that show any damage to the bituminous coating must be repaired by recoating the tank. The gauge of the steel must conform with local code requirements. Fiberglass tanks may be used, but they must also meet local code requirements.

Concrete tanks may be precast or built on site. When a concrete tank is built on site, the floor and sidewalls of the tank must be monolithic, except that a construction joint is permitted in the lower 12 inches of the sidewalls of the tank. The construction joint must have a keyway in the lower section of the joint. The width of the keyway must be approximately 30 percent of the thickness of the sidewall with a depth equal to the width. A continuous water stop or baffle at least 56 inches wide must be set vertically in the joint, embedded one-half its width in the concrete below the joint with the remaining width in the concrete above the joint. The water stop or baffle must be copper, neoprene, rubber, or polyvinyl chloride designed for the specific purpose.

Tongue-and-groove or shiplap joints are required for joints between the concrete tank and the tank cover and between the tank cover and the manhole riser. These joints must be sealed watertight with either cement, mortar, or bituminous compound.

Tank capacity, vertical cylindrical gauge thickness, and minimum diameter are determined by the local code.

Manhole collars and extensions must be made of the same material as the tank being served. Covers for manholes must be made of concrete, steel, cast iron, or another approved material.

**Table 5.1 Tank Capacity:** *(Courtesy of International Private Sewage Disposal Code).*

| TANK DESIGN AND CAPACITY | | MINIMUM GAGE THICKNESS | MINIMUM DIAMETER |
|---|---|---|---|
| **Vertical cylindrical** | | | |
| 500 to 1,000 gallons | Bottom and sidewalls | 12 gage | None |
| | Cover | 12 gage | |
| | Baffles | 12 gage | |
| 1,001 to 1,250 gallons | Complete tank | 10 gage | None |
| 1,251 to 1,500 gallons | Complete tank | 7 gage | None |
| **Horizontal cylindrical** | | | |
| 500 to 1,000 gallons | Complete tank | 12 gage | 54-inch diameter |
| 1,001 to 1,500 gallons | Complete tank | 12 gage | 64-inch diameter |
| 1,501 to 2,500 gallons | Complete tank | 10 gage | 76-inch diameter |
| 2,501 to 9,000 gallons | Complete tank | 7 gage | 76-inch diameter |
| 9,001 to 12,000 gallons | Complete tank | $1/4$-inch plate | None |
| over 12,000 gallons | Complete tank | $5/16$ inch | None |

For SI: 1 inch = 25.4 mm, 1 gallon = 3.785 L.

## Pipe

Pipe for a private sewage-disposal system must have a smooth wall and conform to the requirements of the local code. Distribution pipe, the perforated pipe used for distribution systems, must also conform to the local code requirements.

## Joints and Connections

There are many types of joints and connections that may be used in a private sewage system. All joints must confirm to requirements set forth by the local code. Additionally, some types of joints are more regulated than others. Refer to the following list for a quick review of common code requirements:

- ABS joints—general regulations

- Mechanical joints—general regulations; must be made with elastomeric seals and installed in accordance with manufacturer's instructions.

- Solvent cementing—general regulations; joint surfaces must be clean and dry. Joints must be made while the cement is wet.

- Asbestos-cement joints—general regulations; must be made with a sleeve coupling of the same composition as the pipe and sealed with an elastomeric ring.

- Bituminized fiber joints—general regulations; must be made with tapered-type couplings of the same material.

- Cast-iron joints—general regulation

- Caulked joints—general regulations; joints for hub and spigot pipe must be firmly packed with oakum or hemp. Molten lead must be poured in one operation to a depth of not less than 1 inch. The lead must not recede more than 0.125 inch below the rim of the hub, and must be caulked tight. If the joints are to be painted, varnished, or covered with another coating, this process must wait until the joints have been inspected and approved.

Table 5.2 **Private Sewage Disposal System Pipe:** *(Courtesy of International Private Sewage Disposal Code).*

| MATERIAL | STANDARD |
| --- | --- |
| Acrylonitrile butadiene styrene (ABS) plastic pipe | ASTM D 2661; ASTM D 2751; ASTM F 628 |
| Asbestos-cement pipe | ASTM C 428 |
| Bituminized fiber pipe | ASTM D 1861; ASTM D 1862 |
| Cast-iron pipe | ASTM A 74; CISPI 301 |
| Coextruded composite ABS DWV sch. 40 pipe (Solid) | ASTM F 1488; ASTM F 1499 |
| Coextruded composite ABS sch. 40 DWV pipe with cellular core | ASTM F 1488; ASTM F 1499 |
| Coextruded composite ABS sewer pipe with cellular core with pipe stiffness of PS35, PS50, PS100, PS140 and PS200 | ASTM F 1488; ASTM F 1499 |
| Coextruded composite PVC DWV sch. 40 pipe (solid) | ASTM F 1488 |
| Coextruded composite PVC DWV sch. 40 IPS pipe with cellular core | ASTM F 1488 |
| Coextruded composite PVC DWV sch. 40 IPS pipe with cellular core with pipe stiffness of PS1140 and PS200 | ASTM F 1488 |
| Coextruded composite PVC 3.25 OD DWV pipe | ASTM F 1488 |
| Coextruded composite PVC sewer pipe with cellular core with pipe stiffness of PS35, PS50, PS100, PS140 and PS200 | ASTM F 1488 |
| Concrete pipe | ASTM C 14; ASTM C 76; CSA A257.1; CSA A257.2 |
| Copper or copper-alloy tubing (Type K or L) | ASTM B 75; ASTM B 88; ASTM B 251 |
| Polyvinyl chloride (PVC) plastic pipe (Type DWV, SDR26, SDR35, SDR41, PS50 or PS100) | ASTM D 2665; ASTM D 2949; ASTM D 3034; ASTM F 891; CSA CAN/CSA-B182.2; CSA CAN/CSA-B182.4 |
| Vitrified clay pipe | ASTM C 4; ASTM C 700 |

Table 5.3  **Distribution Pipe:** *(Courtesy of International Private Sewage Disposal Code).*

| MATERIAL | STANDARD |
|---|---|
| Polyethylene (PE) plastic pipe | ASTM F 405 |
| Polyvinyl chloride (PVC) plastic pipe | ASTM D 2729 |
| Polyvinyl chloride (PVC) plastic pipe with pipe stiffness of PS35 and PS50 | ASTM F 1488 |

- Mechanical compression joints—general regulations; gaskets must be compressed when the pipe is fully inserted.

- Mechanical joint coupling—general regulations; these joints, when installed aboveground, must use an elastomeric sealing sleeve with a center stop. Underground joints must be made in accordance with the manufacturer's instructions.

- Concrete joints—general regulations; must be made with an elastomeric seal.

- Copper joints—general regulations

- Soldered joints—general regulations; joint surfaces must be cleaned and coated with an approved flux before being soldered in compliance with the local code requirements.

- Polyethylene joints—general regulations

- Heat-fusion joints—general regulations; joints must be clean and dry. All joint surfaces must be heated to the melting temperature and joined. This type of joint may not be disturbed before it has cooled.

- Mechanical joints—general regulations; they must be made in compliance with the manufacturer's installation recommendations.

- PVC joints—general regulations; joint surfaces must be clean and dry. A purple primer must be applied prior to cementing a joint.

- Vitrified clay joints—general regulations; these joints must be made with an elastomeric seal.

- Copper to cast-iron joints—general regulations; these joints must be made with a brass ferrule or compression joint. A suitable solder joint must be made between the copper and the ferrule. The ferrule must be joined to the cast iron by a caulked or mechanical compression joint.

- Plastic to cast-iron joints—general regulations; the plastic pipe joint must be joined where it meets the cast-iron hub with either a caulked or a mechanical joint.

- Dissimilar plastic joints—general regulations

### Prohibited Joints and Connections

Some joints and connections are prohibited. Concrete and cement joints are not allowed, nor are mastic or hot-pour bituminous joints. Any joint made with a fitting that is not approved for the use is prohibited. Solvent-cement joints between dissimilar plastic pipes are prohibited. Any joint made with an elastomeric rolling O-ring on pipes that have different diameters is prohibited.

The code requirements for materials are easy to understand and comply with. Pay attention: it can save you a lot of time, money, frustration, and embarrassment.

Chapter

# 6

# Soil Absorption Systems

Soil-absorptions systems are the most common type of private sewage disposal systems. Sizing this type of system requires the use of language and tables that exist in local code requirements. When a system has a daily effluent application of 5,000 gallons or less, the sizing is fairly simple. It is possible to use two systems of equal size when the daily effluent application exceeds 5,000 gallons. When this is done, each system must have a minimum capacity of 75 percent of the area required for a single system. Dual systems can be considered as one system, but an approved means of alternating waste application must be provided when such a system is used to accommodate a usage of more than 5,000 gallons a day.

Pressure-distribution systems are permitted in place of a conventional or dosing soil absorption system as long as a site is suitable for the conventional system. When a site is unsuitable for conventional treatment, a pressure distribution system can be used as an alternative system. The result is that a pressure distribution system can be used in either case, but must be used when a conventional system is not feasible.

Flow from a septic or treatment tank to a soil absorption system must be by gravity or dosing for systems receiving 1,500 gallons or less of effluent a day. A tank that discharges effluent at a rate of more than 1,500 gallons a day must be equipped to pump the effluent or to create an automatic siphon for the system.

The sizing of a private sewage soil absorption system for residential properties is done with the use of data provided in the local code. This

**Table 6.1  Minimum absorption area for one- and two-family dwellings:** *(Courtesy of International Private Sewage Disposal Code).*

| PERCOLATION CLASS | PERCOLATION RATE (minutes required for water to fall 1 inch) | SEEPAGE TRENCHES OR PITS (square feet per bedroom) | SEEPAGE BEDS (square feet per bedroom) |
|---|---|---|---|
| 1 | 0 to less than 10 | 165 | 205 |
| 2 | 10 to less than 30 | 250 | 315 |
| 3 | 30 to less than 45 | 300 | 375 |
| 4 | 45 to 60 | 330 | 415 |

For SI:  1 minute per inch = 2.4 s/mm, 1 square foot = 0.0929 $m^2$.

is most commonly done with the use of a sizing table. You must know the perk test data and the type of construction for the system in order to size the system. This applies to one- and two-family dwellings.

If you are sizing a system for another type of building, you will need to know the perk rate, the building usage, and the design details. This sizing is done with data from the local code.

## Excavating Seepage Trenches

Seepage trenches must be 1 to 5 feet wide. The trenches must be spaced at least 6 feet apart. Individual trenches must not exceed a length of 100 feet, unless otherwise approved. The absorption area of a seepage trench must be computed using only the bottom of the trench area. You cannot use the bottom excavation area of the distribution header as an absorption area.

**Table 6.2  Minimum absorption area for other than one- and two-family dwellings:** *(Courtesy of International Private Sewage Disposal Code).*

| PERCOLATION CLASS | PERCOLATION RATE (minutes required for water to fall 1 inch) | SEEPAGE TRENCHES OR PITS (square feet per unit) | SEEPAGE BEDS (square feet per unit) |
|---|---|---|---|
| 1 | 0 to less than 10 | 110 | 140 |
| 2 | 10 to less than 30 | 165 | 205 |
| 3 | 30 to less than 45 | 220 | 250 |
| 4 | 45 to 60 | 220 | 280 |

For SI:  1 minute per inch = 2.4 s/mm, 1 square foot = 0.0929 $m^2$.

Soil Absorption Systems 37

**Table 6.3  Conversion factor:** *(Courtesy of International Private Sewage Disposal Code).*

| BUILDING CLASSIFICATION | UNITS | FACTOR |
|---|---|---|
| Apartment building | 1 per bedroom | 1.5 |
| Assembly hall—no kitchen | 1 per person | 0.02 |
| Auto washer (service buildings, etc.) | 1 per machine | 6.0 |
| Bar and cocktail lounge | 1 per patron space | 0.2 |
| Beauty salon | 1 per station | 2.4 |
| Bowling center | 1 per bowling lane | 2.5 |
| Bowling center with bar | 1 per bowling lane | 4.5 |
| Camp, day and night | 1 per person | 0.45 |
| Camp, day use only | 1 per person | 0.2 |
| Campground and camping resort | 1 per camping space | 0.9 |
| Campground and sanitary dump station | 1 per camping space | 0.085 |
| Car wash | 1 per car | 1.0 |
| Catch basin—garages, service stations, etc. | 1 per basin | 2.0 |
| Catch basin—truck wash | 1 per truck | 5.0 |
| Church—no kitchen | 1 per person | 0.04 |
| Church—with kitchen | 1 per person | 0.09 |
| Condominium | 1 per bedroom | 1.5 |
| Dance hall | 1 per person | 0.06 |
| Dining hall—kitchen and toilet | 1 per meal served | 0.2 |
| Dining hall—kitchen and toilet waste with dishwasher or food waste grinder or both | 1 per meal served | 0.25 |
| Dining hall—kitchen only | 1 per meal served | 0.06 |
| Drive-in restaurant, inside seating | 1 per seat | 0.3 |
| Drive-in restaurant, without inside seating | 1 per car space | 0.3 |
| Drive-in theater | 1 per car space | 0.1 |
| Employees—in all buildings | 1 per person | 0.4 |
| Floor drain | 1 per drain | 1.0 |
| Hospital | 1 per bed space | 2.0 |
| Hotel or motel and tourist rooming house | 1 per room | 0.9 |
| Labor camp—central bathhouse | 1 per employee | 0.25 |
| Medical office buildings, clinics and dental offices  Doctors, nurses and medical staff  Office personnel  Patients | 1 per person  1 per person  1 per person | 0.8  0.25  0.15 |
| Mobile home park | 1 per mobile home site | 3.0 |
| Nursing or group homes | 1 per bed space | 1.0 |
| Outdoor sports facility—toilet waste only | 1 per person | 0.35 |
| Park—showers and toilets | 1 per acre | 8.0 |
| Park—toilet waste only | 1 per acre | 4.0 |
| Restaurant—dishwasher or food waste grinder or both | 1 per seating space | 0.15 |
| Restaurant—kitchen and toilet | 1 per seating space | 0.6 |
| Restaurant—kitchen waste only | 1 per seating space | 0.18 |

| BUILDING CLASSIFICATION | UNITS | FACTOR |
|---|---|---|
| Restaurant—toilet waste only | 1 per seating space | 0.42 |
| Restaurant—(24-hour) kitchen and toilet | 1 per seating space | 1.2 |
| Restaurant—(24-hour) with dishwasher or food waste grinder or both | 1 per seating space | 1.5 |
| Retail store | 1 per customer | 0.03 |
| School—meals and showers | 1 per classroom | 8.0 |
| School—meals served or showers | 1 per classroom | 6.7 |
| School—no meals, no showers | 1 per classroom | 5.0 |
| Self-service laundry—toilet wastes only | 1 per machine | 1.0 |
| Service station | 1 per car served | 0.15 |
| Showers—public | 1 per shower | 0.3 |
| Swimming pool bathhouse | 1 per person | 0.2 |

## Seepage Beds

The excavations for seepage beds must be a minimum of 5 feet wide and have more than one distribution pipe. The bottom of the trench is used to compute the absorption area. Distribution pipes must be spaced evenly with a minimum distance of 3 feet and a maximum distance of 5 feet between the pipes. Spacing between distribution pipes and the sidewall or headwall of a seepage bed shall be set at a maximum of 3 feet and a minimum of 1 foot.

## Seepage Pits

Seepage pits require a minimum inside diameter of 5 feet. They must consist of a chamber walled up with perforated precast concrete rigs, concrete block, brick, or other approved material that allows effluent to percolate in the surround soil. The bed of a seepage bed is to be left open to the soil.

Aggregate of 0.5 to 2.5 inches in size must be placed into a 6-inch minimum annular space separating the outside wall of the chamber and the sidewall excavation. The depth of the annular space is measured from the inlet pipe to the bottom of the chamber. Every seepage pit must be provided with a 24-inch manhole that extends to within 56 inches of the round surface and a 4-inch diameter fresh air inlet.

Seepage pits must be located a minimum of 5 feet apart. The effective area of a seepage pit is the vertical wall area of the walled-up chamber for the depth below the inlet for all strata in which the perk rates are less than 30 minutes per inch. The 6-inch annular opening outside the vertical wall area can be included in determining the effective area. Sizing data should be available in your local code book.

## Rules of Excavation and Construction

There are rules that must be obeyed when excavation and construction of soil-absorption systems is required. The basic rules are as follows:

- The bottom of a trench or bed must be level.

- Excavation is not allowed if the soil is so wet that it creates a soil wire when rolled between a person's hands.

**Table 6.4 Effective square-foot absorption area for seepage pits:** *(Courtesy of International Private Sewage Disposal Code).*

| INSIDE DIAMETER OF CHAMBER IN FEET PLUS 1 FOOT FOR WALL THICKNESS PLUS 1 FOOT FOR ANNULAR SPACE | DEPTH IN FEET OF PERMEABLE STRATA BELOW INLET | | | | | |
|---|---|---|---|---|---|---|
| | 3 | 4 | 5 | 6 | 7 | 8 |
| 7 | 47 | 88 | 110 | 132 | 154 | 176 |
| 8 | 75 | 101 | 126 | 151 | 176 | 201 |
| 9 | 85 | 113 | 142 | 170 | 198 | 226 |
| 10 | 94 | 126 | 157 | 188 | 220 | 251 |
| 11 | 104 | 138 | 173 | 208 | 242 | 277 |
| 13 | 123 | 163 | 204 | 245 | 286 | 327 |

For SI: 1 foot = 304.8 mm.

- Sidewalls or bottoms that suffer from smearing or compaction must be scarified.

- The bottom area must be scarified and loose material must be removed.

### Aggregate and Backfill

A minimum of 6 inches of aggregate ranging in size from 0.5 to 2.5 inches must be laid into a trench or bed below the distribution pipe elevation. The aggregate must be evenly distributed a minimum of 2 inches over the top of the distribution pipe. The aggregate must be covered with approved synthetic materials or 9 inches of uncompacted marsh hay or stray. Building paper is not an approved means of covering. A minimum of 18 inches of soil backfill must be installed on top of the covering. No covering that will prevent the evaporation of effluent should be installed.

### Piping Requirements

Requirements for distribution pipes are not overly complicated. Here are the basics:

- Distribution pipes for gravity systems must have a minimum diameter of 4 inches.

- Distribution heads must be made with solid-wall pipe.

- The top of a distribution pipe must not be less than 8 inches below the original surface in continuous straight or curved lines.

- Grade on a distribution pipe must be 2 to 4 inches per a 100-foot run.

- Effluent must be distributed to all distribution pipes.

- The distribution of effluent to a seepage trench on a sloping site must be accomplished by using a drop box design or some other approved method.

- If dosing is required, the siphon or pump must discharge a dose of minimum capacity equal to 75 percent of the combined volume of the distribution piping in the absorption system.

**Observation Pipes**

Observation pipes are required, and they must have a minimum diameter of 4 inches. The observation pipe must extend at least 12 inches above the final grade. Termination points for observation pipes must be equipped with an approved vent cap.

The bottom 12 inches of an observation pipe should be perforated and extend to the bottom of the aggregate. Observation pipes must be located at least 25 feet from any window, door, or air intake of any building used for human occupancy. No more than four distribution trenches can be served by one common 4-inch observation pipe when interconnected by a common header pipe.

Under special circumstances, and where approved, an observation pipe's location can be permanently recorded and the pipe can be installed not more than 2 inches below the finished grade.

**Weather**

Weather can play a part in the installation of a soil-absorption system. Unless otherwise approved, a system cannot be installed during adverse

## Soil Absorption Systems 41

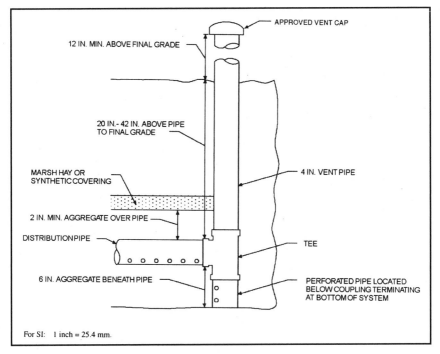

**Figure 6.1** Observation pipe: *(Courtesy of International Private Sewage Disposal).*

weather conditions. If the installation area is frozen, a system cannot be installed. Any existing snow must be removed prior to installing a system. The snow must be moved to avoid any water from ponding as the snow melts.

Backfill material must be protected from freezing. Frozen soil cannot be used as backfill material. Special inspection of the beds or trenches may be required during winter weather conditions. If you are working in cold weather, it is a good idea to contact the local code officer to see if any special inspections will be required before you place your gravel and backfill materials.

# Chapter 7

# Pressure Distribution Systems

Pressure distribution systems are allowed on any site that is suitable for a conventional private sewage-disposal system. There must be at least 6 inches from the original grade to the top of the distribution piping. What is the required minimum depth from original grade? You will have to check your local code requirements. The data should be available to you, probably in tabular form. In general, the depth ranges from 49 to 53 inches.

The estimated daily wastewater flow and the design loading rate based on the perk rate for a site are used to determine the required size of the absorption area. Once again, data in your local code requirements will help you to define a proper size. Normally, you will need an absorption area that equals the wastewater flow divided by the design loading rate that is provided in your local code. The design load factor depends on the perk rate. For example, my local code says that a perk rate of 10 to less than 30 minutes will give me a load factor of 0.8. Since local jurisdictions adopt a code and have the right to alter it, so you must check your current codes to ensure that you remain in compliance.

A rule-of-thumb for computing wastewater flow is to consider that a residence will use 150 gallons per bedroom per day. You can find more data of this type in your local codebook.

## System Design

Any pressure distribution system must discharge effluent into trenches or beds. Every pipe that is connected to an outlet of a manifold is

**Table 7.1 Soil required:** *(Courtesy of International Private Sewage Disposal Code).*

| DISTRIBUTION PIPE (inches) | SUITABLE SOIL (inches) |
|---|---|
| 1 | 49 |
| 2 | 50 |
| 3 | 52 |
| 4 | 53 |

For SI: 1 inch = 25.4 mm.

**Table 7.2 Design loading rate:** *(Courtesy of International Private Sewage Disposal Code).*

| PERCOLATION RATE (minutes per inch) | DESIGN LOADING FACTOR (gallons per square foot per day) |
|---|---|
| 0 to less than 10 | 1.2 |
| 10 to less than 30 | 0.8 |
| 30 to less than 45 | 0.72 |
| 45 to 60 | 0.4 |

For SI: 1 minute per inch = 2.4 s/mm, 1 gallon per square foot = 0.025 L/m$^2$.

counted as a separate distribution pipe. Spacing for horizontal distribution pipes must be between 30 to 72 inches.

Distribution pipe size, hole diameter, and hole spacing have to be selected. The hole diameter and spacing must be equal for each manifold segment. Distribution pipe size is not required to be the same for each segment.

To do a full sizing of piping, you will have to use data from your local code. This will involve the Hazen-Williams Friction Factor in most cases. The math is more than some plumbers want to do. However, most sizing is done by engineers, so few plumbers or septic installers have to learn the finer details of the math required to size a system.

The size of a force main between a pump and a manifold must be based on the friction loss and velocity of effluent through the pipe. Velocity of the effluent in a force main must not be more than 5 feet per second.

## Pumps

The pumps used with a pressure system are sized based on the discharge rate and total dynamic head of the pump performance curve. Total dynamic head must be equal to the difference in feet of elevation

**Table 7.3  Estimated volume for various diameter pipes:**
*(Courtesy of International Private Sewage Disposal Code).*

| DIAMETER (inches) | VOLUME (gallons per foot length) |
|---|---|
| 1 | 0.041 |
| $1^{1}/_{4}$ | 0.064 |
| $1^{1}/_{2}$ | 0.092 |
| 2 | 0.164 |
| 3 | 0.368 |
| 4 | 0.655 |
| 5 | 1.47 |

For SI:  1 inch = 25.4 mm, 1 gallon per foot = 80.5 L/mm.

between the pump and the distribution pipe invert, plus the friction loss and a minimum of 2.5 feet where utilizing low-pressure distribution in the delivery and network pipes.

Control systems for a pumping chamber consist of a control for operating the pump and an alarm system to detect a pump. Start and stop depth controls must be adjustable. Pump and alarm controls must be of an approved type. All switches must be resistant to sewage corrosion.

An alarm system for a pumping station must consist of a bell or light, mounted in a structure, and must be located to be easily seen or heard. A high-water sensing device must be installed approximately 2 inches above the depth set for the "on" pump control but below the bottom of the inlet to the pumping chamber. Alarm systems have to be installed on a separate circuit from the electrical service. All electrical connections must be located outside the pumping chamber.

## Dosing

Dosing frequency must be a maximum of four times a day. The volume per dose is established by dividing the daily wastewater flow by the dosing frequency. The dosing volume must be a minimum of ten times the capacity of the distribution pipe volume.

Your local codebook should have all the data you need to size a pressure distribution system. Most installation contractors don't design and size their own systems. This work is usually done by engineers. However, if you want to size your own system, you should be able to find all the information in your local codebook.

# Chapter 8

# Tanks

A septic tank can be made of welded steel, monolithic concrete, fiberglass, or any other approved material. All septic tanks have to be watertight and constructed as an individual structure, and they must be designed and built to withstand anticipated loads. Precast tanks and tanks built on site must be authorized by the appropriate approval agency.

All septic tanks must have two compartments. Code requires the inlet compartment to be not less than two-thirds of the total capacity of the tank, and the capacity cannot be less than 500 gallons. This means that a minimum size for a modern septic tank is 750 gallons. The secondary compartment must not be less than 250 gallons. If a tank has a capacity of 1,500 gallons, the secondary compartment must not be less than 5 feet long.

The liquid depth of a tank must not be less than 30 inches and a maximum average of 6 feet is required. The total depth cannot be less than 8 inches greater than the liquid depth.

Cylindrical tanks must not be less than 48 inches in diameter. Rectangular tanks must be constructed so that the longest dimensions are parallel to the direction of flow.

## Inlets and Outlets

Open-end coated sanitary tees or baffles made of approved materials constructed to distribute flow and retain scum in a tank or compartment

**Table 8.1  Septic tank capacity for one- and two-family dwellings:** *(Courtesy of International Private Sewage Disposal Code).*

| NUMBER OF BEDROOMS | SEPTIC TANK (gallons) |
|---|---|
| 1 | 750 |
| 2 | 750 |
| 3 | 1,000 |
| 4 | 1,200 |
| 5 | 1,425 |
| 6 | 1,650 |
| 7 | 1,875 |
| 8 | 2,100 |

For SI:   1 gallon = 3.785 L.

are required on the inlet and outlet of all tanks. Inlets and outlets must be equipped to prevent a sewer from entering beyond the inside wall of a tank. Tees or baffles must extend at least 6 inches above and 9 inches below the liquid level, but cannot exceed one-third the liquid depth. A minimum of 2 inches of clear space must be provided over the top of tees or baffles. The bottom of an outlet opening must be a minimum of 2 inches lower than the bottom of the inlet.

### Manholes

At least one manhole opening must be located over the inlet or outlet opening of each tank compartment. The openings must not be less than 24 inches square or 24 inches in diameter.

If the inlet compartment of a septic tank exceeds 12 feet in length, an additional manhole must be provided over the baffle wall. Manholes have to terminate a maximum of 6 inches below the ground surface and be of the same material as the tank. Steel tanks must not have less than a 2-inch collar for the manhole extensions permanently welded to the tank. When a fiberglass tank is used, the manhole extension must be made of fiberglass and must be an integral part of the tank. The collar shaft for a fiberglass tank cannot be less than 2 inches in height.

## Manhole Covers

Manhole covers must be watertight and made of concrete, steel, cast iron, or some other approved material that is capable of withstanding all anticipated loads. Every manhole riser must be fitted with an approved manhole cover. When manhole covers terminate above grade, the covers must be equipped with an approved locking device.

## Inspection Openings

Inspection openings must be provided over either the inlet or outlet baffle of every treatment tank. This type of opening must not be less than 4 inches in diameter and fitted with a tight cover. When inspection pipes terminate above ground, they cannot be less than 6 inches above finished grade. If inspection pipes are authorized to terminate below grade, the pipes must not be more than 2 inches below grade. Inspection pipes terminating below grade must be permanently recorded.

## Sizing Tanks

The sizing of tanks is based on the number of persons using a building or the volume and type of waste that a building produces, whichever is greater. What is the minimum tank size? A septic tank is not allowed to have a capacity of less than 750 gallons. No more than four septic tanks may be installed in a series.

A one- or two-bedroom home can usually be served by a 750-gallon tank. Three-bedroom homes require a 1,000-gallon tank. A home with four bedrooms will need a 1,200-gallon tank. Check your local codebook for data used in your area.

Many factors can come into play when sizing a suitable septic tank. For example, when sizing a tank for a building that is not a one- or two-family dwelling, you will have to consider kitchen and laundry waste. This type of data should be found in your local codebook.

## Installation

Septic tanks have to be located with a specific horizontal distance as noted in your local codebook. We have discussed examples of this earlier.

**Table 8.2  Additional capacity for other buildings:** *(Courtesy of International Private Sewage Disposal Code).*

| BUILDING CLASSIFICATION | CAPACITY (gallons) |
|---|---|
| Apartment buildings (per bedroom—includes automatic clothes washer) | 150 |
| Assembly halls (per person—no kitchen) | 2 |
| Bars and cocktail lounges (per patron space) | 9 |
| Beauty salons (per station—includes customers) | 140 |
| Bowling centers (per lane) | 125 |
| Bowling centers with bar (per lane) | 225 |
| Camp, day use only—no meals served (per person) | 15 |
| Campgrounds and camping resorts (per camp space) | 100 |
| Campground sanitary dump stations (per camp space) (omit camp spaces with sewer connection) | 5 |
| Camps, day and night (per person) | 40 |
| Car washes (per car handwash) | 50 |
| Catch basins—garages, service stations, etc. (per basin) | 100 |
| Catch basins—truck washing (per truck) | 100 |
| Churches—no kitchen (per person) | 3 |
| Churches—with kitchen (per person) | 7.5 |
| Condominiums (per bedroom—includes automatic clothes washer) | 150 |
| Dance halls (per person) | 3 |
| Dining halls—kitchen and toilet waste—with dishwasher, food waste grinder or both (per meal served) | 11 |
| Dining halls—kitchen waste only (per meal served) | 3 |
| Drive-in restaurants—all paper service (per car space) | 15 |
| Drive-in restaurants—all paper service, inside seating (per seat) | 15 |
| Drive-in theaters (per car space) | 5 |
| Employees—in all buildings, per employee—total all shifts | 20 |
| Floor drains (per drain) | 50 |
| Hospitals (per bed space) | 200 |
| Hotels or motels and tourist rooming houses | 100 |
| Labor camps, central bathhouses (per employee) | 30 |
| Medical office buildings, clinics and dental offices<br>  Doctors, nurses, medical staff (per person)<br>  Office personnel (per person)<br>  Patients (per person) | <br>75<br>20<br>10 |
| Mobile home parks, homes with bathroom groups (per site) | 300 |
| Nursing and rest homes—without laundry (per bed space) | 100 |
| Outdoor sports facilities (toilet waste only—per person) | 5 |
| Parks, toilet wastes (per person—75 persons per acre) | 5 |
| Parks, with showers and toilet wastes (per person—75 persons per acre) | 10 |
| Restaurants—dishwasher or food waste grinder or both (per seat) | 3 |
| Restaurants—kitchen and toilet wastes (per seating space) | 30 |
| Restaurants—kitchen waste only—without dishwasher and food waste grinder (per seat) | 9 |
| Restaurants—toilet waste only (per seat) | 21 |
| Restaurants (24-hour)—dishwasher or food waste grinder (per seat) | 6 |
| Restaurants (24-hour)—kitchen and toilet wastes (per seating space) | 60 |

| BUILDING CLASSIFICATION | CAPACITY (gallons) |
|---|---|
| Retail stores—customers | 1.5 |
| Schools (per classroom—25 pupils per classroom) | 450 |
| Schools with meals served (per classroom—25 pupils per classroom) | 600 |
| Schools with meals served and showers provided (per classroom) | 750 |
| Self-service laundries (toilet waste only, per machine)<br>  Automatic clothes washers (apartments, service buildings, etc.—per machine) | 50<br>300 |
| Service stations (per car) | 10 |
| Showers—public (per shower taken) | 15 |
| Swimming pool bathhouses (per person) | 10 |

For SI:  1 gallon = 3.785 L.

Any tank installed in groundwater must be securely anchored. A 3-inch-thick compacted bedding must be provided for all septic installations. The bedding material can be sand, gravel, granite, lime rock, or other noncorrosive materials of such size that the material passes through a 0.5-inch screen.

## Backfill Material

Backfill material for steel and fiberglass tanks must be specified for bedding and shall be tamped into place without damaging the coating on the tank. Concrete tanks can be backfilled with soil material that will pass a 4-inch screen. This fill must be tamped into place.

## Riser Joints

Riser joints on concrete risers and manhole covers have to be tongue-and-groove or shiplap type and sealed watertight with neat cement, mortar, or bituminous compound. Steel risers have to be welded or flanged and bolted watertight. Bituminous compound must be used to

Table 8.3 **Minimum horizontal separation distances for treatment tanks:** *(Courtesy of International Private Sewage Disposal Code).*

| ELEMENT | DISTANCE (feet) |
|---|---|
| Building | 5 |
| Cistern | 25 |
| Foundation wall | 5 |
| Lake, high water mark | 25 |
| Lot line | 2 |
| Pond | 25 |
| Reservoir | 25 |
| Spring | 50 |
| Stream or watercourse | 25 |
| Swimming pool | 15 |
| Water service | 5 |
| Well | 25 |

coat steel risers both inside and out. Fiberglass risers must be watertight and attached in an approved manner.

### Dosing or Pumping Chambers

Dosing or pumping chambers must be made of welded steel, monolithic concrete, glass-fiber-reinforced polyester, or other approved materials. Manholes for these chambers must terminate no less than 4 inches above the ground surface. All chambers are required to be watertight.

Each dosing or pumping chamber has to be sized to permit automatic discharge of the total daily sewage flow, with discharge occurring not more than four times in a 24-hour period. A dosing chamber is required to have a minimum capacity of 500 gallons. A space is required between the bottom of the pump and the floor of the dosing or pumping chamber.

A dosing chamber must have a 1-day holding capacity located above the high-water alarm for one- and two-family dwellings, based on usage of 100 gallons per day per bedroom. In the case of other buildings, other sizing requirements must be met.

If distribution piping exceeds 1,000 feet in developed length, the dosing or pumping chamber must be equipped with two siphons or pumps that dose alternately and serve half of the soil absorption system.

### Treatment Tanks

Designs for other treatment tanks are approved on an individual basis. The capacity, sizing, and installation of the tank must be in compliance

Table 8.4 **Pump chamber sizes:** *(Courtesy of International Private Sewage Disposal Code).*

| NUMBER OF BEDROOMS | MINIMUM PUMPING CHAMBER SIZE (gallons) |
|---|---|
| 1 | 500 |
| 2 | 500 |
| 3 | 750 |
| 4 | 750 |
| 5 | 1,000 |

For SI: 1 gallon = 3.785 L.

with the local code. On occasions when a treatment tank is preceded by a conventional septic tank, credit shall be given for the capacity of the septic tank.

## Maintenance

Septic tanks and treatment tanks must be cleaned of their sludge when the sludge and scum consume one-third of the tank's capacity. Disposal of all septage must be done in an approved manner at an approved location.

## Chemical Restoration

Chemical restoration is a violation of code requirements, unless special approval is given by appropriate authorities.

## Holding Tanks

Holding tanks for septic use are not approved for sites where another type of private sewage system is installed. If a holding tank is allowed, a pumping and maintenance schedule must be established for each tank and submitted to a code official.

Table 8.5 Minimum liquid capacity of holding tanks:
*(Courtesy of International Private Sewage Disposal Code).*

| NUMBER OF BEDROOMS | TANK CAPACITY (gallons) |
|---|---|
| 1 | 2,000 |
| 2 | 2,000 |
| 3 | 2,000 |
| 4 | 2,500 |
| 5 | 3,000 |
| 6 | 3,500 |
| 7 | 4,000 |
| 8 | 4,500 |

For SI: 1 gallon = 3.785 L.

The capacity of holding tanks is required to be much greater than that of septic tanks. For example, a one-bedroom home will require a 2,000-gallon holding tank. Tank capacity is determined by data in local code regulations.

A holding tank may be made of welded steel, monolithic concrete, glass-fiber-reinforced polyester, or other approved materials. Holding tanks must not be installed within 20 feet of any part of a building. The servicing manhole for a holding tank must not be more than 10 feet from an all-weather access road or drive.

A high-water warning device must be installed to activate 1 foot below the inlet pipe. The device must either be audible or equipped with an approved illuminated alarm. An electrical junction box, including warning equipment junctions, must be located outside the holding tank or housed in a waterproof, explosion-proof enclosure. Electrical relays or controls must be located outside a holding tank.

A holding tank must be equipped with a manhole that is not less than 24 inches square or 24 inches in inside diameter. The manhole must not extend less than 4 inches above grade level. Surrounding earth in the manhole area must be sloped to prevent water from running to the manhole. Locking devices are required on manhole covers. Service ports in manhole covers must be at least 8 inches in diameter and must be 4 inches above finished grade. A service port must have an effective locking cover or a brass cleanout plug.

If a septic tank is converted to a holding tank, the septic tank must have its outlet sealed. Removal of the inlet or outlet baffle is prohibited.

Every holding tank is required to be fitted with a vent that has a minimum diameter of 2 inches. The vent must extend at least 12 inches above the finished grade. When the vent is terminated in open air, the pipe must be fitted with either an approved vent cap or a return bend fitting.

If a holding tank is installed in a flood plain, the tank must be adequately anchored to counter buoyant forces in the event of a regional flood. The vent termination and service manhole must be no less than 2 feet above the regional flood elevation.

# Chapter 9

# Mound Systems

Mound systems are prohibited on sites that have shallow soils. In most cases, ground cover with a depth of 2 feet is enough for a mound system. However, impermeable rock strata require a depth of 5 feet. Mound systems are not allowed in flood plains or filled areas. Additionally, mound systems cannot be installed in compacted soil or over a failing conventional system.

Perk tests for mound systems must be conducted at a depth of 20 to 24 inches from existing grade. If a slowly permeable horizon exists at less than 20 to 24 inches, the perk test must be conducted within that horizon. An acceptable perk rate for a mound system is set at a rate of more than 60 minutes per inch and less than, or equal to, 120 minutes per inch.

When shallow permeable soils are encountered over bedrock, a perk test must be conducted at a depth of 12 to 18 inches below existing grade. If a slowly permeable horizon is found within 12 to 18 inches, the perk test must be conducted within that horizon. A mound system in this type of situation is allowed a perk rate between 3 minutes per inch and 60 minutes per inch.

High ground water in permeable soils requires a perk test at 20 to 24 inches. When a slowly permeable horizon is found at less than 20 to 24 inches, a perk test is required in that horizon. Mound systems for this type of site must have a perk rate of 0 minutes per inch to 60 minutes per inch.

Mound systems will not be allowed unless there are at least 2 feet of unsaturated natural soil over creviced or porous bedrock. There must

Table 9.1 Minimum soil depths for mound system installation:
*(Courtesy of International Private Sewage Disposal Code).*

| RESTRICTING FACTOR | MINIMUM SOIL DEPTH TO RESTRICTION (inches) |
|---|---|
| High ground water | 24 |
| Impermeable rock strata | 60 |
| Pervious rock | 24 |
| Rock fragments (50-percent volume) | 24 |

For SI: 1 inch = 25.4 mm.

be a minimum of 2 feet of unsaturated natural soil over high ground water as indicated by soil mottling or direct observation.

If you encounter a slope that is greater than 6 percent with a perk rate of 30 to 120 minutes per inch, you cannot install a mound system. The maximum slope allowed is 12 percent where there is a complex slope.

Mounds must be located so that the longest dimension of the mound and the distribution lines are perpendicular to the slope. The mound must be placed upslope and not at the base of a slope. All mounds built on complex slopes must be situated so that effluent is not concentrated in one direction. By complex slope, I am referring to a slope that has two directions. All surface water runoff must be diverted around the mound.

A minimum of 60 inches of soil must be present over uncreviced, impermeable bedrock. If the soil contains 50 percent coarse fragments by volume in the upper 24 inches, a mound must not be installed unless there are at least 24 inches of permeable, unsaturated soil with less than 50 percent coarse fragments located beneath this layer.

## System Design

Design of mound systems can be complicated. As I have said previously, most system designs are done by engineers and experts in the design field. I am not going to attempt to give you complete design criteria here. If you want full design requirements, check in your local codebook; all the data you need should be there. I will, however, hit some of the major design issues.

One set of rules applies to one- and two-family dwellings and buildings with estimated wastewater flows of less than 600 gallons per day. Mound dimensions for these systems are established with the use of tables and data provided in the code.

## Mound Systems

**Table 9.2** Design criteria for a mount for a one-bedroom home on a 0- to 6-percent slope with loading rates of 150 gallons per day for slowly permeable soil: *(Courtesy of International Private Sewage Disposal Code).*

|   | DESIGN PARAMETER | SLOPE (percent) | | | |
|---|---|---|---|---|---|
|   |   | 0 | 2 | 4 | 6 |
| A | Trench width, feet | 3 | 3 | 3 | 3 |
| B | Trench length, feet | 42 | 42 | 42 | 42 |
|   | Number of trenches | 1 | 1 | 1 | 1 |
| D | Mound height, inches | 12 | 12 | 12 | 12 |
| F | Mound height, inches | 9 | 9 | 9 | 9 |
| G | Mound height, inches | 12 | 12 | 12 | 12 |
| H | Mound height, inches | 18 | 18 | 18 | 18 |
| I | Mound width[a], feet | 15 | 15 | 15 | 15 |
| J | Mound width, feet | 11[a] | 8 | 8 | 8 |
| K | Mound length, feet | 10 | 10 | 10 | 10 |
| L | Mound length, feet | 62 | 62 | 62 | 62 |
| P | Distribution pipe length, feet | 20 | 20 | 20 | 20 |
|   | Distribution pipe diameter, inches | 1 | 1 | 1 | 1 |
|   | Number of holes per distribution pipes[b] | 9 | 9 | 9 | 9 |
|   | Hole spacing[b], inches | 30 | 30 | 30 | 30 |
|   | Hole diameter[b], inches | 0.25 | 0.25 | 0.25 | 0.25 |
| W | Mound width, feet | 25 | 26 | 26 | 26 |

For SI: 1 inch = 25.4 mm, 1 foot = 304.8 mm, 1 gallon = 3.785 L.

a. Additional width to obtain required basal area.
b. Last hole is located at the end of the distribution pipe, which is 15 inches from the other hole.

Table 9.3 Design criteria for a mount for a two-bedroom home on a 0- to 6-percent slope with loading rates of 300 gallons per day for slowly permeable soil: *(Courtesy of International Private Sewage Disposal Code).*

| | DESIGN PARAMETER | SLOPE (percent) | | | |
|---|---|---|---|---|---|
| | | 0 | 2 | 4 | 6 |
| A | Trench width, feet | 3 | 3 | 3 | 3 |
| B | Trench length, feet | 42 | 42 | 42 | 42 |
| | Number of trenches | 2 | 2 | 2 | 2 |
| C | Trench spacing, feet | 15 | 15 | 15 | 15 |
| D | Mound height, inches | 12 | 12 | 12 | 12 |
| E | Mound height, inches | 12 | 17 | 25 | 25 |
| F | Mound height, inches | 9 | 9 | 9 | 9 |
| G | Mound height, inches | 12 | 12 | 12 | 12 |
| H | Mound height, inches | 18 | 18 | 18 | 18 |
| I | Mound width, feet[a] | 12 | 20 | 20 | 20 |
| J | Mound width, feet | 12 | 8 | 8 | 8 |
| K | Mound length, feet | 10 | 10 | 10 | 10 |
| L | Mound length, feet | 62 | 62 | 62 | 62 |
| P | Distribution pipe length, feet | 20 | 20 | 20 | 20 |
| | Distribution pipe diameter, inches | 1 | 1 | 1 | 1 |
| | Number of holes per distribution pipe[b] | 9 | 9 | 9 | 9 |
| | Hole spacing[b], inches | 30 | 30 | 30 | 30 |
| | Hole diameter, inches | 0.25 | 0.25 | 0.25 | 0.25 |
| R | Manifold length, feet | 15 | 15 | 15 | 15 |
| | Manifold diameter[c], inches | 2 | 2 | 2 | 2 |
| W | Mound width, feet | 42 | 46 | 46 | 46 |

For SI: 1 inch = 25.4 mm, 1 foot = 304.8 mm, 1 gallon = 3.785 L.

a. Additional width to obtain required basal area.
b. Last hole is located at the end of the distribution pipe, which is 15 inches from the other hole.
c. Diameter dependent on the size of pipe from pump and inlet position.

Mound Systems 59

Table 9.4 Design criteria for a mound for a three-bedroom home on a 0- to 6-percent slope with loading rates of 450 gallons per day for slowly permeable soil: *(Courtesy of International Private Sewage Disposal Code).*

|   | DESIGN PARAMETER | SLOPE (percent) 0 | 2 | 4 |
|---|---|---|---|---|
| A | Trench width, feet | 3 | 3 | 3 |
| B | Trench length, feet<br>Number of trenches | 63<br>2 | 63<br>2 | 63<br>2 |
| C | Trench spacing, feet | 15 | 15 | 15 |
| D | Mound height, inches | 12 | 12 | 12 |
| E | Mound height, inches | 12 | 17 | 20 |
| F | Mound height, inches | 9 | 9 | 9 |
| G | Mound height, inches | 12 | 12 | 12 |
| H | Mound height, inches | 18 | 18 | 18 |
| I | Mound width[a], feet | 12 | 20 | 20 |
| J | Mound width[a], feet | 12[a] | 8 | 8 |
| K | Mound length, feet | 10 | 10 | 10 |
| L | Mound length, feet | 62 | 62 | 62 |
| P | Distribution pipe length, feet<br>Distribution pipe diameter, inches<br>Number of holes per distribution pipe[b]<br>Hole spacing[b], inches<br>Hole diameter, inches | 31<br>1¼<br>13<br>30<br>0.25 | 31<br>1¼<br>13<br>30<br>0.25 | 31<br>1¼<br>13<br>30<br>0.25 |
| R | Manifold length, feet<br>Manifold diameter[c], inches | 15<br>2 | 15<br>2 | 15<br>2 |
| W | Mound width, feet | 42 | 46 | 46 |

For SI: 1 inch = 25.4 mm, 1 foot = 304.8 mm, 1 gallon = 3.785 L.
a. Additional width to obtain required basal area.
b. First hole is located 12 inches from the manifold.
c. Diameter dependent on the size of pipe from pump and inlet position.

**Table 9.5  Design criteria for a mount for a four-bedroom home on a 0- to 6-percent slope with loading rates of 600 gallons per day for slowly permeable soil:** *(Courtesy of International Private Sewage Disposal Code).*

| | DESIGN PARAMETER | SLOPE (percent) | | | |
|---|---|---|---|---|---|
| | | 0 | 2 | 4 | 6 |
| A | Trench width, feet | 3 | 3 | 3 | 3 |
| B | Trench length, feet<br>Number of trenches | 56<br>3 | 56<br>3 | 56<br>3 | 56<br>3 |
| C | Trench spacing, feet | 15 | 15 | 15 | 15 |
| D | Mound height, inches | 12 | 12 | 12 | 12 |
| E | Mound height, inches | 12 | 20 | 28 | 36 |
| F | Mound height, inches | 9 | 9 | 9 | 9 |
| G | Mound height, inches | 12 | 12 | 12 | 12 |
| H | Mound height, inches | 24 | 24 | 24 | 24 |
| I | Mound width[a], feet | 12 | 20 | 20 | 20 |
| J | Mound width, feet | 12[a] | 8 | 8 | 8 |
| K | Mound length, feet | 12 | 12 | 12 | 14 |
| L | Mound length, feet | 80 | 80 | 80 | 84 |
| P | Distribution pipe length, feet<br>Distribution pipe diameter, inches<br>Number of holes per distribution pipe[b]<br>Hole spacing[b], inches<br>Hole diameter, inches | 27.5<br>1¼<br>12<br>30<br>0.25 | 27.5<br>1¼<br>12<br>30<br>0.25 | 27.5<br>1¼<br>12<br>30<br>0.25 | 27.5<br>1¼<br>12<br>30<br>0.25 |
| R | Manifold length, feet<br>Manifold diameter[c], inches | 30<br>2 | 30<br>2 | 30<br>2 | 30<br>2 |
| W | Mound width, feet | 57 | 61 | 61 | 61 |

For SI: 1 inch = 25.4 mm, 1 foot = 304.8 mm, 1 gallon = 3.785 L.
a. Additional width to obtain required basal area.
b. Last hole is located at the end of the distribution pipe, which is 15 inches from the previous hole.
c. Diameter dependent on the size of pipe from pump and inlet position.

Mound Systems 61

Table 9.6  Design criteria for a mount for a one-bedroom home on a 0- to 12-percent slope with loading rates of 150 gallons per day for shallow permeable soil over creviced bedrock: (Courtesy of International Private Sewage Disposal Code).

| | DESIGN PARAMETER | PERCOLATION RATE (minutes per inch) SLOPE (percent) | | | | | | | | | |
|---|---|---|---|---|---|---|---|---|---|---|---|
| | | 3 to 60 | | | | | | 3 to less than 30 | | | |
| | | 0 | 2 | 4 | 6 | 8 | 10[a] | 12[a] | | | |
| A | Bed width[b], feet | 10 | 10 | 10 | 10 | 10 | 10 | 10 | | | |
| B | Bed length, feet | 13 | 13 | 13 | 13 | 13 | 13 | 13 | | | |
| D | Mound height, inches | 24 | 24 | 24 | 24 | 24 | 24 | 24 | | | |
| E | Mound height, inches | 24 | 26 | 29 | 31 | 34 | 36 | 38 | | | |
| F | Mound height, inches | 9 | 9 | 9 | 9 | 9 | 9 | 9 | | | |
| G | Mound height, inches | 12 | 12 | 12 | 12 | 12 | 12 | 12 | | | |
| H | Mound height, inches | 18 | 18 | 18 | 18 | 18 | 18 | 18 | | | |
| I | Mound width, feet | 12 | 13 | 14 | 17 | 18 | 21 | 26 | | | |
| J | Mound width, feet | 12 | 11 | 10 | 10 | 9 | 9 | 9 | | | |
| K | Mound length, feet | 12 | 12 | 12 | 13 | 13 | 13 | 15 | | | |
| L | Mound length, feet | 37 | 37 | 37 | 39 | 39 | 39 | 43 | | | |
| P | Distribution pipe length[c], feet | 12.5 | 12.5 | 12.5 | 12.5 | 12.5 | 12.5 | 12.5 | | | |
| | Distribution pipe diameter, inches | 1 | 1 | 1 | 1 | 1 | 1 | 1 | | | |
| | Number of distribution pipes | 6 | 6 | 6 | 6 | 6 | 6 | 6 | | | |
| R | Manifold length, feet | 6 | 6 | 6 | 6 | 6 | 6 | 6 | | | |
| | Manifold diameter, inches | 2 | 2 | 2 | 2 | 2 | 2 | 2 | | | |
| S | Distribution pipe spacing, feet | 3 | 3 | 3 | 3 | 3 | 3 | 3 | | | |
| | Number of holes per distribution pipe[d] | 6 | 6 | 6 | 6 | 6 | 6 | 6 | | | |
| | Hole spacing[d], inches | 30 | 30 | 30 | 30 | 30 | 30 | 30 | | | |
| | Hole diameter, inches | 0.25 | 0.25 | 0.25 | 0.25 | 0.25 | 0.25 | 0.25 | | | |
| W | Mound width, feet | 34 | 34 | 34 | 37 | 37 | 41 | 45 | | | |

For SI:  1 inch = 25.4 mm, 1 foot = 304.8 mm, 1 gallon = 3.785 L, 1 minute per inch = 2.4 s/mm.
a. On sites with a 10- to 12-percent slope, the fill depth (D) shall be reduced to a minimum of 1.5 feet or the bed width shall be reduced to decrease E.
b. Bed widths shall not be limited.
c. Use a manifold with distribution pipes only on one side.
d. Last hole is located at the end of the distribution pipe, which is 15 inches from the previous hole.

**Table 9.7  Design criteria for a mount for a two-bedroom home on a 0- to 12-percent slope with loading rates of 300 gallons per day for shallow permeable soil over creviced bedrock:** *(Courtesy of International Private Sewage Disposal Code).*

| | DESIGN PARAMETER | PERCOLATION RATE (minutes per inch) SLOPE (percent) | | | | | | | | | |
|---|---|---|---|---|---|---|---|---|---|---|---|
| | | 3 to 60 | | | | | | | 3 to less than 30 | | |
| | | 0 | 2 | 4 | 6 | 8 | 10[a] | 12[a] | 8 | 10[a] | 12[a] |
| A | Bed width[b], feet | 10 | 10 | 10 | 10 | 10 | 10 | 10 | 10 | 10 | 10 |
| B | Bed length, feet | 25 | 25 | 25 | 25 | 25 | 25 | 25 | 25 | 25 | 25 |
| D | Mound height, inches | 24 | 24 | 24 | 24 | 24 | 24 | 24 | 24 | 24 | 24 |
| E | Mound height, inches | 24 | 26 | 29 | 31 | 34 | 36 | 38 | 34 | 36 | 38 |
| F | Mound height, inches | 9 | 9 | 9 | 9 | 9 | 9 | 9 | 9 | 9 | 9 |
| G | Mound height, inches | 12 | 12 | 12 | 12 | 12 | 12 | 12 | 12 | 12 | 12 |
| H | Mound height, inches | 18 | 18 | 18 | 18 | 18 | 18 | 18 | 18 | 18 | 18 |
| I | Mound width, feet | 12 | 13 | 14 | 17 | 18 | 21 | 26 | 18 | 21 | 26 |
| J | Mound width, feet | 12 | 11 | 10 | 10 | 9 | 9 | 9 | 9 | 9 | 9 |
| K | Mound length, feet | 12 | 12 | 12 | 13 | 13 | 13 | 15 | 13 | 13 | 15 |
| L | Mound length, feet | 49 | 49 | 49 | 51 | 51 | 51 | 55 | 51 | 51 | 55 |
| P | Distribution pipe length[c], feet<br>Distribution pipe diameter, inches<br>Number of distribution pipes | 12<br>1<br>6 | 12<br>1<br>6 | 12<br>1<br>6 | 12<br>1<br>6 | 12<br>1<br>6 | 12<br>1<br>6 | 12<br>1<br>6 | 12<br>1<br>6 | 12<br>1<br>6 | 12<br>1<br>6 |
| R | Manifold length, feet<br>Manifold diameter, inches | 6<br>2 | 6<br>2 | 6<br>2 | 6<br>2 | 6<br>2 | 6<br>2 | 6<br>2 | 6<br>2 | 6<br>2 | 6<br>2 |
| S | Distribution pipe spacing, feet<br>Number of holes per distribution pipe[d]<br>Hole spacing[d], inches<br>Hole diameter, inches | 3<br>5<br>30<br>0.25 | 3<br>5<br>30<br>0.25 | 3<br>5<br>30<br>0.25 | 3<br>5<br>30<br>0.25 | 3<br>5<br>30<br>0.25 | 3<br>5<br>30<br>0.25 | 3<br>5<br>30<br>0.25 | 3<br>5<br>30<br>0.25 | 3<br>5<br>30<br>0.25 | 3<br>5<br>30<br>0.25 |
| W | Mound width, feet | 34 | 34 | 34 | 37 | 37 | 41 | 45 | 37 | 41 | 45 |

For SI: 1 inch = 25.4 mm, 1 foot = 304.8 mm, 1 gallon = 3.785 L, 1 minute per inch = 2.4 s/mm.
a. On sites with a 10- to 12-percent slope, the fill depth (D) shall be reduced to a minimum of 1.5 feet or the bed width shall be reduced to decrease E.
b. Bed widths shall not be limited.
c. This design is based on a manifold with distribution pipes on both sides. An alternative design basis is 24-foot distribution pipes, with manifold at the end.
d. Last hole is located 9 inches from the end of the distribution pipe.

**Table 9.8  Design criteria for a mound for a three-bedroom home on a 0- to 12-percent slope with loading rates of 450 gallons per day for shallow permeable soil over creviced bedrock:** *(Courtesy of International Private Sewage Disposal Code).*

| | DESIGN PARAMETER | PERCOLATION RATE (minutes per inch) SLOPE (percent) | | | | | | |
|---|---|---|---|---|---|---|---|---|
| | | 3 to 60 | | | | | 3 to less than 30 | |
| | | 0 | 2 | 4 | 6 | 8 | 10[a] | 12[a] |
| A | Bed width[b], feet | 10 | 10 | 10 | 10 | 10 | 10 | 10 |
| B | Bed length, feet | 38 | 38 | 38 | 38 | 38 | 38 | 38 |
| D | Mound height, inches | 24 | 24 | 24 | 24 | 24 | 24 | 24 |
| E | Mound height, inches | 24 | 26 | 29 | 31 | 34 | 36 | 38 |
| F | Mound height, inches | 9 | 9 | 9 | 9 | 9 | 9 | 9 |
| G | Mound height, inches | 12 | 12 | 12 | 12 | 12 | 12 | 12 |
| H | Mound height, inches | 18 | 18 | 18 | 18 | 18 | 18 | 18 |
| I | Mound width, feet | 12 | 13 | 14 | 17 | 18 | 21 | 26 |
| J | Mound width, feet | 12 | 11 | 10 | 10 | 9 | 9 | 9 |
| K | Mound length, feet | 12 | 12 | 12 | 13 | 13 | 13 | 15 |
| L | Mound length, feet | 62 | 62 | 62 | 64 | 64 | 64 | 68 |
| P | Distribution pipe length[c], feet / Distribution pipe diameter, inches / Number of distribution pipes | 18.5 / 1 / 6 | 18.5 / 1 / 6 | 18.5 / 1 / 6 | 18.5 / 1 / 6 | 18.5 / 1 / 6 | 18.5 / 1 / 6 | 18.5 / 1 / 6 |
| R | Manifold length, feet / Manifold diameter, inches | 6 / 2 | 6 / 2 | 6 / 2 | 6 / 2 | 6 / 2 | 6 / 2 | 6 / 2 |
| S | Distribution pipe spacing, feet / Number of holes per distribution pipe[d] / Hole spacing[d], inches / Hole diameter, inches | 3 / 8 / 30 / 0.25 | 3 / 8 / 30 / 0.25 | 3 / 8 / 30 / 0.25 | 3 / 8 / 30 / 0.25 | 3 / 8 / 30 / 0.25 | 3 / 8 / 30 / 0.25 | 3 / 8 / 30 / 0.25 |
| W | Mound width, feet | 34 | 34 | 34 | 37 | 37 | 41 | 45 |

For SI:  1 inch = 25.4 mm, 1 foot = 304.8 mm, 1 gallon = 3.785 L, 1 minute per inch = 2.4 s/mm.
a. On sites with a 10- to 12-percent slope, the fill depth ($D$) shall be reduced to a minimum of 1.5 feet or the bed width shall be reduced to decrease $E$.
b. Bed widths shall not be limited.
c. Use a manifold with distribution pipes only on one side.
d. Last hole is located at the end of the distribution pipe, which is 27 inches from the previous hole.

**Table 9.9  Design criteria for a mount for a four-bedroom home on a 0- to 12-percent slope with loading rates of 600 gallons per day for shallow permeable soil over creviced bedrock:** *(Courtesy of International Private Sewage Disposal Code).*

| | DESIGN PARAMETER | PERCOLATION RATE (minutes per inch) SLOPE (percent) | | | | | | | | |
|---|---|---|---|---|---|---|---|---|---|---|
| | | 3 to 60 | | | | | 3 to less than 30 | | | |
| | | 0 | 2 | 4 | 6 | 8 | 10[a] | 12[a] | | |
| A | Bed width[b], feet | 10 | 10 | 10 | 10 | 10 | 10 | 10 | | |
| B | Bed length, feet | 50 | 50 | 50 | 50 | 50 | 50 | 50 | | |
| D | Mound height, inches | 24 | 24 | 24 | 24 | 24 | 24 | 24 | | |
| E | Mound height, inches | 24 | 26 | 29 | 31 | 34 | 36 | 38 | | |
| F | Mound height, inches | 9 | 9 | 9 | 9 | 9 | 9 | 9 | | |
| G | Mound height, inches | 12 | 12 | 12 | 12 | 12 | 12 | 12 | | |
| H | Mound height, inches | 18 | 18 | 18 | 18 | 18 | 18 | 18 | | |
| I | Mound width, feet | 12 | 13 | 14 | 17 | 18 | 21 | 26 | | |
| J | Mound width, feet | 12 | 11 | 10 | 10 | 9 | 9 | 9 | | |
| K | Mound length, feet | 12 | 12 | 12 | 13 | 13 | 13 | 15 | | |
| L | Mound length, feet | 74 | 74 | 74 | 76 | 76 | 76 | 78 | | |
| P | Distribution pipe length[c], feet | 24.5 | 24.5 | 24.5 | 24.5 | 24.5 | 24.5 | 24.5 | | |
| | Distribution pipe diameter, inches | 1 | 1 | 1 | 1 | 1 | 1 | 1 | | |
| | Number of distribution pipes | 6 | 6 | 6 | 6 | 6 | 6 | 6 | | |
| R | Manifold length, feet | 6 | 6 | 6 | 6 | 6 | 6 | 6 | | |
| | Manifold diameter, inches | 2 | 2 | 2 | 2 | 2 | 2 | 2 | | |
| S | Distribution pipe spacing, feet | 3 | 3 | 3 | 3 | 3 | 3 | 3 | | |
| | Number of holes per distribution pipe[d] | 10 | 10 | 10 | 10 | 10 | 10 | 10 | | |
| | Hole spacing[c], inches | 30 | 30 | 30 | 30 | 30 | 30 | 30 | | |
| | Hole diameter, inches | 0.25 | 0.25 | 0.25 | 0.25 | 0.25 | 0.25 | 0.25 | | |
| W | Mound width, feet | 34 | 34 | 34 | 37 | 37 | 41 | 45 | | |

For SI: 1 inch 25.4 mm, 1 foot = 304.8 mm, 1 gallon = 3.785 L, 1 minute per inch = 2.4 s/mm.
a. On sites with a 10- to 12-percent slope, the fill depth (D) shall be reduced to a minimum of 1.5 feet or the bed width shall be reduced to decrease E.
b. Bed widths shall not be limited.
c. Use a manifold with distribution pipes only on one side.
d. Last hole is located 9 inches from the end of the distribution pipe.

Table 9.10 Design criteria for a mount for a one-bedroom home on a 0- to 12-percent slope with loading rates of 150 gallons per day for permeable soil with a high water table: *(Courtesy of International Private Sewage Disposal Code).*

| | DESIGN PARAMETER | PERCOLATION RATE (minutes per inch) SLOPE (percent) | | | | | | |
|---|---|---|---|---|---|---|---|---|
| | | 0 to 60 | | | | | 0 to less than 30 | |
| | | 0 | 2 | 4 | 6 | 8 | 10 | 12 |
| A | Bed width, feet | 4 | 4 | 4 | 4 | 4 | 4 | 4 |
| B | Bed length, feet | 32 | 32 | 32 | 32 | 32 | 32 | 32 |
| D | Mound height, inches | 12 | 12 | 12 | 12 | 12 | 12 | 12 |
| E | Mound height, inches | 12 | 13 | 14 | 14 | 16 | 17 | 18 |
| F | Mound height, inches | 9 | 9 | 9 | 9 | 9 | 9 | 9 |
| G | Mound height, inches | 12 | 12 | 12 | 12 | 12 | 12 | 12 |
| H | Mound height, inches | 18 | 18 | 18 | 18 | 18 | 18 | 18 |
| I | Mound width, feet | 9 | 10 | 11 | 12 | 13 | 14 | 15 |
| J | Mound width, feet | 9 | 9 | 8 | 8 | 7 | 7 | 6 |
| K | Mound length, feet | 10 | 10 | 10 | 10 | 10 | 11 | 11 |
| L | Mound length, feet | 52 | 52 | 52 | 52 | 52 | 53 | 53 |
| P | Distribution pipe length | 15.5 | 15.5 | 15.5 | 15.5 | 15.5 | 15.5 | 15.5 |
|   | Distribution pipe diameter, inches | 1 | 1 | 1 | 1 | 1 | 1 | 1 |
|   | Number of distribution pipes | 2 | 2 | 2 | 2 | 2 | 2 | 2 |
|   | Number of holes per distribution pipe[a] | 7 | 7 | 7 | 7 | 7 | 7 | 7 |
|   | Hole spacing[a], inches | 30 | 30 | 30 | 30 | 30 | 30 | 30 |
|   | Hole diameter, inches | 0.25 | 0.25 | 0.25 | 0.25 | 0.25 | 0.25 | 0.25 |
| W | Mound width, feet | 22 | 23 | 23 | 24 | 24 | 25 | 25 |

For SI: 1 inch = 25.4 mm, 1 foot = 304.8 mm, 1 gallon = 3.785 L, 1 minute per inch = 2.4 s/mm.
a. Last hole is located at the end of the distribution pipe, which is 21 inches from the previous hole.

**Table 9.11** Design criteria for a mount for a two-bedroom home on a 0- to 12-percent slope with loading rates of 300 gallons per day for permeable soil with a high water table: *(Courtesy of International Private Sewage Disposal Code).*

| | DESIGN PARAMETER | PERCOLATION RATE (minutes per inch) SLOPE (percent) | | | | | | | | | | |
|---|---|---|---|---|---|---|---|---|---|---|---|---|
| | | 0 to 60 | | | | | | 0 to less than 30 | | | | |
| | | 0 | 2 | 4 | 6 | 8 | 10 | 12 | | | | |
| | | | | | | | | | 0 | 6 | 8 | 10 | 12 |
| A | Bed width, feet | 6 | 6 | 6 | 6 | 6 | 6 | 6 | | | | |
| B | Bed length, feet | 42 | 42 | 42 | 42 | 42 | 42 | 42 | | | | |
| D | Mound height, inches | 12 | 12 | 12 | 12 | 12 | 12 | 12 | | | | |
| E | Mound height, inches | 12 | 13 | 14 | 17 | 18 | 19 | 22 | | | | |
| F | Mound height, inches | 9 | 9 | 9 | 9 | 9 | 9 | 9 | | | | |
| G | Mound height, inches | 12 | 12 | 12 | 12 | 12 | 12 | 12 | | | | |
| H | Mound height, inches | 18 | 18 | 18 | 18 | 18 | 18 | 18 | | | | |
| I | Mound width, feet | 9 | 10 | 11 | 12 | 13 | 15 | 16 | | | | |
| J | Mound width, feet | 9 | 9 | 8 | 8 | 7 | 7 | 6 | | | | |
| K | Mound length, feet | 10 | 10 | 10 | 10 | 10 | 11 | 11 | | | | |
| L | Mound length, feet | 62 | 62 | 62 | 62 | 62 | 64 | 64 | | | | |
| P | Distribution pipe length[a], feet | 20 | 20 | 20 | 20 | 20 | 20 | 20 | | | | |
| | Distribution pipe diameter, inches | 1 | 1 | 1 | 1 | 1 | 1 | 1 | | | | |
| | Number of distribution pipes | 4 | 4 | 4 | 4 | 4 | 4 | 4 | | | | |
| R | Manifold length, feet | 3 | 3 | 3 | 3 | 3 | 3 | 3 | | | | |
| | Manifold diameter, inches | 2 | 2 | 2 | 2 | 2 | 2 | 2 | | | | |
| S | Distribution pipe spacing, feet | 3 | 3 | 3 | 3 | 3 | 3 | 3 | | | | |
| | Number of holes per distribution pipe[b] | 9 | 9 | 9 | 9 | 9 | 9 | 9 | | | | |
| | Hole spacing[b], inches | 30 | 30 | 30 | 30 | 30 | 30 | 30 | | | | |
| | Hole diameter, inches | 0.25 | 0.25 | 0.25 | 0.25 | 0.25 | 0.25 | 0.25 | | | | |
| W | Mound width, feet | 24 | 25 | 25 | 26 | 26 | 28 | 29 | | | | |

For SI: 1 inch = 25.4 mm, 1 foot = 304.8 mm, 1 gallon = 3.785 L, 1 minute per inch = 2.4 s/mm.

a. Use a manifold with distribution pipes only on one side.
b. Last hole is located at the end of the distribution pipe, which is 15 inches from the previous hole.

## Mound Systems

**Table 9.12  Design criteria for a mount for a three-bedroom home on a 0- to 12-percent slope with loading rates of 450 gallons per day for permeable soil with a high water table:** *(Courtesy of International Private Sewage Disposal Code).*

| | DESIGN PARAMETER | PERCOLATION RATE (minutes per inch) SLOPE (percent) | | | | | | | | |
|---|---|---|---|---|---|---|---|---|---|---|
| | | 0 to 60 | | | | | | 0 to less than 30 | | |
| | | 0 | 2 | 4 | 6 | 8 | 10 | 12 |
| A | Bed width, feet | 8 | 8 | 8 | 8 | 8 | 8 | 8 |
| B | Bed length, feet | 47 | 47 | 47 | 47 | 47 | 47 | 47 |
| D | Mound height, inches | 12 | 12 | 12 | 12 | 12 | 12 | 12 |
| E | Mound height, inches | 12 | 12 | 16 | 18 | 19 | 22 | 24 |
| F | Mound height, inches | 9 | 9 | 9 | 9 | 9 | 9 | 9 |
| G | Mound height, inches | 12 | 12 | 12 | 12 | 12 | 12 | 12 |
| H | Mound height, inches | 18 | 18 | 18 | 18 | 18 | 18 | 18 |
| I | Mound width, feet | 9 | 11 | 12 | 13 | 15 | 17 | 18 |
| J | Mound width, feet | 9 | 9 | 8 | 8 | 7 | 7 | 6 |
| K | Mound length, feet | 10 | 10 | 10 | 10 | 10 | 11 | 12 |
| L | Mound length, feet | 67 | 67 | 67 | 67 | 69 | 69 | 71 |
| P | Distribution pipe length, feet<br>Distribution pipe diameter, inches<br>Number of distribution pipes | 23<br>1<br>6 | 23<br>1<br>6 | 23<br>1<br>6 | 23<br>1<br>6 | 23<br>1<br>6 | 23<br>1<br>6 | 23<br>1<br>6 |
| R | Manifold length, feet<br>Manifold diameter, inches | 64<br>2 | 64<br>2 | 64<br>2 | 64<br>2 | 64<br>2 | 64<br>2 | 64<br>2 |
| S | Distribution pipe spacing, feet<br>Number of holes per distribution pipe[a]<br>Hole spacing[a], inches<br>Hole diameter, inches | 32<br>10<br>30<br>0.25 | 32<br>10<br>30<br>0.25 | 32<br>10<br>30<br>0.25 | 32<br>10<br>30<br>0.25 | 32<br>10<br>30<br>0.25 | 32<br>10<br>30<br>0.25 | 32<br>10<br>30<br>0.25 |
| W | Mound width, feet | 26 | 28 | 28 | 29 | 30 | 32 | 32 |

For SI: 1 inch = 25.4 mm, 1 foot = 304.8 mm, 1 gallon = 3.785 L, 1 minute per inch = 2.4 s/mm.
a. Last hole is located at the end of the distribution pipe, which is 21 inches from the previous hole.

**Table 9.12b  Design criteria for a mount for a four-bedroom home on a 0- to 12-percent slope with loading rates of 600 gallons per day for permeable soil with a high water table:** *(Courtesy of International Private Sewage Disposal Code).*

| | DESIGN PARAMETER | PERCOLATION RATE (minutes per inch) SLOPE (percent) | | | | | | | | |
|---|---|---|---|---|---|---|---|---|---|---|
| | | 0 to 60 | | | | | 0 to less than 30 | | | |
| | | 0 | 2 | 4 | 6 | 8 | 10 | 12 |
| A | Bed width, feet | 10 | 10 | 10 | 10 | 10 | 10 | 10 |
| B | Bed length, feet | 50 | 50 | 50 | 50 | 50 | 50 | 50 |
| D | Mound height, inches | 12 | 12 | 12 | 12 | 12 | 12 | 12 |
| E | Mound height, inches | 12 | 14 | 17 | 19 | 22 | 24 | 26 |
| F | Mound height, inches | 9 | 9 | 9 | 9 | 9 | 9 | 9 |
| G | Mound height, inches | 12 | 12 | 12 | 12 | 12 | 12 | 12 |
| H | Mound height, inches | 18 | 18 | 18 | 18 | 18 | 18 | 18 |
| I | Mound width, feet | 9 | 11 | 13 | 14 | 17 | 18 | 19 |
| J | Mound width, feet | 9 | 9 | 8 | 8 | 7 | 7 | 6 |
| K | Mound length, feet | 10 | 10 | 10 | 10 | 11 | 11 | 12 |
| L | Mound length, feet | 70 | 70 | 70 | 70 | 72 | 72 | 74 |
| P | Distribution pipe length, feet | 24.5 | 24.5 | 24.5 | 24.5 | 24.5 | 24.5 | 24.5 |
| | Distribution pipe diameter, inches | 1 | 1 | 1 | 1 | 1 | 1 | 1 |
| | Number of distribution pipes | 6 | 6 | 6 | 6 | 6 | 6 | 6 |
| R | Manifold length, feet | 6 | 6 | 6 | 6 | 6 | 6 | 6 |
| | Manifold diameter, inches | 2 | 2 | 2 | 2 | 2 | 2 | 2 |
| S | Distribution pipe spacing, feet | 3 | 3 | 3 | 3 | 3 | 3 | 3 |
| | Number of holes per distribution pipe[a] | 10 | 10 | 10 | 10 | 10 | 10 | 10 |
| | Hole spacing[a], inches | 30 | 30 | 30 | 30 | 30 | 30 | 30 |
| | Hole diameter, inches | 0.25 | 0.25 | 0.25 | 0.25 | 0.25 | 0.25 | 0.25 |
| W | Mound width, feet | 28 | 29 | 31 | 32 | 34 | 35 | 36 |

For SI: 1 inch = 25.4 mm, 1 foot = 304.8 mm, 1 gallon = 3.785 L, 1 minute per inch = 2.4 s/mm.
a. Last hole is 9 inches from the end of the distribution pipe.

For buildings with wastewater flow in excess of 600 gallons a day, different sizing requirements come into play.

Daily wastewater flow should be estimated at 150 gallons per day per bedroom for one- and two-family dwellings. Tables and data in the codebook are required to rate the wastewater flow of other types of buildings.

## Trenches

Mound systems must be equipped with trenches that distribute effluent for slowly permeable soils with or without high groundwater. The length of a trench is selected by determining the longest dimension perpendicular to any slope on the site. Width requirements for trenches are subject to specific site conditions. The same is true of trench spacing. Trenches are usually required to have a width of 2 to 4 feet. No trench may be longer than 100 feet. When multiple trenches are required, they must share a uniform length. There cannot be more than three trenches in one mound.

## Beds

A long, narrow bed design is required for permeable soils with high water tables. The bed must be square or rectangular for shallow permeable soils over bedrock. Bed length is established once you know the longest dimension available and perpendicular to any slope on the site.

Mound height consists of the fill depth, the bed or trench depth, the cap, and the topsoil depth. Fill depth must not be less than 1 foot for slowly permeable soils and permeable soils with high water tables, and not less than 2 feet of fill are required for shallow permeable soils over bedrock. Additional fill will be needed on the downslope end of a bed or trench if the site is not level. This is to make the bottom of the bed or trench level.

Bed and trench depth must be at least 9 inches. A minimum of 6 inches of aggregate are needed under the distribution pipes, and not less than 2 inches of aggregate should be placed over the top of the distribution pipes.

The cap and topsoil depth at the center of a mound must not be less than 18 inches, which includes 1 foot of subsoil and 6 inches of topsoil. The outer edges of a mound and the minimum cap and topsoil depth must not be less than 1 foot, which includes 6 inches of subsoil and 6 inches of topsoil. Caps must be made of topsoil or finer textured soil.

Table 9.13 **Downslope and upslope width corrections for mounds on sloping sites:** *(Courtesy of International Private Sewage Disposal Code).*

| SLOPE (percent) | DOWNSLOPE CORRECTION FACTOR ($S_D$) | UPSLOPE CORRECTION FACTOR ($S_U$) |
|---|---|---|
| 0 | 1 | 1 |
| 1 | 1.03 | 0.97 |
| 2 | 1.06 | 0.94 |
| 3 | 1.10 | 0.915 |
| 4 | 1.14 | 0.89 |
| 5 | 1.18 | 0.875 |
| 6 | 1.22 | 0.86 |
| 7 | 1.27 | 0.83 |
| 8 | 1.32 | 0.80 |
| 9 | 1.38 | 0.785 |
| 10 | 1.44 | 0.77 |
| 11 | 1.51 | 0.75 |
| 12 | 1.57 | 0.73 |

## Construction Methods

As with other types of systems, mound systems cannot be built when the soil is frozen or so wet that it can be rolled in a person's hand to create a soil wire. Excess vegetation and small trees must be removed from the site. The force main from the pumping chamber must be installed before the mound site is plowed. The force main must be sloped uniformly toward the pumping chamber, so that the main drains after each dose.

### Plowing

Plowing is done with a moldboard plow or chisel plow. A depth of 7 to 8 inches is required. Plowing must be perpendicular to the slope. Rototillers are not allowed for plowing purposes. Sand fill must be placed immediately after plowing. Once a site is plowed, all foot and vehicular traffic must be prohibited from the plowed area.

## Fill

Fill material must be medium sand texture defined as 25 percent or more very coarse, coarse, and medium sand and a maximum of 50 percent fine sand, very fine sand, silt, and clay. The percentage of silt plus one and one-half times the percentage of clay must not exceed 15 percent. A fill material with a higher content of silt and clay is not acceptable.

Fill is put in place for the upslope and side edges of the plowed area. Vehicular traffic is not allowed within 25 feet beyond the downslope edge of the mound. Fill must be put in place with a track-type tractor, and not less than 6 inches of sand should be kept beneath the tracks at all times.

## Absorption Area

Beds and trenches must be formed with sand fill. The bottoms of beds and trenches must be level. A minimum of 6 inches of coarse aggregate ranging in size from 0.5 inch to 2.5 inches must be placed in the bed or trench excavation. The top of the aggregate must be level.

## Distribution System

A distribution system is placed on top of the aggregate, with the holes located on the bottom of the distribution pipes. Clear markings that indicate each end of the distribution pipes must be placed so that they are visible from the surface. An observation pipe must be placed at the bottom of each bed or trench.

## Cover

The top of the beds or trenches must be covered with not less than 2 inches of aggregate ranging in size from 0.5 inch to 2.5 inches. Next, a minimum of 4 to 5 inches of uncompacted straw or marsh hay or an approved synthetic fabric is placed over the aggregate. The cap and topsoil covers must be in place and the mound must be seeded immediately and protected from erosion.

Table 9.14  **Infiltrative capacity of natural soil:** *(Courtesy of International Private Sewage Disposal Code).*

| PERCOLATION RATE (minutes per inch) | INFILTRATIVE CAPACITY (gallons per foot per day) |
|---|---|
| 0 to less than 30 | 1.2 |
| 30 to 60 | 0.74 |
| greater than 60 to 120 | 0.24 |

For SI: 1 gallon per foot per day = 80.5 L/mm/day,
1 minute per inch = 2.4 s/mm.

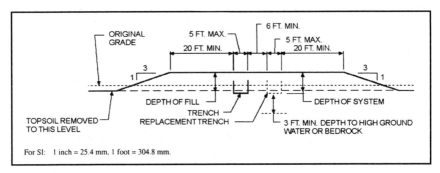

For SI: 1 inch = 25.4 mm, 1 foot = 304.8 mm.

**Figure 9.1**  Design of filled area system: *(Courtesy of International Private Sewage Disposal Code).*

## Maintenance

Pump chambers are to be checked each time that a septic tank is pumped out. Any solids found in the pump chamber are to be removed. Excess traffic in the mound area must be avoided.

Mound systems are expensive, but they are sometimes the only reasonable way to make a lot buildable. These systems are fairly common, so you should make yourself aware of their existence and value.

# Chapter 10

# Cesspools

Normally, cesspools are prohibited. Special permission is required from a code officer before a cesspool can be installed in compliance with the code. When a cesspool is used, it is to be considered a temporary expedient pending the construction of a public sewer. A cesspool might be considered an overflow facility. Another potential authorization for a cesspool would be for a means of sewage disposal for limited, minor, or temporary applications.

Cesspools have to conform to local code requirements that are set forth for seepage pits. The seepage pit must have a minimum sidewall of 20 feet below the inlet opening. Where a stratum of gravel or equally pervious material of 4 feet or more in thickness is found, the sidewall need not be more than 10 feet below the inlet.

# Chapter 11

# Inspections

Inspections are required for private sewage disposal systems. The installation of these systems must take place after construction is complete and before backfilling is done. If backfilling is done prior to a successful inspection, a contractor can be required by a code officer to uncover the work. Code officers can require any inspection that they feel is warranted. It is a contractor's responsibility to notify code officials when an installation is ready for inspection. Installers must make arrangements to enable code officials to inspect all parts of a system. Inspections are needed for additions, alterations, or modifications of existing systems. If an inspection reveals defective materials or workmanship, the nonconforming parts of a system must be removed, replaced, and reinspected.

Again, we have a short chapter, due to little language on the subject in the code. Let's move on to the next chapter.

# Chapter 12

# Referenced Standards

Various standards are often used in the plumbing industry. The Private Sewage Disposal Code uses such standards. Many of the standards used in the code are listed here.

**Table 12.1  Referenced standards:** *(Courtesy of International Private Sewage Disposal Code).*

**ASTM**
American Society for Testing and Materials
100 Barr Harbor Drive
West Conshohocken, PA 19428-2959

| Standard reference number | Title | Referenced in code section number |
|---|---|---|
| A 74—98 | Specification for Cast Iron Soil Pipe and Fittings | Table 505.1 |
| B 32—96 | Specification for Solder, Metal | 505.8.2 |
| B 75—97 | Specification for Seamless Copper Tube | Table 505.1 |
| B 88—96 | Specification for Seamless Copper Water Tube | Table 505.1 |
| B 251—97 | Specification for General Requirements for Wrought Seamless Copper and Copper-Alloy Tube | Table 505.1 |
| B 813—93 | Standard Specification for Liquid and Paste Fluxes for Soldering Applications of Copper and Copper Alloy Tube | 505.8.2 |
| C 4—98 | Specification for Clay Drain Tile | Table 505.1 |
| C 14—95 | Specification for Concrete Sewer, Storm Drain, and Culvert Pipe | Table 505.1 |
| C 76—98 | Specification for Reinforced Concrete Culvert, Storm Drain, and Sewer Pipe | Table 505.1 |
| C 425—98 | Specification for Compression Joints for Vitrified Clay Pipe and Fittings | 505.11, 505.12 |
| C 428—97 | Specification for Asbestos-Cement Nonpressure Sewer Pipe | Table 505.1 |
| C 443—98 | Specification for Joints for Circular Concrete Sewer and Culvert Pipe, Using Rubber Gaskets | 505.7, 505.12 |
| C 564—97 | Specification for Rubber Gaskets for Cast Iron Soil Pipe and Fittings | 505.6.2, 505.6.3, 505.12 |
| C 700—97 | Specification for Vitrified Clay Pipe, Extra Strength, Standard Strength, and Perforated | Table 505.1 |
| C 913—96 | Specification for Precast Concrete Water and Wastewater Structures | 504.2 |
| C 1173—97 | Specification for Flexible Transition Couplings for Underground Systems | 505.3.1, 505.7, 505.10.1, 505.11, 505.12 |
| D 1861—88 | Specification for Homogeneous Bituminized Fiber Drain and Sewer Pipe | Table 505.1 |
| D 1862—88 | Specification for Laminated-Wall Bituminized Fiber Drain and Sewer Pipe | Table 505.1 |
| D 1869—95 | Specification for Rubber Rings for Asbestos-Cement Pipe | 505.4, 505.12 |
| D 2235—96a | Specification for Solvent Cement for Acrylonitrile-Butadiene-Styrene (ABS) Plastic Pipe and Fittings | 505.3.2 |
| D 2564—96a | Specification for Solvent Cements for Poly (Vinyl Chloride) (PVC) Plastic Pipe and Fittings | 505.10.2 |
| D 2657—97 | Standard Practice for Heat-Joining Polyolefin Pipe and Fittings | 505.9.1 |
| D 2661—97a | Specification for Acrylonitrile-Butadiene-Styrene (ABS) Schedule 40 Plastic Drain, Waste, and Vent Pipe and Fittings | Table 505.1, 505.3.2 |
| D 2665—98 | Specification for Poly (Vinyl Chloride) (PVC) Plastic Drain, Waste, and Vent Pipe and Fittings | Table 505.1 |
| D 2729—96a | Specification for Poly (Vinyl Chloride) (PVC) Sewer Pipe and Fittings | Table 505.1.1 |
| D 2751—96a | Specification for Acrylonitrile-Butadiene-Styrene (ABS) Sewer Pipe and Fittings | Table 505.1 |

*(Continued)*

77

## Table 12.1 (Continued) Referenced standards: *(Courtesy of International Private Sewage Disposal Code).*

| | | |
|---|---|---|
| D 2855—96 | Standard Practice for Making Solvent-Cemented Joints with Poly (Vinyl Chloride) (PVC) Pipe and Fittings | 505.10.2 |
| D 2949—98 | Specification for 3.25-In Outside Diameter Poly (Vinyl Chloride) (PVC) Plastic Drain, Waste, and Vent Pipe and Fittings | Table 505.1 |
| D 3034—97 | Specification for Type PSM Poly (Vinyl Chloride) (PVC) Sewer Pipe and Fittings | Table 505.1 |
| D 3212—96a | Specification for Joints for Drain and Sewer Plastic Pipes Using Flexible Elastomeric Seals | 505.3.1, 505.10.1 |
| D 4021—92 | Specification for Glass-Fiber-Reinforced Polyester Underground Petroleum Storage Tanks | 504.4 |
| F 405—97 | Specification for Corrugated Polyethylene (PE) Tubing and Fittings | Table 505.1.1 |
| F 477—96a | Specification for Elastomeric Seals (Gaskets) for Joining Plastic Pipe | 505.12 |
| F 628—97a | Specification for Acrylonitrile-Butadiene-Styrene (ABS) Schedule 40 Plastic Drain, Waste, and Vent Pipe and Fittings | Table 505.1, 505.3.2 |
| F 656—89a | Specification for Primers for Use in Solvent Cement Joints of Poly (Vinyl Chloride) (PVC) Plastic Pipe and Fittings | 505.10.2 |
| F 891—98 | Specification for Coextruded Poly (Vinyl Chloride) (PVC) Plastic Pipe with a Cellular Core | Table 505.1 |
| F1488—98 | Standard Specification for Coextruded Composite Pipe | Table 505.1, Table 505.1.1 |
| F1499—95 | Standard Specification for Coextruded Composite Drain Waste and Vent (DWV) Pipe | Table 505.1 |

## Table 12.2 Referenced standards: *(Courtesy of International Private Sewage Disposal Code).*

**CSA**
Canadian Standards Association
178 Rexdale Blvd.
Rexdale (Toronto), Ontario, Canada M9W 1R3

| Standard reference number | Title | Referenced in code section number |
|---|---|---|
| A257.1—92 | Circular Concrete Culvert, Storm Drain, Sewer Pipe and Fittings | Table 505.1 |
| A257.2—92 | Reinforced Concrete Culvert, Storm Drain, Sewer Pipe and Fittings | Table 505.1 |
| A257.3—92 | Joints for Circular Concrete Sewer and Culvert Pipe, Manhole Sections, and Fittings Using Rubber Gaskets | 505.7, 505.12 |
| CAN/CSA-B137.3—93 | Rigid Poly (Vinyl Chloride) (PVC) Pipe for Pressure Applications—with Revisions through September 1992 | 505.10.2 |
| CAN/CSA-B181.1—96 | ABS Drain, Waste, and Vent Pipe and Fittings | 505.3.2 |
| CAN/CSA-B181.2—96 | PVC Drain, Waste, and Vent Pipe and Pipe Fittings | 505.10.2 |
| CAN/CSA-B182.1—96 | Plastic Drain and Sewer Pipe and Pipe Fittings | 505.10.2 |
| CAN/CSA-B182.2—95 | PVC Sewer Pipe and Fittings (PSM Type)—with Revisions through November 1992 | Table 505.1 |
| CAN/CSA-B182.4—97 | Profile PVC Sewer Pipe and Fittings | Table 505.1 |
| CAN/CSA-B602—90 | Mechanical Couplings for Drain, Waste, and Vent Pipe and Sewer Pipe | 505.3.1, 505.6.3, 505.7, 505.10.1, 505.11, 505.12 |

**CISPI**
Cast Iron Soil Pipe Institute
5959 Shallowford Road, Suite 419
Chattanooga, TN 37421

| Standard reference number | Title | Referenced in code section number |
|---|---|---|
| 301—97 | Specification for Hubless Cast Iron Soil Pipe and Fittings for Sanitary and Storm Drain, Waste and Vent Piping Applications | Table 505.1 |

**ICC**
International Code Council
5203 Leesburg Pike, Suite 708
Falls Church, VA 22041

| Standard reference number | Title | Referenced in code section number |
|---|---|---|
| IBC—2000 | International Building Code® | 201.3 |
| IPC—2000 | International Plumbing Code® | 201.3, 505.13 |

**NSF**
National Sanitation Foundation
3475 Plymouth Road
P.O. Box 130140
Ann Arbor, MI 48113-0140

| Standard reference number | Title | Referenced in code section number |
|---|---|---|
| 40—96 | Residential Wastewater Treatment Systems | 1102.1 |
| 41—98 | Non-Liquid Saturated Systems | 1301.2 |

**UL**
Underwriters Laboratories Inc.
333 Pfingsten Road
Northbrook, IL 60062-2096

| Standard reference number | Title | Referenced in code section number |
|---|---|---|
| 70—79 | Bituminous-Coated Metal Septic Tanks | 504.3 |

# Chapter 13

# Site Evaluation for Water Wells

On-site evaluations for water wells are often taken for granted by builders. I have known many builders who have submitted bids for work without ever seeing the building lot. There is a lot of risk in doing this. A builder cannot bid a job competitively without knowing what type of well will be used. If a bid goes in for a bored well and it turns out that a drilled well is needed, the bidder will lose money. If you estimate for the most expensive type of well in lieu of examining the site, you can overbid and lose the job. Failure to do a site inspection can be a very big mistake.

What can you really tell by walking a piece of land? It depends on your past experience and skill level. Some conditions point to obvious solutions. For example, when I built my last home, I was able to see bedrock sticking up in places. The type of humps in the land indicated rock close to the surface. A little scratching and digging proved that bedrock was right at the ground surface in some spots and not more than a couple of feet down in most locations. This automatically told me that a drilled well would be needed. When bedrock is present, drilling is the only sensible option.

You can't always see what's likely to be under the ground by looking at the surface. Knowledgeable builders want to know what they will get into when drilling wells and digging footings. Many experienced builders require customers or landowners to provide them with soil studies before giving a firm bid. When such studies have not been done, some builders do their own. I'm one of these builders.

It is not uncommon to see me, or one of my people, out digging holes on potential building sites. A post-hole digger can reveal a lot about

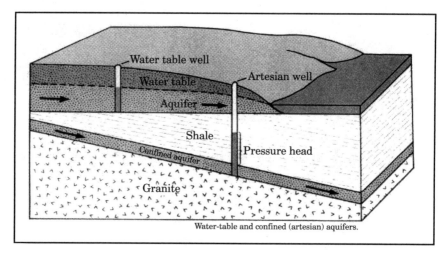

**Figure 13.1** Cross-section of underground conditions.

what conditions exist below the topsoil. Augers and probe rods can also provide some insight into what is likely to be encountered. A probe rod will tell you if a lot of rock is present. But, to see the soils, you need a hole. An auger or post-hole digger is the best way to get these samples. Augers are often easier to use. A power auger is ideal, if you happen to have one.

When you can create some test holes, you have a lot more information to base your bid decisions on. There are only a few ways to bid a job where a well will be installed. You can guess what will be needed, but this is very risky. Digging test holes will give you a very good idea of what types of wells might be suitable. Interviewing well owners on surrounding property can provide a lot of data that can help you with your decision. Hiring soil-testing companies is a great but expensive way to find out what you are getting into. Having a few well installers walk the land with you so that they can provide you with solid bids is another good way to protect yourself.

Since soil tests are required for land where a septic system will be used, the results from these tests can be reviewed to aid in the evaluation of a well type. This, however, is not always a safe way to proceed. Since septic locations will typically be at least 100 feet from a well location, ground conditions might be very different. If you are going to rely on tests, the testing should be done at the proposed well site.

Since most builders are not experts on the subject of wells, I think it is wise to have well installers make site inspections and solid bids. If you have three to five bids from reputable well installers, you can feel secure in the fact that a bid you present to a customer is safe. This is the easy way out, but it's a good way to go.

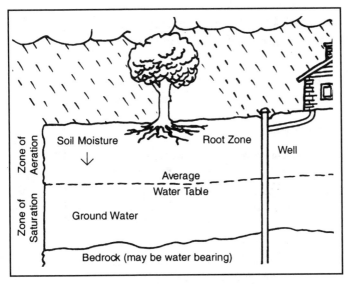

**Figure 13.2** Ground conditions that affect well placement.

When you have firm quotes from well installers, you should be able to depend on them. Having only one bid is risky. The well installer might be too busy to do your job when the work is needed. It's possible that the installer who gave you a bid will be out of business before you can request service from the company. As long as you have multiple bids from reputable installers, you should have very little to fear in terms of your well price.

### Location

The location of a well is important. Choosing a location is not always easy. Many factors can influence the location of a well. The most obvious might be the location of a house. It is not common to place a well beneath a home, so most people will choose a location outside of the foundation area of a home. Septic fields are another prime concern. Wells are required to be kept a certain distance from septic systems. The distance can vary from jurisdiction to jurisdiction and because of topography. Access to a location with a well-drilling rig is also a big factor. These large trucks aren't as maneuverable as a pickup truck. Picking a place for a well must be done with access in mind.

Appearance can be a factor in choosing a well location. Wells are not pretty, so they usually aren't welcome in locations where their presence is obvious. A drilled well is easier to hide than a bored well. The

## Recommended Minimum Separation Distances From Potential Sources of Contamination to Private Wells

**Well** — 2 ft. • Pump house floor drain draining to ground surface.

10 ft. • Watertight sewer line (cast iron or approved plastic).

25 ft. • Farm silo.

50 ft.
- Watertight sewer line (clay tile, orangeburg, etc.).
- Septic tank, subsurface drain field or lagoon (downhill from the well).
- House foundation (downhill from well).
- Other potential or known sources of pollution, such as oil and gas wells, manure piles, landfills, etc. (downhill from the well).

75 ft.
- Septic tank, subsurface drain field or lagoon (same elevation as the well).
- House foundation (same elevation as well).
- Other potential or known sources of pollution, such as oil and gas wells, manure piles, landfills, etc. (same elevation as the well).

100 ft.
- Septic tank, subsurface drain field or lagoon (uphill from the well).
- Septic tank, subsurface disposal field (soil percolation greater than or equal to one inch in less than five minutes).
- Seepage pit.
- Barnyard or feedlot.
- House foundation (uphill from the well).
- Other potential or known sources of pollution, such as oil and gas wells, manure piles, landfills, etc. (uphill from the well).

300 ft. • Privies, cesspools, or other known sources of pollution.*

* For sources not addressed, provide as much separation as practical from the well. If the well is on a hillside or at the foot of a hill where pollutant sources are located, the corresponding separation value is a horizontal distance.

These distances constitute the minimum separation and should be increased in areas of fractured rock or limestone, or where the direction of ground water movement is from sources of contamination toward the well. These are distances in the well standards. Local waste storage ordinances may recommend or require different separation distances. (Source: OAC 785:35-7.)

Figure 13.3  Suggestions for well locations.

difference between a 6-inch well casing and a 3-foot well casing is considerable.

The location of a well also depends on where an expert believes water will be found. Few people know for sure where water will be found, but some people have a knack for being right more often than not. This brings us to the question of prospecting for water. Is it possible to predict where water will be found? A lot of people think so. Let's talk about this awhile.

## Reading Signs To Find Water

Have you ever heard of reading signs to find water? Well, you won't find billboard signs with arrows pointing down and the words "Water Is Here" painted on them. However, there are some natural signs that an experienced eye can detect that can reveal data on potential underground water.

## Maps

Maps can give you a lot of guidance on where water might be found. Some regional authorities maintain records on wells already in existence. Reviewing this historical data can definitely help you pinpoint your well type and location. Unfortunately, there is never any guarantee that water will be where you think it is. A neighboring landowner might have a well that is 75 feet deep while your well turns out to be 150 deep. It is, however, likely that wells drilled in close proximity will average about the same depth. I've seen houses in subdivisions where one house has great well water and the next-door neighbors' water suffers from unpleasant sulfur content. Fifty feet can make quite a difference in the depth, quantity, and quality of a well.

Topographical maps show land elevations. You can look at these maps and put many things into perspective. If there is a river or stream in the area, you can plot the position of your property in respect to the surface water. Does this help you? Not necessarily.

My property has a good deal of river frontage. My well location is probably 50 feet, or so above the river. Yet my well is over 400 feet deep. Why? I'm not sure. I think it has to do with the bedrock that my well is drilled into. Water doesn't run through solid rock very well.

Some maps are helpful, especially the ones that indicate depths and types of wells already in existence. Water does run through rock, but it

Figure 13.4 Risk factors to consider when choosing a well location.

| | 1. Low Risk / Safest Situation | 2. Medium Risk / Potential Hazard | 3. High Risk / Unsafe Situation | Your Risk |
|---|---|---|---|---|
| **Position of well in relation to pollution sources** | Well is uphill from all pollution sources. Surface water flows away from well. | Well is uphill from most pollution sources. Some surface depressions can store water near well. | Well located downhill from pollution sources or in a pit or depression. Surface water accumulates near well. | |
| **Separation distances between wells and pollution sources** | Exceeds all state minimum required distances. | Meets minimum distance requirements. | Does not meet minimum separation distances for most or all potential sources. | |
| **Soil type** | Fine-textured soils, such as clay loams and silty clay. | Medium-textured soils, such as sandy clay loams and loams. | Coarse-textured soils, such as loamy sands and sands. | |
| **Subsurface conditions** | Water table deeper than 20 feet and protected by tight bedrock or clay. | Water table 10 to 12 feet deep. No clay or bedrock. | Water table or impermeable layer shallower than 10 feet. | |

needs cracks or some other form of access to get into and out of it. This makes it difficult to predict what you will find when you attempt to install a well.

## Plants

Plants can be very good indicators of whether and how much water is available beneath the ground's surface. Trees and plants require water. The fact that plants and trees need water might not seem to provide much insight into underground water. Take cattails as an example. If you see cattails growing nearby, you can count on water being close by. It's suggested that the depth of water in the earth can be predicted to some extent by the types of trees and plants in the area.

Cane and reeds are believed to indicate that water is within 10 feet of the ground's surface. Arrow weed means that water is within 20 feet of the surface. There are many other rules of thumb for various types of plants and trees in regard to finding water. From a well-drilling point-of-view, I'm not sure how accurate these predictions are. I know that cattails and ferns indicate water is close by, but I can't say that it will be potable water or how deep it will be found in the ground. I suspect that there are some very good ways to predict water with plants and trees.

Since I am not an expert in plants, trees, or finding water, I won't attempt to pump you up with ideas of how to find water at a certain depth just because some particular plant grows in the area. It is my belief that with enough knowledge and research, a person can probably predict water depths with good accuracy in many cases. If water can be predicted with plants and trees, can it be found with a forked stick?

## Dowsing

Dowsing is a subject that can bring a lot of controversy. There are people who swear they can find water with nothing more than a forked stick. Can they do it? Sometimes they can. People have been successful in locating water with what some people call witching sticks, but how hard is it to do this? There is a huge amount of underground water in the United States. Finding water might be just as possible with a fireplace poker as it is with a twig from a willow tree. I'm not in a position to make a judgment call on this. Dowsing is an activity that I have no first-hand experience with.

I have known builders who hired dowsers to come to job sites and pick locations for wells. They probably always hit water, but then, so have I, and I've never used the services of a dowser. Rumor has it that the type of wood used for a divining rod (forked stick) is important. Favorite wood species include peach, willow, hazel, and witch hazel. It's said that since these trees require a lot of water to grow, their wood is ideal for locating water.

I've studied the use of divining rods to a minor extent. People use metal rods instead of forked sticks to try finding gold, water, and other underground treasures. As an amateur treasure hunter, I'm often captivated by new ways of locating hidden bounty. One treasure-hunting show I saw on television conducted a controlled test of self-proclaimed dowsing experts. If my memory is correct, the results of their tests were much better in identifying water than they were in finding gold. The test was controlled over a man-made test bed. Did the testing prove anything? I don't know, but it was interesting to watch.

In my experience as a builder, I've never found a need for hiring a professional dowser. You can try their services if you like, but I don't feel it's necessary. An experienced well installer who knows the area and has access to maps is probably just as good a bet and maybe a better one.

### High-Tech Stuff

Some professional water hunters have high-tech stuff to work with. I assume their sophisticated scientific gear works, but I don't know this to be true. If you were looking for an underground water source to serve a small community, it might be worthwhile to enlist the services of some of these professionals. For a single house, I can see no reason to go to the trouble or expense. Hire a well driller on a flat-rate basis who will guarantee water and you are probably money ahead.

### It Doesn't Take Very Long

It doesn't take very long to perform a site inspection. An hour or so is the most time that is normally needed. Can you afford to spend thousands of dollars to save an hour or so? Maybe you can, but I can't. Site inspections are important. Whether it is you or your well installer who makes a decision on what type of well will be needed, someone is assuming a lot of responsibility. Planning to use a bored well, and then finding that a 400-foot drilled well is needed, is going to cost somebody a lot of money. Customers will expect you to guarantee them a firm price. You should expect the same from your well installer. While you might win more often than you lose by playing the odds, the losses can be very expensive. Someone should definitely perform a site inspection before commitments are made on wells.

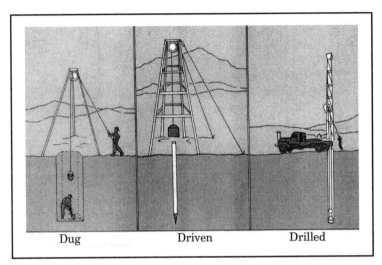

**Figure 13.5.** Types of wells.

**Chapter**

# 14

# Testing and Protecting Potable Water

Private water supplies need to be tested before they are used for human consumption. Periodic testing is recommended as the private water source continues to be used as potable water. Local code requirements generally cover the installation of new wells for the initial testing. Lenders who provide money to buy homes with private water systems generally require a satisfactory water test before they will allow the closing of a real-estate transfer. But, there is no one to insist that water wells be tested every so often once they are in use and the property is not changing hands.

Just because a well tests fine when it is installed doesn't mean that the water is safe to drink two years later. Water tables shift. All sort of changes take place over time. A well could be affected by an influx of surface water that is not supposed to be able to get into the well. Pesticides can become a problem over time. Periodic testing is not very expensive, but it is the intelligent thing to do.

Private water supplies can suffer from a variety of unwanted contaminants. Not all of the pollutants and contaminants are deadly, but this doesn't mean that they are desirable. In some cases the problem with well water may be taste or clarity. These problems don't usually pose a health risk, but they can make drinking the well water unpleasant.

Fixing well problems doesn't have to be expensive. In fact, it is often very inexpensive. Let's discuss this matter in more depth.

Table 14.1  Water Testing Facts

| Contaminant Problem | Definition | Health Effects | Treatment Methods to Use |
|---|---|---|---|
| Acidic water | Acidic water has a pH level of less than 7. | Corrosive or acidic water can leach metals from pipes into drinking water. Lead leached into drinking water can cause developmental delays in children. In adults, lead can cause kidney problems and high blood pressure. Copper can cause gastrointestinal distress, or liver and kidney damage. | ▪ Neutralize the water by correcting the pH level. |
| Chloride | An element that can originate naturally in rock, or may be caused by municipal sewage, industrial wastes, or road salt. | No known health effects. | ▪ De-ionization<br>▪ Reverse osmosis |
| Coliform Bacteria | Live organisms found in water that have been contaminated by human or animal waste. | Consuming water with pathogenic organisms may cause typhoid, dysentery, cholera, or gastrointestinal disorders. | ▪ Chlorine disinfection |
| Fluoride | An element that occurs naturally in water from erosion of natural deposits. It is also found in some fertilizers and as by-product of aluminum factories. | There is a debate over the optimal level of fluoride in water. The standard is currently set at less than 4ppm. Fluoride can promote strong teeth. However, fluoride may cause bone disease and contribute to mottled teeth in children. | ▪ Ion-exchange<br>▪ Reverse osmosis |

## Confusion

There is often a lot of confusion about the quality of private water systems. For example, a person might think that since water coming from a well looks, smells, and tastes fine that the water is safe to drink. This may not be the case. Testing is needed to confirm the quality of a water

Table 14.2  Water Testing Facts

| Contaminant Problem | Definition | Health Effects | Treatment Methods to use |
|---|---|---|---|
| Radon | A colorless, odorless, and tasteless radioactive gas. Formed from the decay of naturally occurring minerals containing radioactive elements. | Radon becomes radon gas when vaporized. Radon gas has been linked to increased rates of lung cancer. Scientists remain unclear of the effects of radon in drinking water. | • Aeration |
| Taste and Odor | Most sources of odor and taste-bearing substances in water are generally harmless organic materials. | No known health effects. | • Carbon filtration and/or oxidation<br>• Potassium permanganate<br>• Ozonation<br>• Aeration |
| Turbidity and Color | Visual haziness in water. Discolored water may be caused by organic compounds or metallic ions. | Turbidity has no health effect, per se, but it can interfere with disinfectants and provide a medium for bacteria to grow. | • Settling<br>• Filtration |

source. And the water tested should be collected at the point of use. Taking water directly from a well for testing is okay, but the true test will be when water is taken from the point of use, such as from a kitchen faucet. There could be problems in the plumbing between the well and the faucet that would not show up if the water is taken directly from the well.

Trying to test water for every conceivable risk is not practical. It's possible, but not feasible. It is generally recommended to test for the following:

- Coliform bacteria

- Nitrate

Table 14.3  Water Testing Facts

| Contaminant Problem | Definition | Health Effects | Treatment Methods to Use |
|---|---|---|---|
| Iron | An element that can leach into ground water from rocks and soils. Iron in water may be caused by the corrosion of metal pipes, pumps and fixtures. | No known health effects. | - Water softeners<br>- Catalytic oxidizing filters<br>- Oxidation-filtration systems<br>- Chlorination, or ozonation if iron bacteria are present |
| Manganese | An element found in rock, soil, and organic materials. | No known health effects. | - Water softener<br>- If level exceeds 2.0 mg/L oxidizing filters or oxidation-filtration may be required. |
| Nitrates | Are naturally occurring forms of nitrogen. Formed when fertilizers, decaying plants, manure, and/or other organic residue is broken down. They are also a by-product of septic system waste. | When consumed in high quantities nitrates can be absorbed by the blood, converting hemoglobin to methemoglobin. Methemoglobin does not carry oxygen effectively. Babies are especially susceptible to this condition causing "blue-baby syndrome." | - Ion exchange<br>- Reverse osmosis |

- Lead

- Sulfate

Many people feel that the earth will purify well water. This is largely true, but it is not a guarantee of safe drinking water. If surface water gets through a well casing, the surface water could contaminate a well.

Testing and Protecting Potable Water 93

Table 14.4 Risk factors associated with private water wells

| | 1. Low Risk / Safest Situation | 2. Medium Risk / Potential Hazard | 3. High Risk / Unsafe Situation | Your Risk |
|---|---|---|---|---|
| Well age | Less than 20 years old. | 20 to 50 years old. | More than 50 years old. | |
| Well type | Drilled well, drilled deep through protective layer. | Shallow drilled well, less than 100 feet. | Dug well or driven well. | |
| Casing height above land surface | More than 8 inches above the surface. | At surface or up to eight inches above. | Casing at or below surface or in pit or basement. | |
| Condition of casing and well cap (seal) | No holes or cracks. Cap tightly attached. Screened vent. | No holes or cracks visible. Cap loose. Condition below ground unknown. | Holes or cracks visible. Casing loose or poor condition suspected. Cap loose or missing. Running water can be heard. | |
| Condition of concrete slab | Two-foot concrete slab in good condition. | Concrete slab cracked or deteriorated. | No concrete slab. | |
| Casing depth | Casing extends into deep bedrock or clay formation below land surface. | Casing extends 10 feet below minimum water level during pumping. | Casing extends less than 10 feet below minimum water level during pumping. | |
| Backflow prevention | Anti-backflow devices (check valves) installed on all outside faucets. | | No anti-backflow device. | |
| Well inspection and "tune-up" | Well was inspected and water tested within the last 10 years. | Well was inspected and water tested 10 to 20 years ago. | Don't know when well was last inspected, or water tested, or well was inspected more than 20 years ago. | |

Other facilities and buildings, such as gas stations, shopping malls, mining, and other activities, can also affect the ground's ability to purify water.

Will a water filter solve any potential problems? Not really. It is possible to filter out a vast majority of problems, but no single pump can accomplish this task. To be reasonable, you have to define symptoms and problems and work with the known data. In some cases, a filter

## Private Well Protection Checklist*

|  | Yes | No |
|---|---|---|
| Have you had your well water tested within the last two years? | ❏ | ❏ |
| Is your well less than 20 years old? | ❏ | ❏ |
| Is the water table around your well deeper than 30 feet? | ❏ | ❏ |
| Is your well securely capped and does it have intact casing which extends 1 foot above ground level? | ❏ | ❏ |
| Is your well properly grouted to prevent entry of surface water? | ❏ | ❏ |
| Do you mix, load, or store pesticides or fertilizer at least 100 feet from the well? | ❏ | ❏ |
| Do you have backflow prevention devices installed on all faucets used to fill spray tanks? | ❏ | ❏ |
| Are feedlots and lagoons located at least 100 feet away from the well? | ❏ | ❏ |
| Is your leach field at least 100 feet from the well? | ❏ | ❏ |
| Are abandoned wells on your property properly sealed? | ❏ | ❏ |
| Have all underground storage tanks been removed? | ❏ | ❏ |
| Are all fuel tanks at least 100 feet from the well? | ❏ | ❏ |
| Do you dispose of pesticide containers, solvents, lubricants, paints, household cleaning products, and old batteries in a way that will not pose a risk to your drinking water? | ❏ | ❏ |

* If you answer no to any of these questions, your well may be vulnerable to contamination. Consider having your water tested and evaluate how you can improve the safety of your water supply.

**Figure 14.1** Water protection checklist

could make a problem worse. I am not kidding! If your water contains bacteria, a filter may trap it near the outlet point, such as at a faucet. Since the filter may collect and store the bacteria, the end result could be a much higher delivery of bacteria to your drinking glass than if there was no filter at all.

How many sources of contamination can you think of? Well, there are plenty of them. Here is a partial list of what might pollute your drinking water:

- Mining

- Acid rain

- Pesticides

- Gas stations

- Industrial facilities

Table 14.5 Water treatment methods

| Treatment method | Contaminants removed | How it works | Limitations |
|---|---|---|---|
| Cation or anion exchange (water softeners) | Barium, radium, iron, magnesium, calcium | Sodium exchanged for calcium and magnesium | Sodium may cause health problems |
| Activated carbon filters | Volatile organic compounds, some pesticides, radon, mercury, odors | Water filtered through carbon granules | Must replace filters regularly |
| Chlorination | Bacteria and other microbial contaminants | Chlorine metered into water | Chlorine by-products may be harmful or affect taste of water |
| Distillation | Radium, odors, heavy metals, salt, nitrate, off-tastes | Evaporation/condensation | Slow, energy intensive, expensive |
| Reverse osmosis | Radium, sulfate, nitrate, calcium, magnesium, slats, some pesticides and VOCs | Membrane filters dissolved impurities | Expensive, slow, wastes water |
| Mechanical filtration | Dirt, sediments, scale, insoluble iron and magnesium | Sand or other filtering material strains impurities | Does not remove dissolved contaminants |
| Ultraviolet radiation | Bacteria and other microbial contaminants | Water passes under a special UV light | No residual effect; may not work in cloudy water, slow |
| Ozonation | Bacteria and other microbes | Water exposed to ozone gas | Equipment is expensive; no residual effect |
| Oxidizing filters | Iron, manganese, hydrogen sulfide | Contaminants removed through filtering and chemical reactions | Potassium permanganate is caustic |

- Septic systems

- Lead

- Trash dumps

- Oil spills

There are, of course, many other sources of pollution, but the list above will give you a hint of fairly common sources that could ruin your water. There are four basic types of pollutants. They are:

- Microbial pollutants

- Radiological pollutants

- Organic chemical pollutants

- Inorganic chemical pollutants

**Testing Recommendations**

What are the testing recommendations for private water supplies? They vary, depending upon whom you ask. However, there are some general, rule-of-thumb recommendations that most experts agree with. Depending upon whom you ask, you may be told to test your water once a year. This is not bad advice. Some people will tell you to test the water once every two years or when you expect a problem may exist. This is a personal matter that only you can decide.

Primary testing should be done for coliform bacteria, nitrate, sulfate, and lead. Older homes may contain lead plumbing pipes or solder joints. When this is the case, testing should be done for pH, lead, copper, cadmium, zinc, and a corrosion index.

Many homebuyers want a fairly full battery of tests run before a home is purchased. Some of the most common elements tested for include:

Table 14.6  Water treatment recommendations

| Contaminant | Activated Alumina Filters | Activated Carbon Filters[7] | Air Stripping | Anion Exchange | Cation Exchange/Water Softener | Chlorination | Distillation | Mechanical Filtration | Oxidizing Filters | Ozonation | Reverse Osmosis | Ultraviolet Radiation |
|---|---|---|---|---|---|---|---|---|---|---|---|---|
| Arsenic | X | | X | | | | X | | | | X | |
| Asbestos | | X | | | | | X | | | | X | |
| Atrazine | | X | | | | | X | | | | X | |
| Benzene | | X | X | | | | X | | | | X | |
| Chlorine | | X | | | | | | | | | | |
| Coliform bacteria | | | | | | X | X | | | X | | X |
| Color | | X | | X | X | | | | X | X | | |
| Flouride | X | | | | | | X | | | | X | |
| Hardness | | | | | X | | | | | | | |
| Hydrogen sulfide | | X | X | | | X[1] | | | X | X[1] | | |
| Inorganics, minerals (some) | | | | | | | X | | | | X | |
| Iron/manganese — dissolved | | | | | X[2] | X[1] | | | X | X[1] | | |
| Iron/manganese — insoluble | | | | | | | | X | X | | | |
| Lead | | | | | | | X | | | | X | |
| Mercury | | X | | X | | | X | | | | X | |
| Nitrate | | | | X | | | X | | | | X | |
| Odor and taste | X | X | X | X | X | X | | | X | X | X | |
| Pesticides (some) | | X | X | | | | X | | | | X | X |
| Radium | | | | | X[5] | | X[4] | | | | X[4] | |
| Radon gas | | X[6] | X | | | | | | | | | |
| Salt | | | | | | | X | | | | X | |
| Sand, silt, clay (turbidity) | | | | | | | | X | | | | |
| Volatile organic chemicals (some) | | X | X | | | | X[3] | | | | X | |

1 When followed by mechanical filtration or an activated carbon filter
2 When present in low concentrations
3 Only for volatile organic chemicals with high boiling points
4 Other water quality problems may interfere with treatment
5 With zeolite softening
6 Often requires pretreatment system
7 There are several different types of activated carbon filters (e.g. granular, block, powder, etc), not all types work on all substances listed.

Table adapted from the Water Quality Association and from 1996. Rick Weinzierl, et.al (1996). "57 Ways to Protect Your Home Environment (and Yourself)," University of Illinois at Urbana-Champaign, North Central Regional Extension, Publication 583.

Table 14.7  Testing for potential health concerns

| Problem or concern | Test |
|---|---|
| Family or guests become ill | total coliform bacteria,* nitrate* |
| Water supply used by an infant less than six months old | nitrate |
| Water supply used by young children | lead |

* Coliform bacteria and nitrate serve as indicator tests for possible contamination.

- Lead

- Iron

- Coliform bacteria

- Nitrate

- Sulfate

- pH

- Hardness

How important could pH be? If you have ever kept fish in an aquarium, you probably have some experience in testing the pH of the water. Acidity can be important. I clearly remember a lady from many years ago who suffered terribly with stomach discomfort. She sought many remedies and found none of them to be satisfactory. Then I did a water test on her home and found the pH level to be extremely acidic. With the installation of an acid neutralizer, her stomach problems disappeared. This is not a common occurrence, but it is a fine example of how something as simple as the pH rating can affect a person's quality of life.

Iron and manganese can be a problem even for water softeners. Water stains on plumbing fixtures can indicate a problem with iron, manganese, or copper. Then there are the elements that contribute to awful smells and water that is not tasteful. Any of the following may be responsible for foul odors and taste:

- Sodium

- Hydrogen sulfide

- pH

- Copper

- Lead

- Iron

- Zinc

- Sodium

- Chloride

Cloudy water is often a problem when drinking well water. This is usually turbid water. When pipes indicate corrosion, such as pinholes in copper plumbing pipes, look for copper, acid, iron, manganese, lead, and the corrosion index. When a water pump wears out quickly, check for pH and the corrosion index. Soap scum is a common problem. This is almost always the result of hard water.

## New Wells

New wells are generally disinfected prior to testing. This is a simple procedure that uses bleach. You should check with local health officials to obtain the proper ratio of bleach to introduce into a well. But as an example, you might use three cups of liquid bleach in a well that is 100 feet deep. Ideally, you want to pour the bleach along the sides of the well casing. Then flush the bleach further into the well. This is easy to do with pressurized water from a hose, but it may be necessary to dump clean water into the well with buckets.

Table 14.8  Potential problems and solutions for water treatment

| Problem | Recommended Corrective Action |
|---|---|
| The well is newly constructed, or maintenance or repair was recently done. | Go to step two below. |
| The pump was primed with impure water. | |
| There is standing water around the well or water draining toward the well. | Re-grade around the well so the ground slopes away from your well. |
| The concrete well pad is cracked or separated from the well casing. | Re-pour pad or fix and seal all cracks and gaps. |
| The well is not completely sealed against surface water, insects, or other foreign matter. | Replace any missing plugs, cap any open pipes, and seal any openings, gaps or cracks. |
| | *Contact a licensed well contractor to replace or install a new wellhead gasket. |
| The storage tank is dirty or unprotected. | Contact a water system contractor to clean and seal. |
| There are cross-connections in the plumbing system. | Make sure that your plumbing is not connected to another source of water that may be contaminated (e.g. a defunct community water system). |
| There is not adequate back-flow protection. | Install a back-flow prevention device on every outdoor faucet (available at most hardware and plumbing supply stores). |
| | *Contact a licensed well contractor to ensure that there is proper back-flow protection within the well. |
| There are dead-end or unused water lines connected to your plumbing system. | Flush lines regularly or |
| | Remove any unused lines or sections of the water system. |
| The well casing is corroded. | *Contact a licensed well contractor to assess and repair. |
| There is sediment at the bottom of the well. | |
| The well casing is perforated too high or the sanitary seal is not adequate. | *Contact a licensed well contractor to drill a new well and to properly destroy the old well. |

Make sure that the water supply to the building is turned off. Let the well set with the bleach in it for 8 to 24 hours. After the waiting period, turn the water service cutoff on and flush the plumbing pipes and well. This is done best by running water into a bathtub. You should smell the bleach when the water is turned on. Run the water until there is no further odor of bleach. Some people attempt to run a well dry before stopping the flushing process. The bottom line is to get all of the bleach out of the well and then wait until the well recovers its static level.

Once the well has recovered, go to a potable water outlet, such as a faucet. Remove any aerator that may be present. Use a cigarette lighter or other small flame to sterilize the outlet of the faucet. Turn the water on and let it run for a few minutes. Do a "clean catch" in a sterile test bottle. Now you are ready to deliver the water for a laboratory test.

### Who Tests The Water?

Who tests the water? Many county extension offices will conduct water tests for homeowners and contractors. There are numerous private, professional labs that conduct water tests. Even some plumbing sup-

## Testing and Protecting Potable Water

Table 14.9  Troubleshooting water quality

| Problem | Possible Cause | Health Risk Category* |
|---|---|---|
| Water is orange or reddish brown | This may be due to high levels of iron (Fe). | ❶ |
| Porcelain fixtures or laundry are stained brown or black | This is commonly a result of high manganese (Mn) and/or iron (Fe) levels. As little as 50 parts per billion (ppb) manganese and 300 ppb iron can cause staining. | ❶ |
| White spots on the dishes or white encrustation around fixtures | High levels of calcium (Ca) and manganese (Mn) can cause hard water, which leaves spots. Hardness can also be measured directly. | ❶ |
| Water is blue | Blue water or blue deposits may be due to high levels of copper (Cu), especially if coupled with corrosive water. | ❷ |
| Water smells like rotten eggs | This is most likely caused by hydrogen sulfide (H₂S). | ❶ |
| Water heater is corroding | Water can be corrosive, neutral, or noncorrosive. Water that is very corrosive can damage metal pipes and water heaters. The lab can calculate the corrosivity of your water by measuring calcium, pH, total dissolved solids (TDS), and alkalinity. | ❶ |
| Water appears cloudy, frothy or colored | Suspended particulates, measured directly or as turbidity, can cause the water to appear cloudy, frothy or colored. Detergents and/or sewage waste may also be the culprit. | ❷ |
| Home's plumbing system has lead pipes, fittings, or solder joints | Corrosive water can cause lead (ppb), copper (Cu), cadmium (Cd), and/or zinc (Zn) to be leached from lead pipes, fittings, and solder joints. | ❷ |
| Water has a turpentine odor | This may be due to methyl tertiary butyl ether (MTBE) | ❷ |
| Water has a chemical smell or taste | This may be due to volatile or semivolatile organic compounds (VOCs) or pesticides. | ❷ |

*Are You Concerned That A Nearby Activity May Be Contaminating Your Well?*
Here are some land uses and possible contaminants to test for.

| Land Use | Possible Contaminants | Health Risk Category* |
|---|---|---|
| Landfill, industry, or dry cleaning operation | Consider testing for volatile organic compounds (VOCs), pH, total dissolved solids (TDS), chloride (Cl), sulfate (SO₄), and metals. | ❷ |
| Agricultural crop production | Consider testing for pesticides commonly used near the well (consult the farmer or Department of Agriculture for a list), nitrate (NO₃), pH, and total dissolved solids (TDS). | ❷ |
| Livestock enclosure, manure, or compost storage area | Consider testing for bacteria, nitrate (NO₃), and total dissolved solids (TDS). | ❷ |
| Gas station or automobile repair shop | Consider testing for total petroleum hydrocarbons (TPHg), total oil and grease (TOG), benzene, toluene, ethylbenzene, xylenes (BTEX), MTBE, ethylene dibromide (EDB). | ❷ |

*❶ No known health risk at commonly found concentrations
❷ Some of the possible causes can have a detrimental effect on health even if present in low concentrations

pliers are equipped to help their customers obtain water tests. The test should be done by an outside agency. It is not wise for contractors to test the water quality in systems that they install. This practice is fine for a preliminary test, but the final and official test results should come from a recognized independent testing facility.

What can you expect water to be tested for by a professional lab? It depends on the lab, where you are located, local regulations, and what is requested of the lab. If a comprehensive test is done, it can cover a multitude of test factors. Here are some of what might be tested for:

## Chapter Fourteen

Table 14.10 Water testing recommendations

| Test | Recommended Test | | | Interpreting Your Results |
|---|---|---|---|---|
| | Recommended Frequency | Cost | If the lab report shows: | Then you may want to consider one or more of the following options: |
| Total Coliform Bacteria | Twice per year: Wet season Dry season | $20 – 50 | Present | Eliminate cause, disinfect and retest (see page 13). <br><br> Increase testing frequency <br><br> Install a treatment system such as distillation, chlorination, ozonation, or ultraviolet radiation. Consult a water treatment professional for more advice. |
| Nitrate | Annually | $25 – 45 | ≥ 45 mg/l $NO_3$ or ≥ 10 mg/l $NO_3$-N | Install a treatment system or find an alternate water supply. Reverse osmosis, distillation, or anion exchange, will remove some of the nitrate. Consult a water treatment professional for more advice. <br><br> Increase testing frequency |
| Electrical Conductance (EC) | Annually | $12 – 20 | ≥ 1600 μmhos/cm or significantly different from previous year result | Conduct further testing, such as nitrate and/or minerals to determine the cause of the high EC, or the change in EC. |
| MINERALS <br><br> Aluminum (Al) Arsenic (As) Barium (Ba) Cadmium (Cd) Chromium (Cr) Fluoride (F) Iron (Fe) Lead (Pb) Manganese (Mn) Mercury (Hg) Selenium (Se) Silver (Ag) | Every 5-10 years, or If EC changes significantly, or If taste, color, odor or surrounding land use change | Package $250 – 300 <br><br> Individual $20 – 30 <br><br> Mercury $30 – 40 | Al ≥ 0.2 mg/l <br> As ≥ 0.05 mg/l <br> Ba ≥ 1.0 mg/l <br> Cd ≥ 0.005 mg/l <br> Cr ≥ 0.05 mg/l <br> F ≥ 2.0 mg/l <br> Fe ≥ 0.3 mg/l <br> Pb ≥ 0.015 mg/l <br> Mn ≥ 0.05 mg/l <br> Hg ≥ 0.002 mg/l <br> Se ≥ 0.05 mg/l <br> Ag ≥ 0.1 mg/l | Compare to previous results <br><br> Install a treatment system or find an alternate water supply. The appropriate treatment system is dependent on your overall water chemistry and what constituents you would like to remove. Consult a water treatment professional for more advice. |

≥ is greater than or equal to
mg/l is milligrams per liter. 1 mg/l = 1 part per million (ppm). 1 mg/l = 1000 microgram per liter (μg/l). 1μg/l = 1 part per billion (ppb)

Arsenic

Barium

Cadmium

Chloride

Coliform Bacteria

Chromium

Color

Copper

Fluoride

Iron

Lead

Manganese

## When should you test your water?

| Situation | Test |
|---|---|
| Family members or house guests have recurrent incidents of gastrointestinal illness. | Test for coliform bacteria, nitrate, and sulfate. |
| Household water plumbing contains lead pipes, fittings, or solder joints. | Test for pH, corrosion index, lead, copper, cadmium, and zinc. |
| You are buying a home and wish to assess the safety and quality of the existing water supply. | Test for coliform bacteria, nitrate, lead, iron, hardness, pH, sulfate, total dissolved solids (TDS), corrosion index, and other parameters depending on proximity to potential sources of contamination. |
| You need a water softener to treat hard water. | Before purchase and installation, test for iron and manganese, which decrease the efficiency of cation exchange softeners. |
| You wish to monitor the efficiency and performance of home water treatment equipment. | Test for the specific water problem being treated upon installation, at regular intervals after installation, and if water quality changes. |
| Water stains plumbing fixtures and laundry. | Test for iron, manganese, and copper. |
| Water has an objectionable taste or smell. | Test for hydrogen sulfide, pH, corrosion index, copper, lead, iron, zinc, sodium, chloride and TDS. |
| Water appears cloudy, frothy or colored. | Test for color, turbidity, and detergents. |
| Pipes or plumbing show signs of corrosion. | Test for corrosion index, pH, lead, iron, manganese, copper, and zinc. |
| Water leaves scaly residues and soap scum and decreases the cleaning action of soaps and detergents. | Test for hardness. |
| Water supply equipment (pump, chlorinators, etc.) wears rapidly. | Test for pH, corrosion index. |

Figure 14.2  Suggestions for when a water supply should be tested.

Mercury

Nitrate-nitrogen

Odor

PH

Selenium

Silver

Sodium

Sulfate

Zinc

### Pesticides in Potable Water

Pesticides in potable water can be a very serious problem. Protecting well water from all forms of contamination is important. Understanding pesticide pollution and preventing it are essential parts of a private water supply. There are five primary types of pesticides to be aware of. The list is comprised of herbicides, insecticides, fungicides, nematocides, and rodenticides.

What do pesticides do to drinking water? They can make it lethal. Several factors come into play when pesticides are introduced into a water supply. The active ingredient in the pesticide formulation is a primary cause. Then there are the contaminants that exist as impurities in the active ingredient of the pesticide. Any additives that may be present in the pesticide can attack the water supply. Degradation of the water quality from the chemical, microbial, or photochemical agent is also possible.

Primary factors in assessing the toxicity of pesticides in water include the following:

- Toxicity

- Persistence

- Degradates

**Health Effects**

What are the health effects on humans who come into contact with pesticides? Cancer can be a big one. Pulmonary problems are also a prime risk. Hematological morbidity and inborn deformities are also possible. Immune-system deficiencies round out the top of the list of potential problems.

Humans may be affected by pesticides through skin contact, inhalation, and ingestion. In terms of potable water, ingestion is the primary concern.

Surface water is a major source of pesticide contamination in private water supplies. Runoff water can carry significant amounts of agricultural pesticides for long distances. This makes the protection of water wells from the infiltration of surface water a key concern.

**Protecting Private Wells**

Protecting private wells is not a complicated issue. Most of the work is common sense. The first and most important element of protection is a watertight well casing that extends down to bedrock. A watertight cover is also essential. What else should an installer be concerned with?

**Table 14.11  Testing for nuisance problems**

| Problem or concern | Recommended test |
|---|---|
| **Appearance**<br>  black flakes<br>  brown or yellow water | manganese<br>iron |
| **Stains on fixtures or clothing**<br>  red or brown<br>  black<br>  green or blue | iron<br>manganese<br>copper |
| **Odor or taste**<br>  rotten egg<br>  metallic<br>  salty<br>  septic, musty, earthy<br>  gasoline or oil | hydrogen sulfide<br>pH, iron, zinc, copper<br>total dissolved solids, chloride, sodium<br>total coliform bacteria, iron<br>hydrocarbon scan, volatile organic chemicals |
| **White deposits on pots and fixtures, hard-to-lather soap** | hardness |
| **Corrosion of plumbing** | corrosivity, pH, lead, iron, zinc, manganese, copper, sulfates, chloride |

How about where the well pipe or pipes enter the well casing? If a pitless adapter is used, which is normally the case for a submersible pump, a watertight gasket is intended to seal the hole made for the adapter. This gasket must be effective to protect the water supply. What happens in a shallow well when a hole is knocked into the side of a concrete casing and a pipe is inserted through it? Well, this is a bit tougher to seal. The open space around the pipe where it enters the casing and the space between the pipe and any sleeving material should be packed solid. At this point, a good mortar mix is applied to seal the opening. In all cases, the pipe penetration areas must be watertight and secure.

Okay, you have the casing, the cap, and the pipe penetrations secure. Is there anything else needed? Legally, you are probably good to go. However, there are some added security measures that can add to the protection of a well. For example, you could have the finished grade around a well sloped in such a way that ground water would not be able to reach the well casing or cap under any normal runoff conditions. This doesn't cost a lot of money, and it is good protection.

The location of a well is, of course, a major part of protecting the water source. Local regulations generally enforce the potential placement of wells. An installer should never install a well downstream of a potential contamination source. Minimum distance requirements exist in local codes, as we discussed much earlier in this book. Water is an essential element of life, so test it and protect it properly.

Figure 14.3  Special water testing considerations

| Situation | Test |
|---|---|
| Your well is in an area of intensive agricultural use. | Test for pesticides commonly used in the area, coliform bacteria, nitrate, pH, and TDS. |
| You live near a mining operation. | Test for iron, lead, arsenic, manganese, aluminum, pH, and corrosion index. |
| Your well is near a gas drilling operation. | Test for chloride, sodium, barium, and strontium. |
| Your water smells of gasoline or fuel oil and your well is located near an operating or abandoned gas station or near buried fuel storage tanks. | Test for fuel components or volatile organic compounds (VOC). |
| Your well is near a dump, junkyard, landfill, factory or dry cleaning operation. | Test for volatile organic chemicals (such as gasoline components and cleaning solvents), pH, TDS, chloride, sulfate, and metals. |
| Your well is near a road salt storage site or a heavily salted roadway and the water tastes salty or corrosion appears on pipes. | Test for chloride, TDS, and sodium. |
| You are concerned about radon where you live. | Test for radon. |

Chapter

# 15

# Drilled Wells

Drilled wells are the most dependable individual water source I know of. These wells extend deep into the earth. They reach water sources that other types of wells can't come close to tapping. Since drilled wells take advantage of water that is found deep in the ground, it is very unusual for drilled wells to run dry. This accounts for their dependability. However, dependability can be expensive.

Of all the common well types, drilled wells are the most expensive to install. The difference in price between a dug well and a drilled well can be thousands of dollars. But the money is usually well spent. Dug wells can dry up during hot summer months. Contamination of water is also more likely in a dug well. When all the factors are weighed, drilled wells are worth their price.

A house is a big investment. Like a house, a well can be looked upon as an investment. Nobody likes to spend more money than they have to. Builders don't want to run the prices of their projects up unnecessarily. Homebuyers prefer not to pay outrageous prices for the properties they purchase. In an effort to keep costs down, some builders look for the cheapest water source they can find. This can be a big mistake. How good is a house without water? Is it really a bargain to install a cheap well only to have it fail six months later?

I don't want to paint a picture that indicates the only type of well worth using is a drilled well. Dug wells can perform very well. Even driven wells can provide a suitable supply of potable water. Some houses get their water from springs and other natural water sources. Your selection of a water supply for the houses you build will depend on many factors. Drilled wells will not always be practical. However, deep wells with submersible pumps are hard to beat when you want to guarantee a good supply of drinking water.

When I built houses in Virginia, most of the wells used were dug wells. In Maine, drilled wells are much more common than dug wells. Geographic location makes a difference in the type of water source used for a house.

## Depth

The depth of a drilled well can vary a great deal. In my experience, drilled wells are usually at least 100 feet deep. Some drilled wells extend 500 feet or more into the earth. My personal well is a little over 400 feet deep. Based on my experience as both a builder and plumber, I've found most drilled wells to range between 125 to 250 feet deep. When you think about it, that's pretty deep.

It's hard for some people to envision drilling several hundred feet into the earth. I've had many homebuyers ask me how I was going to give them such a deep well. Many people have asked me what will happen if the well driller hits rock. Getting through bedrock is not a problem for the right well-drilling rig. While we are on the subject of drilling rigs, let's talk about the different ways to drill a well.

## Well-Drilling Equipment

Well-drilling equipment is available in various forms. While one type of rig may be the most common, all types of well rigs have advantages and disadvantages. As a builder, it can be helpful to know what your options are for drilling wells on different types of sites. There are two basic types of drilling equipment.

Rotary drilling equipment is very common in my area. This type of rig uses a bit to auger its way into the earth. The bit is attached to a drill pipe. Extra lengths of pipe can be added as the bit cuts deeper into the ground. The well hole is constantly cleaned out with air, water, or mud under pressure.

Percussion cable tool rigs make up the second type of drilling rig. These drilling machines use a bit that is attached to a wire cable. The cable is raised and dropped repeatedly to create a hole. A bailer is used to remove debris from the hole. Between rotary equipment and percus-

Drilled Wells 109

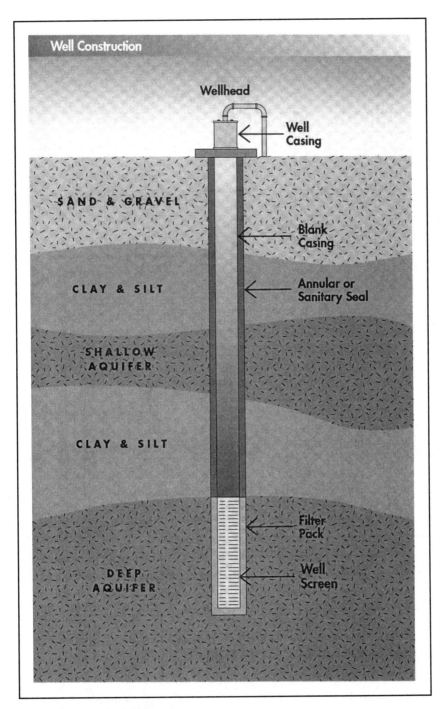

**Figure 15.1** Sample of well construction

## 110 Chapter Fifteen

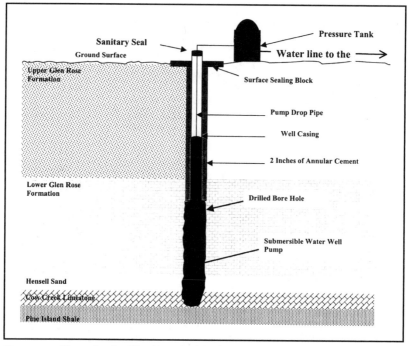

**Figure 15.2** Example of a Glen Rose Aquifer

sion cable equipment, there are a number of variations in the specific types of drilling rigs.

We could go into a lengthy discussion of all the various types of well rigs available. But since you are a builder and probably have no desire to become a well driller, there seems to be little point in delving into all the details of drilling a well. What you should know, however, is that there are several types of drilling rigs in existence. Your regional location may affect the types of rigs being used. A few phone calls to professional drillers will make you aware of what your well options are. For my money, I've also contracted the services of rotary drillers. There are times when other types of rigs could be a better choice. My best advice to you is to check with a number of drilling companies in your area and see what they recommend.

### The Basics

Let's go over the basics of what is involved, from a builder's point of view, when it comes to drilling a well. Your first step must be deciding on a well location. The site of a proposed house will, of course, have

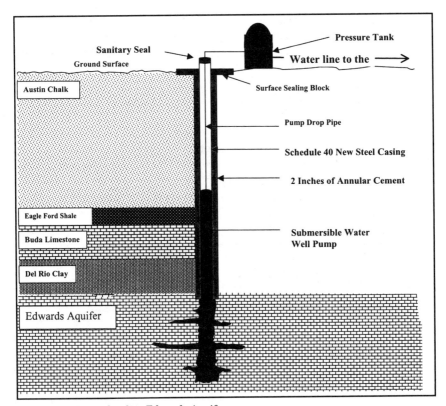

**Figure 15.3** Example of an Edwards Aquifer

some bearing on where you want the well to be drilled. Local code requirements will address issues pertaining to water wells. For example, the well will have to be kept at some minimum distance from a septic field, assuming that one is to be used.

Who should decide on where a well will go? Once local code requirements are observed, the decision for a well location can be made by a builder, a homebuyer, a well driller, or just about anyone else. If you're building spec houses, the decision will be up to you and your driller. Buyers of custom homes may want to take an active role in choosing a suitable well site. As an experienced builder, I recommend that you consult with your customers on where they would like their well. Some people are adamant about where they do and do not want a well placed.

I often allow my customers to choose preferential well sites. Many of the homebuyers leave the decision on well placement up to me. Sometimes a customer will choose a site that is not practical. For example, a well-drilling rig might not be able to get access to the location easily. When a customer makes what I perceive to be a bad decision, I offer recommendations. Working out an agreeable location is rarely a problem.

I never leave well location to the discretion of well drillers. Some drillers will take the path of least resistance when installing a well. This can result in some very unpleasant well sites. Drilled wells are not as obtrusive as dug wells, but they still don't make good lawn ornaments. If you don't want to arrive on your job site to find a freshly drilled well in the middle of the front lawn, don't allow well drillers to pick their sites at random.

An experienced driller can provide you with a lot of advice when it comes to picking a spot for a well. Don't overlook this opportunity. It always pays to listen to experienced people. Once you or your customer have chosen a well site, talk the decision over with your driller. You may find that there is a good reason to move from the intended site to a more suitable one. While I feel that you, or a trusted supervisor, should

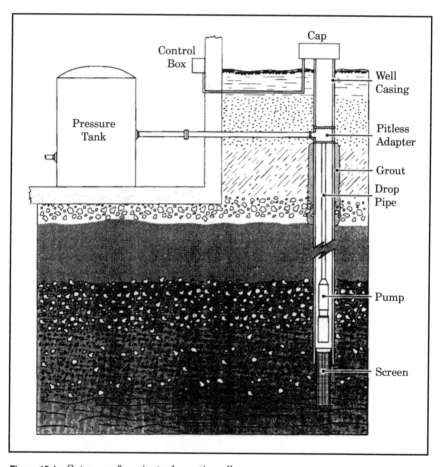

**Figure 15.4** Cutaway of a private domestic well

take an active part in picking a well location, I don't think that you should go against the recommendations of a seasoned well driller. If a reputable driller advises you to choose a new location, you should seriously consider doing so.

## Access

Access is one of the biggest concerns a builder has in the well-drilling process. It is a builder's responsibility to make access available to a well driller. Drilling rigs require a lot of room to maneuver. A narrow, private drive with overhanging trees may not be suitable for a well rig.

Having enough width and height to get a well truck into a location is not the only consideration. Well rigs are heavy—very heavy, in fact. The ground that these rigs drive over must be solid. New construction usually requires building roads or driveways. If you can arrange to have a well installed along the roadway, your problems with access are reduced.

It is not always desirable to install a well alongside a driveway. This may not cause any additional concern, but it could. If the ground where you are working is dry and solid, a well rig can drive right over it. But, if the ground is wet and muddy or too sandy, a big truck won't be able to cross it. You must consider this possibility when planning a well installation.

Don't build obstacles for yourself. Just as the old joke goes about painting yourself into a corner, you can build yourself right out of room. Some builders put wells in before they build houses. Others wait until the last minute to install a well. Why do they wait? They do it to avoid spending the money for a well before it's needed. This reduces the interest they pay on construction loans and keeps their operating capital as high as possible. When the well is installed last, it can often be paid for out of the closing proceeds from the sale of a house.

I have frequently waited until the end of a job to install wells. There is some risk to this method. It's possible, I suppose, that a house could be built on land where no water could be found with a well. This would truly be a mess. A more likely risk is that the house construction will block the path of a well rig.

When you are confirming the location for a well, make sure that you will be able to get drilling equipment to it when you need to. Going to the buyers of a custom home and informing them that their beautiful shade trees will have to be cut down to get a well rig into the site is not a job I would enjoy having. It is safer to install wells before foundations are put in.

## Working On Your Site

Once you have a well driller working on your site, there's not much for you to do but wait. The drilling process can sometimes be completed in a single day. There are times, however, when the rig will be working for longer periods of time. Depending on the type of drilling rig being used, a pile of debris will be left behind. This doesn't usually amount to much, and your site contractor can take care of the pile when preparing for finish grading.

The well driller will usually drill a hole that is suitable for a six-inch steel casing. The casing will be installed to whatever depth is necessary. Once bedrock is penetrated, the rock becomes the well casing. How much casing is needed will affect the price of your well. Obviously, the less casing that is needed, the lower the price should be.

Your well driller will grout the well casing as needed to prevent ground water from running through the casing and dropping down the well. This is normally a code requirement. It limits the risk of contamination in the well from surface water. A metal cap is then installed on the top edge of the casing, and then the well driller's work is done.

## Pump Installers

Many well drillers offer their services as pump installers. You might prefer to have the driller install the pump, or maybe you would rather have your plumber do the job. A license may be needed in your area for pump installations. Any master plumber will be allowed to make a pump installation, but you should check to see if drillers are required to have special installation licenses. If they are, make sure that any driller you allow to install a pump is properly licensed.

If you want your driller to install a pump setup as soon as the drilling is done, you will need to make arrangements for a trench to run the water service pipe in. The cost of this trench is usually not included in prices quoted by pump installers. Watch out for this one, because it is an easy way to lose a few hundred dollars if you don't plan on the cost of the trench.

After your well is installed, you should take some steps to protect it during construction. Heavy equipment could run into the well casing and damage it. I suggest that you surround the well area with some highly visible barrier. Colored warning tape works well, and it can be supported with nothing more than some tree branches stuck in the ground. A lot of builders don't take this safety precaution, but they should. A bulldozer can really do a number on a well, and a well casing can be difficult for an operator to see at times.

How far will the pump installer take the job? Will the water service pipe be run just inside the home's foundation and left for future hookup? Who will run the electrical wires and make the electrical connections? Does the installation price include a pressure tank and all the accessories needed to trim it out? Who will install the pressure tank? You need answers to these questions before you award a job to a subcontractor. Pump systems involve a lot of steps and materials. It's easy for a contractor to come in with a low bid by shaving off some of the work responsibilities in fine print. Be careful, or you may wind up paying a lot more than you planned to for a well system.

We have now covered the installation of a drilled well and a pump system in an abbreviated version. There are some particulars, however, that you should be aware of, so that you can supervise the work more effectively. Let's concentrate on these issues now.

## Quantity And Quality

When it comes to the quantity and quality of water produced by a well, few (if any) well drillers will make commitments. Every well driller I've ever talked to has refused to guarantee the quantity or quality of water. The only guarantee that I've been able to solicit has been one of hitting water. This is an issue that builders have to be aware of.

Customers may ask you, the builder, to specify what the flow rate of their new well will be. It is beyond the ability of a builder to do this. An average, acceptable well could have a 3-gallon-per-minute recovery rate. Another well might replenish itself at a rate of 5 gallons per minute. There are wells with much faster recovery rates, and there are those with slower rates. Anything less than 3 gallons per minute is less than desirable, but it can be made to suffice.

How can you deal with the recovery rate? You really can't. Sometimes by going deeper, a well driller can hit a better aquifer that will produce a higher rate of recovery, but there is no guarantee. So don't make a guarantee to your customers. Show your customers the disclaimers on the quotes from well drillers and use that evidence to back up your point that there is no guarantee of recovery rate in the well business. Now I could be wrong. You might find some driller who will guarantee a rate, but I never have.

Even though you can't know what a recovery rate will be when you start to drill a well, you can determine what it is after water has been hit. Your well driller should be willing to test for and establish the recovery rate for you. Every driller I've ever used has performed this service. You need to know what the recovery rate is in order to size a pump properly. There is an old-fashioned way of determining a recovery rate, but I'd ask my driller to do it for me.

It is also important that you know the depth of the well, and this is something that your driller can certainly tell you. As you gather information, record it for future reference. The depth of the well will also be a factor when selecting and installing a pump. Don't let your driller leave the job until you know what the depth and the recovery rate are.

If you're the type of person who likes to do a little engineering on your own, you can refer to information provided by pump manufacturers to calculate your pump needs.

### Quality

The quality of water is difficult to determine when a new well is first drilled. It can take days or even weeks for the water in a new well to assume its posture. In other words, the water you test today may offer very different results when tested two weeks from now. Before a true test of water quality can be conducted, it is often necessary to disinfect a new well. Many local codes require disinfection before testing for quality.

Wells are usually treated with chlorine bleach to disinfect them. Local requirements on disinfection vary, so I won't attempt to tell you exactly what to do. In general, a prescribed amount of bleach is poured into a well. It is allowed to sit for some specific amount of time, as regulated by local authorities. Then the well pump is run to deplete the water supply in the well. As a rule of thumb, the pump is run until there is no trace of chlorine odor in the tap water. When the well replenishes itself, the new water in the well should be ready for testing. But, again, check with your local authorities for the correct procedure to use in your area. By the way, builders usually are responsible for conducting the disinfection process.

Once the well is ready to test, a water sample is taken from some faucet in the house. Test bottles and collection instructions are available from independent laboratories. Follow the instructions provided by the testing facility. As a rule of thumb, you should remove any aerator that may be installed on the spout of a faucet before taking your water collection. It is often recommended that a flame be held to the spout to kill bacteria that may be clinging to the faucet and washed into the collection bottle. Don't attempt this step when testing from a plastic faucet! Many plumbers take water tests from outside hose bibbs, and this is fine. A torch can be used to sterilize the end of the hose bibb, and a hose bibb provides almost direct access to water.

When collecting water for a test, water should be run through the faucet for several minutes before catching any for a test. It is best to drain the contents of a pressure tank and have it refill with fresh well water for the test. You can run reserve water out quickly by opening the cold water faucet for a bathtub.

Once a water sample is collected, it is taken or mailed to a lab. Time is of the essence when testing for bacteria, so don't let a water bottle ride around in your truck for a few days before you get around to mailing it. Again, follow the instructions provided by the lab for delivering the water.

There are different types of tests that you can request the lab to perform. A mandatory test will reveal if the water is safe to drink. If you want to know more, you have to ask for additional testing. For example, you might want to have a test done to see if radon is present. Many wells have mineral contents in sufficient quantity to affect the water quality. Acid levels in the well may be too high. The water could be considered hard, which can make washing with soaps difficult and can leave stains in plumbing fixtures. There are a number of potential tests to run.

Drinking water, or potable water as it is often called, is your primary goal when drilling a well. It is rare that a drilled well will not test well enough to meet minimum requirements for safe drinking. In fact, I've never known of a drilled well that wasn't suitable for drinking. Nevertheless, an official statement of acceptability is generally required by code officials and lenders who loan money on houses.

If you get into a discussion on water quality, you may have to pinpoint exactly what it is that you are talking about. Is it only safety? Or does it extend to mineral contents and such? Most professionals look at water quality on an overall basis, which includes mineral contents.

### Supervising the Drilling

Supervising the drilling of a new well is not something that you should have to worry about. Standing around all day, watching a well rig drill, is pretty boring and not very productive. Once you have chosen a well location and have made sure the driller is drilling in the proposed area, you can pretty much ignore the drilling process until it's done. Near the end of the job, you might want to make an appearance to see what height the well casing is being set at and to obtain information about the well depth and flow rate.

## Trenching

Trenching will be needed for the installation of a water service and for the electrical wires running out to a submersible pump. It is possible to pump water from a deep well with a two-pipe jet pump, but submersible pumps are, in my opinion, far superior. When a submersible pump is used, it hangs in the well water. Electrical wires must be run to the well casing and down into the well. The wires and the water service pipe can share the same trench.

As with any digging, you must make sure that there are no underground utilities in the path of your excavation. Most communities offer some type of underground utility identification service. In many places, one phone call will be all that is needed to get all underground utility locations on your work site marked. It may, however, be necessary in some parts of the country to call individual utility companies. I expect that you are familiar with this process; most builders are.

Once you have a clear path to dig, a trench must be dug to a depth that is below the local frost line. The water service pipe will have water in it at all times, so it must be buried deep enough to avoid freezing. How deep is deep enough? It varies from place to place. In Maine, I have to get down to a depth of 4 feet to be safe from freezing. In Virginia, the frost line was set at 18 inches. Your local code office can tell you what the prescribed depth is in your area.

After a trench is opened up, you can arrange for your pump installer. Most installers will want to do all of their work in one trip. This means that you must have enough of the house built to allow an installer to bring the water pipe through the foundation and to set the pressure tank. As a helpful hint, you should install a sleeve in your foundation as it is being poured so that the pump installer will not have to cut a hole through the foundation.

Check with your local plumbing inspector to determine what size sleeve will be needed. Most plumbing codes require a sleeve in a foundation wall to be at least two pipe sizes larger than the pipe being installed. For a typical 1-inch well pipe, this would mean that the sleeve would have to be at least 2 inches in diameter. But again, check with your local code office, because plumbing codes do vary from place to place.

At this point, you are ready to have your pump system installed. The pump might be installed by your well driller or your plumber. Wiring for the pump should be done by a licensed electrician. All of your subcontractors should, of course, carry liability insurance that protects you and the property owner from their mistakes and accidents.

## Installing A Pump System

Installing a pump system for a drilled well is not a difficult procedure. It is, however, a job that is usually required to be done by a licensed professional. Even if you know how to do all of the work, you probably can't do it legally without a license. Since almost everyone uses submersible pumps when dealing with deep wells, I'll base my examples on the assumption that you, too, will be using a submersible pump.

Since you are a builder rather than a pump installer, I will not go into every little detail of a pump installation. I will cover all the key points, but without the step-by-step instructions that I would give someone

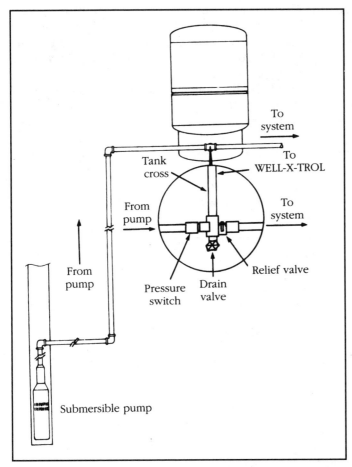

**Figure 15.5** Pressure tank in use with a submersible pump: *(Courtesy of Amtrol, Inc.)*

wishing to learn the trade. In other words, I'm not going to waste your time with instructions to apply pipe dope or how tight to turn a fitting, or how to carry out every other little plumbing process. What I will tell you is how to make sure that the installer you choose is doing a good job.

## At The Well Casing

Let's start our work at the well casing. You should be looking at an empty trench and the side of a steel well casing when a pump installation begins. A hole has to be cut through the side of the well casing to allow what is called a pitless adapter to be installed. A cutting torch can be used to make this hole, but most installers use a metal-cutting hole saw and a drill. The size of the hole needed is determined by the pitless adapter. Keeping the hole at proper tolerances is important. If the pitless adapter doesn't fit well, ground water might leak into the well by getting past the gasket provided with the pitless adapter.

The hole in the side of the well casing should be positioned so that a water pipe lying in the trench will line up with the pitless adapter once it is installed. Once the hole has been cut, the pitless adapter needs to be installed. This is really a two-person job. One person will work down in the trench, while another person will work from above the main opening in the well casing.

A long, threaded pipe is needed to position the piece of the pitless adapter that is installed inside the well. Most plumbers make up a T-

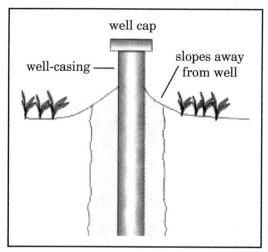

**Figure 15.6** Example of a well casing and cap that is installed properly.

shaped pipe tool to use for this part of the job. The pitless adapter is screwed onto the threads of the T-shaped tool. With this done, one person lowers the pitless adapter down the well and into position. A threaded protrusion on the adapter is intended to poke through the hole in the side of the well casing. When it does, the person in the trench installs a gasket and the other part of the pitless adapter, which is basically a retaining ring. The ring is tightened to create a watertight seal and to hold the pitless adapter in place. At this point, the T-shaped tool is unscrewed from the pitless adapter and laid aside. As a builder, you should check to see that the pitless adapter is tight and in position to prevent contaminated ground water from entering around the hole.

Most installers put their pump rigs together next. This involves one of two things: either a truck that is equipped with a reel system for well pipe or enough room to lay all of the well pipe out in a fairly straight line. With deep wells, this can be a problem. It is important, however, that the well pipe be laid out and not allowed to kink, unless it is being fed into the well from a reel system.

The pipe used for most wells comes in long coils. It is best to avoid joints in well pipe whenever possible. With the long lengths of pipe available, there should be no reason why a joint would be needed to splice two pieces of pipe together. When joints and connections are made with standard well pipe, it is best to use metallic fittings. Nylon fittings are available, but they may break under stress more quickly than metal fittings.

Polyethylene (PE) pipe is the type used most often for well installations. This black plastic pipe is sold in coils and can be used for both the vertical drop pipe in the well and the horizontal water service in the trench. In houses where hot water will be available, PE pipe must not extend more than 5 feet inside the foundation wall. The plumbing code requires the same pipe type to be used for water distribution of hot and cold water. PE pipe is not rated to handle hot water, so it cannot be used as an interior water distribution pipe.

Other types of pipe can be used for well systems, but PE pipe remains the most popular. This pipe has some drawbacks for installers. The material will kink easily. If the pipe kinks, it should not be used. The bend in the pipe weakens it. If you see an installer kink a pipe, make sure that the section of pipe that was kinked is not used. Since couplings are undesirable both in the well and in the trench, a kinked pipe may mean getting a whole new roll of pipe, depending upon where the kink occurs.

Another fault of PE pipe is its tendency to become very hard to work with in cold weather. Fittings, which are an insert type, are difficult to push into the pipe when it's cold. Warming the ends of the pipe with a torch or heat gun will make the material pliable and easy to work with.

However, care must be used to avoid melting the pipe. All connections made with PE pipe should be made with two stainless-steel hose clamps. One clamp is all that is required by code, but a second clamp provides cheap insurance against leaks.

When a coil of PE pipe is unrolled, it will take some work to straighten it out. This is basically a two-person job, although I have done it alone. The pipe should be stretched out as straight as possible, without kinking it, and then the pipe must be manipulated in a looping motion to make it lay flat. This step of the installation process should not be ignored.

Once the well pipe is lying flat, an installer can proceed with putting the pump and accessories together with the pipe. A torque arrestor should be installed on the pipe. This will limit vibration in the well as the pump runs. A male insert adapter (brass please) should be used to connect the pipe to the pump. Another insert adapter will connect the pipe to the pitless adapter fitting that has yet to be installed.

Electrical wires must run from the pump to the top of the well casing. Some plumbers use electrical tape to secure electrical wire to the well pipe. This is a common procedure. Other installers use plastic guides that slide over the well pipe to secure the wires. Either way, someone has to make sure that the wires are secure. It is also very important to make sure that the wires are not damaged as the pump assembly is lowered into the well. Waterproof splice kits can be used to join electrical wires that are extending into a well, and of course, waterproof wire is required.

All of the piping and wiring is connected to a submersible pump before the pump is installed. This work is normally done near the well on the ground. Once the entire assembly is put together, it is lowered into the well. Some companies have special trucks for this part of the job, but a lot of installers do it the old-fashioned way, by hand.

When the pump assembly is put down the well, the work goes much better if two people are involved. Lowering a pump assembly by hand is not difficult. The T-bar tool is once again connected to the part of a pitless adapter that is wedge-shaped and attached to the upper end of the drop pipe. Smart installers attach nylon rope to the submersible pump as a safety rope. I've seen a number of installers skip this step, but I won't allow it on my jobs. I insist on a safety rope being installed.

Submersible pumps are held in the well by only the well pipe unless a safety rope is used. If a fitting pulls loose, an expensive pump can be lost in the well forever. The rope, which is ultimately tied to the top of the well casing, gives you something to retrieve the pump with if anything goes wrong with the piping arrangement. For the low cost of nylon rope, it's senseless not to use it.

The pump is then lowered into the well. During this process, the person working the assembly down into the casing must take care not to scrape the electrical wires along the edge or sides of the casing. If the insulation on the wires is cut by the casing, the pump may fail to operate. Once the pump is in position, the part of the pitless adapter piece that is secured to the well casing is put into place. This is done simply by lining the wedge up with the groove in the stationary piece and tapping it into place. Then the safety rope is adjusted and tied off to the casing. The T-bar is unscrewed and removed. The only step remaining at the well head is the connection of electrical wires and the replacement of the well cap. An experienced crew can put together and install a pump assembly in about two hours or so.

**The Water Service**

The next step in the installation of a pump system is the water service. This pipe might be the same type as is used in the well, or it may be some other type. Copper tubing was used as a water service material for years. It is still used at times. Most installers use PE pipe, and there is no reason not to. It's less expensive than copper, it's not affected by acidic water in the same way that copper is, and PE pipe has a long life span.

PE pipe for a water service is prepared in the same way that it would be for use as a drop pipe. It is laid out and straightened to get the loops out. When this is complete, it is placed in the trench. One end of the pipe connects to the protrusion from the pitless adapter. This connection is made with two hose clamps. The other end of the pipe is placed through a foundation sleeve and extended into a home. The entire length of the pipe should be lying flat on the bottom of the trench. There should not be any rocks or other sharp objects under or around the pipe. This is important, and it is something that not all installers are especially careful about, so you might want to check it yourself. Any sharp objects may puncture the pipe during the backfilling process. It is also possible for rough or sharp objects to cut the pipe months after installation as the ground settles and the pipe moves. Make sure the trench and the backfill material are free of objects which may harm the pipe.

Water service pipes often have to be inspected before they are covered up. This is not directly your responsibility, but you should make sure that an inspection has been done if one is required. Backfilling the trench should be done gradually. If someone takes a backhoe and pushes large piles of dirt into the ditch, the pipe might kink or collapse.

## 124 Chapter Fifteen

Require the backfilling to be done in layers so that there will not be excessive weight dumped on the pipe all at once.

**Inside the Foundation**

The rest of the work will take place inside the foundation. A pressure tank should be used on all jobs. This tank gives a reserve supply of water that allows the house to have good water pressure. The tank also preserves the life of the pump. Without a pressure tank, a water pump would have to cut on every time a faucet was turned on. This short cycling of the pump would wear it out quickly. A pressure tank removes the need for a pump to cut on every time water is needed. Until the tank drops to a certain pressure, the pump is not required to run. When the pump refills the tank, it runs long enough to avoid short cycling.

The size of a pressure tank is normally determined by the number of people and plumbing fixtures expected in a house. Larger tanks require the pump to run less often. In addition to the pressure tank, there are

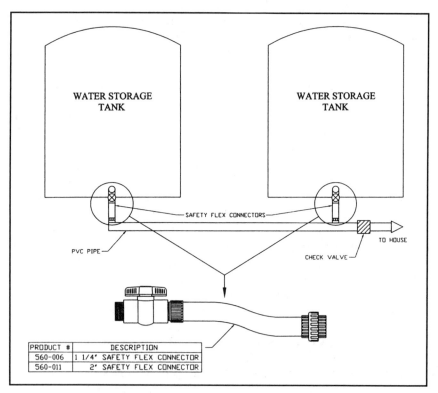

**Figure 15.7** Example of a multiple-tank installation for pressure tanks.

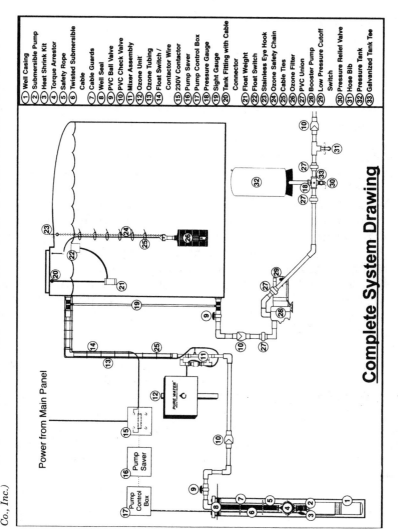

Figure 15.8 Complete installation diagram for a submersible pump system: *(Courtesy of American Tank Co., Inc.)*

many accessories that must be installed. For example, there will be a pressure gauge, a relief valve, a drain valve, a pressure switch, and a tank tee. Electrical regulations may call for a disconnect box at the tank location. Even if your local code doesn't require a disconnect box, it's a good idea to install one.

Under normal circumstances, an experienced installation crew can complete an entire pump installation, including inside work, in less than a day. As a builder, you will have to hang around the job if you want to see that all of the work is being done correctly, because it will happen quickly. However, you can tell a lot about the workmanship by making periodic inspections.

## The Finer Points

We didn't discuss the finer points of pump installations, but you now have a good idea of what goes on with the process. There are a few other issues that we should cover. One is pump selection. This decision is usually left up to the driller or plumber who will supply and install the pump. It's best not to undersize a pump. While it makes no sense to buy an extremely powerful pump that is not required, it is in your customer's best interest to get a pump that exceeds its requirements by some degree. Getting a pump that is too powerful can cause some problems. Let me explain.

Pumps are rated on their output in gallons per minute (GPM). Sizing charts are available to show you what pumps of various sizes are capable of producing. Let's say that you have a well with a recovery rate of 3 GPM. Can you imagine what might happen if you install a pump that is rated at 5 GPM? If you guessed that the well might be pumped dry, you're right. The pump should not produce more water than the well can replenish. Keep this in mind when you are looking over the specifications provided to you by bidding subcontractors. Compare the pump rate with the well rate, and make sure that the pump is not too powerful for the well.

Pumps should not be suspended too close to the bottom of a well. How far from the bottom should a pump be placed? It depends on the static water level in the well. If a pump is placed too close to the bottom of a well, it can pick up gravel, sand, and other debris. Some wells fill in a little over time, and having a pump too close to the bottom in this circumstance is bad news. A pump should be, in my opinion, at least 15 feet above the bottom, and higher when practical.

My personal pump is about 30 feet above the bottom of the well. The depth of my well is a little over 400 feet. The static level of my well water is only about 15 feet from the top of my well casing. This means that I have a column of water that is about 385 feet deep. That's a substantial reserve. I would have to pump hundreds of gallons of water at one time to run the well dry. Since I have so much water, it's easy for me to keep my pump hanging high above the bottom. Not all wells have so much reserve water, and this can force an installer to hang a pump closer to the bottom. The key is to keep the pump far enough from the bottom to avoid problems with debris being sucked into the pump.

### Problems

Problems sometimes arise with well installations. What would you do if the trench for your water service could not be dug deep enough to protect the pipe from freezing? This doesn't happen often, but it does happen. In fact, it happened at my house. Bedrock was between 18 and 24 inches down, so getting to a depth of 4 feet was not feasible. Winters are very cold in Maine, so there was no doubt that my water service would freeze if left at such a shallow depth and unprotected. You might run into a similar situation.

In my case, I took three precautions. I insulated the water service pipe with foam pipe insulation. Then I ran it through a continuous sleeve of 3-inch plastic pipe. The air space between the sleeve and the water service adds to frost protection. My final and most effective step was to install an in-line heat tape. This type of heat tape is expensive, but it's a bargain when you are faced with difficult situations such as mine.

The heat tape is waterproof. It runs the full length of the water pipe. The heating element is inside the water pipe, in the water. A sensor and control are installed in some practical location. Mine is in a crawl space, with the temperature sensor positioned by an air vent. This allows the sensor to feel the outside temperature.

During the winter, I plug in the control unit, which is thermostatically controlled, and the heat tapes runs automatically. It heats the water in the pipe to a point above freezing. I used it all of last winter and experienced no problems. If you can't dig as deep as you would like to, I strongly recommend an in-line heat tape.

Chapter

# 16

# Shallow Wells

Dug wells and bored wells, or shallow wells as they are often called, are very different from drilled wells. While most drilled wells have diameters of 6 inches, a typical dug well will have a diameter of about 3 feet. Concrete usually surrounds a dug well, rather than the steel casing used for a drilled well. While drilled wells often reach depths of 300 feet, a dug well rarely runs deeper than 30 feet. There are, to be sure, many differences between the two types of wells.

In the old days, dug wells were created with picks, shovels, and buckets. It was dangerous work. Today, boring equipment is normally used to create a dug well. I suppose it would be more proper to call these wells bored wells, but most professionals I know refer to them as dug wells. To be specific on the type of well I'm talking about, let me explain the basic makeup.

If you see a concrete cylinder that has a diameter of approximately 3 feet sticking up out of the ground, you are looking at what I call a dug well. The concrete casing will typically be covered with a large, heavy concrete disk. You might want to call this type of well a bored well or a dug well; it's up to you. For my purposes, they are dug wells.

Old dug wells were dug by hand. Many of them were lined with stones. I've crawled down a few of these as a plumber to work on pipes and such, but I don't think I'd do it today. My younger years as a plumber were more adventuresome than I would care to repeat.

Dug wells are always shallow in comparison with drilled wells. Finding a dug well that is 50 feet deep would be similar to finding

**130 Chapter Sixteen**

pirate treasure sitting on top of a sandy beach. Most dug wells that I've seen have been no more than 35 feet deep. Bored wells can run much deeper. It's possible for a bored well to reach a depth of 100 feet or more. The depth is regulated by the ground the well is dug in. Some types of earth allow for a deeper well than others. Still, in practical terms, most dug or bored wells won't exceed 50 feet in depth. Beyond this level, a drilled well is more practical.

Large-diameter wells go by many names. I call them dug wells. Sometimes they are bored wells. A lot of people know them as shallow wells. This is why jet pumps, which are often used with these types of wells, are called shallow-well pumps. Without trying to get extremely technical, I will simply refer to these wells as dug wells. You now know that I might be lumping bored wells into the category, and that both types are considered shallow wells, so we shouldn't have a problem understanding each other.

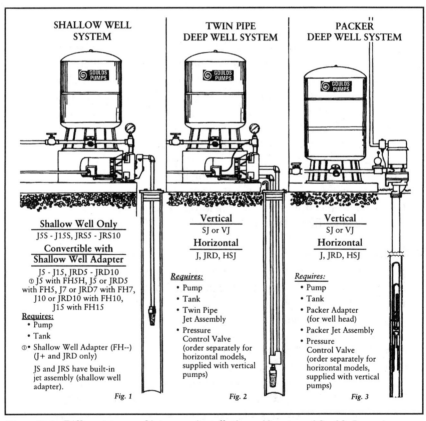

**Figure 16.1** Different types of jet pump installations: (*Courtesy of Goulds Pumps*).

## Shallow Wells

Shallow wells are common in many parts of the country. When the water table is high and reasonably constant, shallow wells work fairly well. They may dry up during some hot, dry times of the year, but this is not always the case. Some shallow wells maintain a good volume of water all through the year.

Most of the homes that I've owned have had shallow wells. My present home has a drilled well, and I feel much more secure about it than I ever did with shallow wells. But I only ever had trouble with one of my shallow wells and that trouble only occurred for one summer. At other times and with my other wells, I never experienced any problems.

Shallow wells are much less expensive to install than drilled wells. This is one good reason for using a shallow well. But there are drawbacks as well. Having a sufficient water quantity is one of these drawbacks. Due to the large diameter of a dug well, a lot of water can be stored in reserve. While the water in a shallow well may be only a few feet deep, it has a lot more surface area than water stored in a deep well. The increased surface area helps to make up for the lack of depth.

Even with a large diameter, dug wells do often run dry for short periods of time. If they don't run completely dry, they may contain such a small quantity of water that rationing is needed. For example, you may have to go to a local laundromat to wash clothes for a few weeks out of the year. This, of course, is more inconvenience than some homeowners are willing to put up with.

### When Should You Use A Shallow Well?

When should you use a shallow well? The ground conditions where the well will be drilled have a lot of influence on this decision. If bedrock is near the ground surface, a dug well is not practical. Drilled wells are the answer when you have to penetrate hard rock. To give you a good idea of what depths are possible and in what types of soil conditions different types of wells can be used, let me break them down into categories.

### True Dug Wells

True dug wells can be created with depths of up to 50 feet. Their diameter may range from a common 3 feet to a massive 20 feet. In terms of geographic formations, clay, silt, sand, gravel, cemented gravel, and

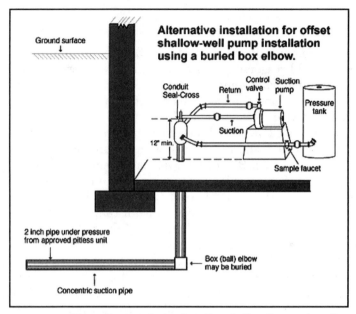

**Figure 16.2** Alternative installation for offset shall-well pump installation using a buried box elbow

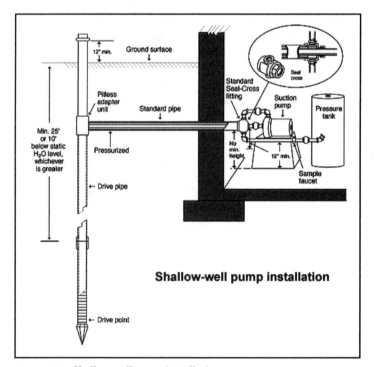

**Figure 16.3** Shallow-well pump installation

even boulders can be dealt with. Sandstone and limestone may allow the use of a dug well, but the material must either be soft or fractured. Dense igneous rock cannot be penetrated with a dug well.

## Bored Wells

Bored wells can reach depths of 100 feet. The diameter of this type of well can be a minuscule 2 inches. Larger diameters in the 30-inch range are also common. Clay, silt, sand, and gravel can all support a bored well. Cemented gravel will stop a bored well, and so will bedrock. Boulders can sometimes be worked around, and sandstone and limestone affect a bored well in the same way that they do a dug well.

## Drilled Wells

Drilled wells can run to depths of 1,000 feet. This is far from being a shallow well. All of the geologic formations mentioned for the well

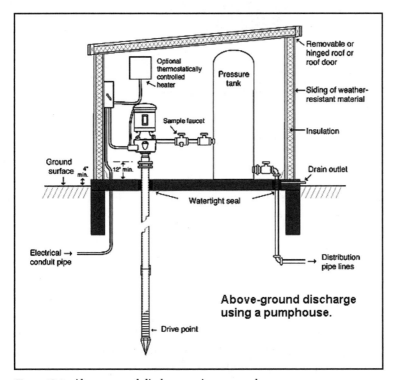

**Figure 16.4** Above-ground discharge using a pumphouse

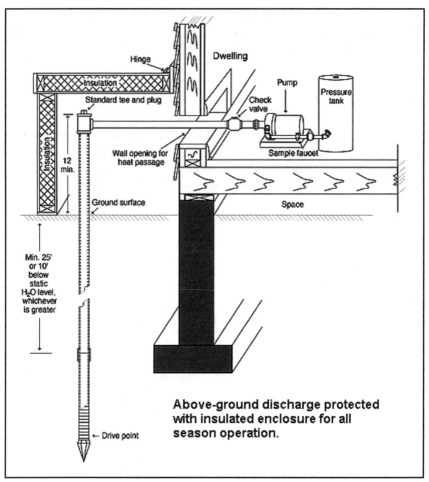

**Figure 16.5** Above-ground discharge protected with insulated enclosure for all season operation

types above can be overcome with a drilled well. Percussion and hydraulic rotary drilling are very similar in their abilities.

**Jetted Wells**

Jetted wells can run up to 100 feet in depth. They cannot be installed in bedrock, limestone, or sandstone. Boulders, cemented gravel, and large, loose gravel all prohibit the use of jetted wells. Clay, sand, and silt are the best types of geologic formations to use a jetted well in. The diameter of a jetted well may be anywhere from two to twelve inches.

Shallow Wells    135

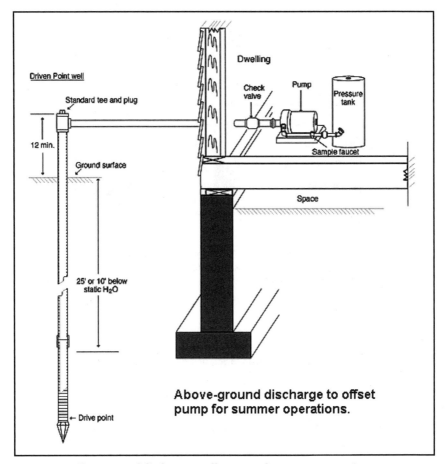

Figure 16.6  Above-ground discharge to offset pump for summer operations

## Characteristics

Let's talk about some of the characteristics of shallow wells. Dug wells can extend only a few feet below a water table. The geologic conditions affect the recovery rate of a well. For example, a dug well that is surrounded by gravel might produce an excellent recovery rate, while one having fine sand surrounding it may produce a poor rate of recovery. Dug wells cannot be considered terrific water producers.

Bored wells are a little different. They can extend deeper into water table. Going 10 feet below the edge of a water table is not uncommon. This added depth gives bored wells an advantage over dug wells. Having an 8-foot head of water with a 3-foot diameter provides a good supply of water, assuming that the recovery rate is good. It would not

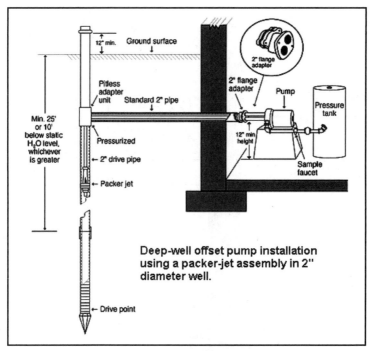

**Figure 16.7** Deep-well offset pump installation using a packer-jet assembly in 2" diameter well

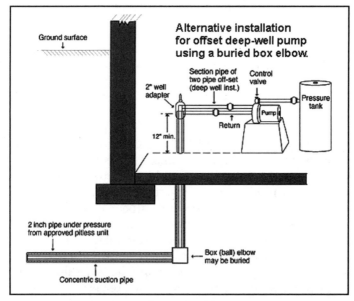

**Figure 16.8** Alternative installation for offset deep-well pump using a buried box elbow

be considered strange or unusual for a bored well to have a recovery rate of 20 GPM. This is about double what would be expected of a good dug well.

Even though many professionals, myself included, call all shallow wells dug wells, there is clearly a difference between a bored well and a true dug well. Most shallow wells installed today are bored wells. If your well installer talks about a dug well, make sure that the installer is actually referring to a bored well.

Some bored wells hit artesian aquifers. When this happens, the static water level in the well rises. For example, a bored well that extends 10 feet past the water table would normally be thought of as holding 10 feet of water. If an artesian aquifer is hit, the actual water reserve could be 20 feet deep or more. I've seen bored wells with such powerful artesian effects that water actually ran out over the top of the well casing. It is certainly possible to obtain plenty of water for a residence with a bored well. But, there is no guarantee of it.

**Look Around**

If you are unsure of what type of well to use, look around at the wells being used for houses near your building site. If you see a lot of concrete casings sticking up out of the ground, it's a good indication that shallow wells work in the area. A host of steel casings indicate a need for a drilled well. Looking at neighboring homes is a good way to get a feel for what type of well is likely to be needed. It is not, however, a sure bet.

If you want more dependable information on which type of well is likely to be suitable for your project, talk to some experts. Well installers are an excellent place to start. Local installers should know a great deal about the pros and cons of various types of wells in the region where you plan to build. But be careful. You may run across an installer who is not equipped to install all types of wells. If this is the case, it's possible that the recommendations made may be in the best interest of the installer rather than you or your customer.

Most rural areas have county extension offices or some similar facility to help with all sorts of issues. Wells fall into this category. A visit to the right local agency may reveal detailed maps that show aquifers and soil types. If you can pinpoint your building location on a map of this type, you can make some sound decisions yourself on the type of well to use.

Combined research is the true key to success. Visit your county extension office and look over the maps. Take an inventory of wells on surrounding properties. Talk to several well installers. Combine all the knowledge and advice you collect to make a wise decision. Don't be afraid to install a bored well if evidence points to it being a good choice.

As a builder in Virginia, I was responsible for the installation of dozens upon dozens of shallow wells. To the best of my knowledge, only one of the wells ever gave anyone any trouble, and that was one on my personal property. When the conditions are right and the job is done properly, a shallow well can give years of quality service at an affordable price.

### The Preliminary Work

The preliminary work for a builder who is having a shallow well installed is very similar to that of a builder who is preparing for the installation of a drilled well. Since we covered the steps for this in the last chapter, I won't repeat them here. The only big difference is the size of a shallow well. A drilled well can be hidden easily with a few strategically placed shrubs. The casing and cap of a shallow well is much harder to hide. Aside from the size of the well, the rest of the steps are about the same.

### After A Well Is In

After a well is in, you are faced with the disinfection process. Just as we discussed in the last chapter, chlorine is normally used to purify a well before testing is done on the water. There are a few extra things to consider, however, when testing a shallow well. For one thing, the well cover is heavy. One person can slide the cover aside, but it is easier with two people. The concrete cover is brittle. Rough handling can cause it to crack or break. Having a concrete cover drop and catch your finger between the lip of the well casing and the cover is extremely painful. Be careful when handling the cover.

The odds of a child or an animal falling into an uncovered drilled well are much lower than they are with a shallow well. Two adults could jump into the opening of a bored well without any problem. Never leave the cover off of a shallow well when you are not right at the well site. Even if you are just going into the house to turn the water on or off, put the cover back on the well. Pets and people can fall into an open well quickly, and the results can be disastrous. Even drilled wells should be covered every time they are left unattended. Curious kids and pets can find themselves in some very dangerous circumstances in the blink of an eye.

When you are purging chlorine from a shallow well, you must monitor the water level closely. While it is uncommon for a drilled well to be pumped dry during a purging, it is not unusual for a dug or bored well to have its water level fall below the foot valve or end of the drop pipe. If this happens, the pump keeps pumping, but it's only getting air. This is bad for the pump and can burn it up. Monitor the water level closely as you empty a shallow well.

You can watch the water level in most shallow wells without any special equipment. Your eyes should be the only tools needed to tell when the water level drops down too far. However, a string with a weight attached to it can be used to gauge the water depth. If you know how long the drop pipe in the well is, and your pump installer should be able to tell you, make the string a foot shorter than the drop pipe. As long as the string hits water each time it is dropped into the well, you know the water level is safe.

## Pump Selection

Pump selection for a shallow well is sometimes a little more complex than it is for a deep well. Almost everyone uses submersible pumps for drilled wells. It's possible to use a two-pipe jet pump for a deep well, but few installers recommend this course of action. Shallow wells, on the other hand, can use single-pipe jet pumps, two-pipe jet pumps, or even submersible pumps. The use of submersible pumps in shallow wells is not common. In wells where the water lift is not more than about 25 feet, single-pipe jet pumps are often used. Deeper bored wells, and some dug wells, see the use of two-pipe jet pumps. If a well of any type has an adequate depth of water to work with, a submersible pump can always be used.

## Single-Pipe Jet Pumps

Single-pipe jet pumps are the least expensive option available for a shallow well. But their lifting capacity is limited. Since single-pipe jet pumps work on a suction-only basis, they must be able to pull a vacuum on the water. Since physics plays a part in how high above sea level a vacuum can be made, jet pumps can only pull water so high. Without looking it up, I can't remember the exact maximum lift under ideal conditions. I know it's around 30 feet. But, for practical purposes, most

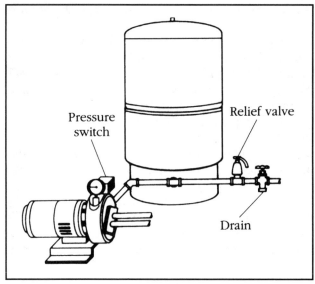

**Figure 16.9** Typical jet-pump set-up: *(Courtesy of Amtrol, Inc.)*

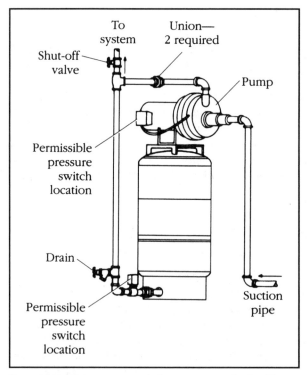

**Figure 16.10** Jet pump mounted on a pressure tank with a pump bracket: *(Courtesy of Amtrol, Inc.)*

professionals agree that a suction-based pump should not be expected to lift water more than 25 feet.

Jet pumps are installed at some location outside of the well. Unlike submersible pumps, which hang in the well water, jet pumps are normally installed in basements, crawl spaces, or pump houses. The pumps and their pipes must be protected from freezing temperatures. A pressure tank should be used with jet pumps, just as they are with submersible pumps. Many jet pumps are made to sit right on top of a pressure tank with the use of a special bracket. The pressure tank must also be protected from freezing temperatures.

**Two-Pipe Jet Pumps**

Two-pipe jet pumps can be used to pump water from deeper wells. These pumps use two pipes. The pump forces water down one pipe to allow it to be sucked up in the other. Since a two-pipe pump is not

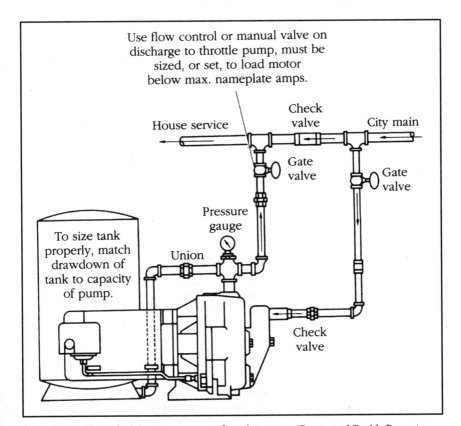

**Figure 16.11** Typical piping arrangement for a jet pump: *(Courtesy of Goulds Pumps)*

dependent solely on suction, it can handle higher lifts. The physical appearance of a two-pipe pump is basically the same as that of a one-pipe pump, except for the extra pipe.

### Submersible Pumps

Submersible pumps push water up out of a well. Since the pump is pushing, rather than sucking, it can be installed at greater depths. In many ways, submersible pumps are far less prone to failure than jet pumps. They are more expensive, but they tend to last longer and there are not as many parts and pieces to fail. This makes submersible pumps a favorable choice when the water depth is sufficient to warrant their use.

Since jet pumps are most often used with shallow wells, we will concentrate our efforts on them. Submersible systems were described in the last chapter if you wish to review them. Since most shallow wells are equipped with a single-pipe jet pump, we will start our installation procedures here.

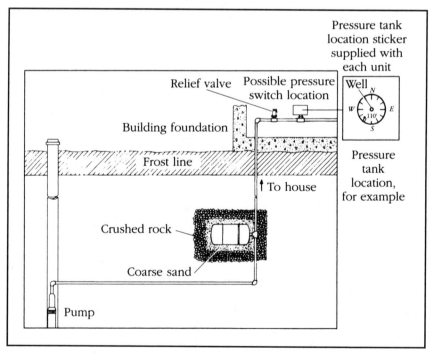

**Figure 16.12** Example of a submersible pump with an underground pressure tank: *(Courtesy of Amtrol, Inc.)*

## Installing a Single-Pipe Jet Pump

Installing a single-pipe jet pump is fairly easy. The well portion of the work is particularly simple. Almost anyone with modest mechanical skills and an ability to read instructions can manage the installation of a jet pump. However, a license to perform this type of work is probably required in your area. This may prohibit you from making your own installations.

The first step in installing a jet pump will be digging a trench between the well casing and the pump location. Since we covered trenching in the last chapter, we will skirt the issue here. Once the trench is dug, you are ready to make a hole in the side of the well casing. Since most shallow wells are cased with concrete, a cold chisel and a heavy hammer are all that's needed to make a hole. The hole will have to be patched to make it watertight after the well piping is installed. Otherwise, ground water will run into the well and may contaminate it.

After a hole is made, you are ready to see the well pipe installed. The pipe will probably be PE pipe. Insert fittings should be made of metal, in my opinion, and all connections should be double-clamped. A foot valve is usually installed on the end of the pipe that will hang in the well water. The foot valve serves two purposes. It acts as a strainer to block out gravel and similar debris that might otherwise be sucked into the pipe and pump. Foot valves also act as check valves to prevent water in the drop pipe from running out into the well. If the water in a drop pipe is not controlled with a check valve or foot valve, the jet pump would lose its prime and fail to pump water.

After the foot valve is installed, an elbow fitting is attached to the other end of the pipe. This fitting will allow the water service pipe, which will be placed in the trench, to connect to the drop pipe. Some form of protection should be provided for the water service pipe where it penetrates the concrete casing. Foam pipe insulation can sometimes be used, and rigid plastic pipe, like the type used for drains and vents in plumbing systems, can always be used. If PE pipe is allowed to rest on the rough edge of concrete, constant vibration of the pipe when the pump is pumping will eventually wear a hole through the pipe.

After the connection is made between the drop pipe in the well and the water service pipe in the trench, the rest of the work will be done at the pump location. Backfilling the trench should be done with caution.

Jet pumps can be installed almost anywhere where they will not get wet or become frozen. Crawl spaces, basements, pump houses, and even closets are all potential installation locations. A pressure tank is usually installed in close proximity to a jet pump. In some cases, the pump will be bolted to a bracket on a pressure tank. Sometimes a small

**AV5 foot valves:** Bronze body with stainless steel strainer.

| Order no. | Pipe size | Outside diameter |
|---|---|---|
| AV5-07 | 3/4" | $1^{11}/_{16}$" |
| AV5-10 | 1" | $1^{15}/_{16}$" |
| AV5-12 | $1^1/_4$" | $2^1/_4$" |
| AV5-15 | $1^1/_2$" | $2^7/_8$" |
| AV5-20 | 2" | $3^1/_2$" |

**Figure 16.13** Foot valve: *(Courtesy of Gould Pumps, Inc.)*

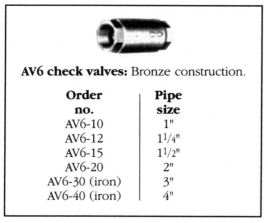

**AV6 check valves:** Bronze construction.

| Order no. | Pipe size |
|---|---|
| AV6-10 | 1" |
| AV6-12 | $1^1/_4$" |
| AV6-15 | $1^1/_2$" |
| AV6-20 | 2" |
| AV6-30 (iron) | 3" |
| AV6-40 (iron) | 4" |

**Figure 16.14** Check valve: *(Courtesy of Gould Pumps, Inc.)*

pressure tank is suspended above a pump. Many tanks stand independently on a solid surface, such as a floor. Part of the decision as to the type of tank used and its location will be based on the space available for the tank and pump. It is not mandatory that a pressure tank be used, but it is highly recommended to prolong the life of a pump. However, a pressure tank does not have to be installed adjacent to a pump. It can be in a remote location. Pressure tanks can even be installed underground, as long as the tank is an underground model.

The piping arrangement for a jet pump is not complicated. There are, however, many accessories that are commonly used, such as an

Shallow Wells    145

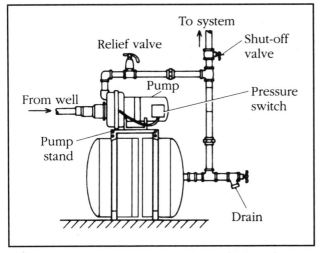

**Figure 16.15** Bracket-mounted jet pump on a horizontal pressure tank: *(Courtesy of Amtrol, Inc.)*

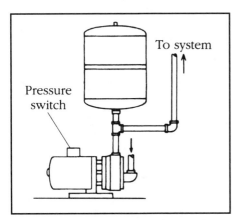

**Figure 16.16** Small, vertical pressure tank installed above pump: *(Courtesy of Amtrol, Inc.)*

air-volume control. Shut-off valves are, of course, installed in the pipes leaving the pump. In the case of a one-pipe jet pump, one large pipe brings water to the pump. A smaller pipe distributes the water to a water distribution system. A pressure switch is needed to control when the pump cuts on and off. Relief valves are needed to protect pressure tanks. Drain valves are typically installed for pressure tanks, and a check valve might be installed. In special situations, a pressure-control valve might be desirable.

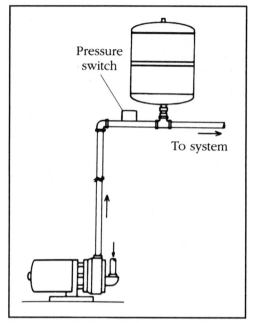

**Figure 16.17** Small, vertical pressure tank installed above pump: *(Courtesy of Amtrol, Inc.)*

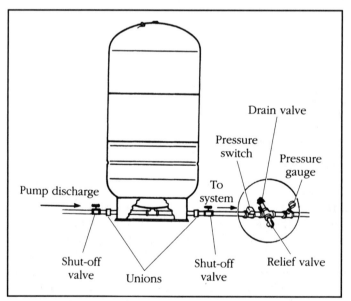

**Figure 16.18** Stand-type pressure tank installed with a straight-through method instead of a tank tee: *(Courtesy of Amtrol, Inc.)*

## Two-Pipe Pumps

Two-pipe pumps look very similar to one-pipe pumps except for the extra pipe involved. While one-pipe jet pumps have only a suction pipe, two-pipe pumps have both a pressure and a suction pipe. There are some differences in the piping arrangement in the well to go along with this most noticeable difference. An ejector is installed to enable the pressure pipe to assist the suction pipe in lifting water from the well.

A pressure tank should still be used with a two-pipe system. In fact, a pressure tank should be installed with all types of well pumps for residential use.

## Pressure Tanks

Pressure tanks should be considered standard equipment with every residential pump installation. The tanks are not very expensive, and they can add years to the life of a pump. They can also provide residents with better water pressure, which is an important aspect to consider. It's very rare today to see a plumbing system that is fed by a well where a pressure tank is not used. I can't imagine a pump installer bidding a job without including the cost of a pressure tank, but make sure that your next well system does provide a pressure tank. With that said, let's talk about the specifics of various sizes and types of pressure tanks.

### Small Tanks

Small tanks are available in sizes that hold no more than 2 gallons of water. This is a very small tank. In my opinion, it is too small for almost any application. A tank that has such a minimum capacity will do almost no good in a routine, residential plumbing system. When you consider that many toilets use more than 2 gallons of water each time they are flushed, you can see that the pump will be running a lot. Think about showers. At a flow rate of 3 gallons per minute, the first minute of a shower will deplete the supply of a super-small pressure tank.

The whole purpose of installing a pressure tank is to take some strain off a pump. While a 2-gallon buffer provides some support to a pump, the help is minimal. A pressure tank should be sized to meet the needs of the house it serves. In other words, the number of people using

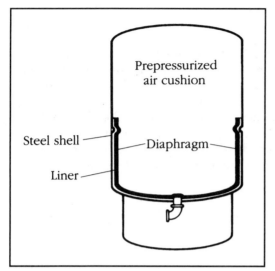

**Figure 16.19** Diaphragm-type pressure tank: *(Courtesy of Goulds Pumps)*

the plumbing system and the types of fixtures available should be taken into consideration when choosing a pressure tank.

### Large Tanks

Large tanks can take up a substantial amount of room. This is not normally a problem in houses where a basement, cellar, or crawl space is available. But if the pump system and tank must be installed within the primary living space of a home, tank size can become an issue. Residential pressure tanks can be purchased with a capacity of 100 gallons of water or more. In most circumstances, this is overkill. A tank that holds between 20 and 40 gallons should perform well under average residential conditions.

### In-Line Tanks

In-line tanks are designed to be installed right off a water pipe. They may be suspended below the pipe, or they may rise above the pipe. The size of in-line tanks vary. A brand that I use offers in-line tanks with capacities of 2 gallons, about 4 gallons, about 8 gallons, a little over 10 gallons, and 14 gallons. This range of sizes can be adequate for small homes.

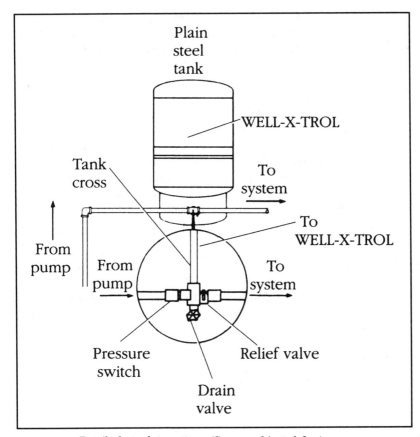

**Figure 16.20** Detail of a tank-tee set-up: *(Courtesy of Amtrol, Inc.)*

An in-line tank is not normally mounted on a bracket or set on a floor. It typically hangs from a pipe or sticks up above the pipe. When floor space is at a premium, an in-line tank is desirable. As a builder, I would try to keep the size up in the 10-gallon range as a minimum. However, I recently installed a smaller tank for a summer cottage. It was one of the 4-gallon models. Since the cottage had only one bathroom and rarely accommodated more than two people, the small tank was deemed adequate. You must match your tank to your needs.

**Stand Models**

Stand models are the type of pressure tank most often installed in homes. These units sit on a floor in a freestanding mode. Piping is run from the pump to the tank and then from the tank to the water

## 150  Chapter Sixteen

distribution system. With capacities ranging from about 10 gallons up to 119 gallons, this type of tank can meet any residential need.

I have a stand model in my home. Its capacity is around 40 gallons. This tank cost more to buy than a smaller tank would, but its size allows my pump to run less often. This saves electricity and prolongs the life of my pump. I think the trade-off is worthwhile.

When a stand model is used, it is easiest to install with a tank tee, or a tank cross as they are often called. This is a fitting that is designed for use with a pressure tank. The tank tee screws into the pressure tank and provides both an inlet and outlet opening. In addition to these openings, the fitting is tapped for accessories, such as pressure gauges and relief valves. The use of this fitting not only makes a job easier to install; it gives the appearance of a more professional installation.

### Pump-Stand Models

Pressure tanks are available in pump-stand models. These tanks are frequently used in conjunction with jet pumps. They are not needed

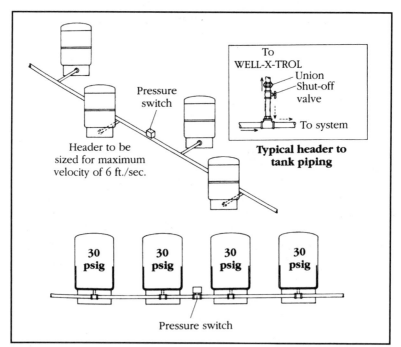

**Figure 16.21** Diagram of multiple pressure tanks being installed together: *(Courtesy of Amtrol, Inc.)*

with submersible pumps, since submersibles are hung in a well. A pump-stand model is designed to sit horizontally. It has a bracket on top of it so that a pump can be bolted down to it.

Using a pump-stand model is one way to conserve floor space. Since a pump attaches to the top of the stand, there is only one object sitting on the floor. With a typical stand model, both the pressure tank and the pump would be installed on the floor. The manufacturer I deal with provides two sizes of pump-stand tanks. One is about 8 gallons, and the other is 14 gallons. I would opt for the latter.

### Underground Tanks

Although I have never installed one, there are underground tanks available. Their size ranges from 14 gallons to 62 gallons when purchased from the manufacturer of my choice. Personally, I can't think of a time when an underground tank would have helped me, but I'm sure there must be occasions when they are desirable.

### Diaphragm Tanks

Diaphragm pressure tanks are common in today's plumbing systems. This was not always the case. Older houses often have plain galvanized storage tanks. These standard tanks can be a real pain to live with. They frequently become waterlogged. By this, I mean that they lose their air content and fill with water. When this happens, the tank must be drained, pumped up with air, and refilled with water. A waterlogged tank will make a pump run every time water is called for at a faucet, thus eliminating the advantage of having a pressure tank.

Modern pressure tanks come precharged with air, and they have a diaphragm system which eliminates waterlogging. Standard tanks are still available, but they are rarely used. I suggest that you specify a diaphragm tank in your well specs.

Another problem with a standard tank is rust. After some period of time, a metal holding tank will begin to rust. Air will leak out, and eventually water will leak. This means a patch or replacement will have to be made. With modern pressure tanks, liners are used so that water never comes into contact with the metal housing of the tank. This, of course, eliminates the possibility of interior rust. Bacteria growth and rust in drinking water are not nearly as likely with a lined tank as they are with a traditional metal tank.

## Sizing

Sizing a pressure tank is an important step in designing and installing a good pump system. The pressure tank protects the pump. I mentioned earlier that you should take the number of people and plumbing fixtures into consideration when choosing a size for your pressure tank. My comment was meant to get you thinking about how little use it takes to drain down a small tank. If you want to seriously size a tank, you should work from the specifications for the pump that will be installed. In other words, the ideal sizing comes from carefully matching your tank to your pump. The relevance of a number of people or plumbing fixtures plays only a small role, if any, in determining tank size.

Pressure tanks are designed to work with pressure switches. These switches can be set for various cut-in and cutout pressures. It has long been common for a pump to cut on when tank pressure drops to 20 pounds per square inch (PSI). At this cut-in rate, a typical cutout rate is 40 PSI. For years, this has been something of a standard in well systems. However, times change, and so do standards.

It is not at all uncommon for cut-in pressures to be set at 30 PSI today. A corresponding cutout pressure is 50 PSI. Some homes have their well systems set up with cut-in pressures of 40 PSI and cutout pressures of 60 PSI. The increase in pressure is due to many factors. People tend to enjoy increased water pressure at some of their plumbing fixtures, such as showers. Some plumbing fixtures and devices require higher working pressures than what was common in the past. This all contributes to the trend for higher pressures.

Before you can size a pressure tank, you must determine what your cut-in and cutout pressures will be. You must also know the gallons-per-minute (GPM) rating of the pump. How many start-ups can the pump be subjected to in a 24-hour period? A time period must be established as to the minimum run time for the pump. This is usually established by the manufacturer, so check your pump paperwork to see what the minimum run time should be.

Once you know all the variablesß mentioned above, you can select a tank of the proper size. However, you need some type of sizing chart to do this math. Most manufacturers of pressure tanks will be happy to provide you with sizing tables. You can see for yourself how easy it is to make sure that the homes you build are provided with suitable pressure tanks.

## Installation Procedures

Installation procedures for pressure tanks can take many forms. You've already seen that there are numerous types of tanks available. The installation of these tanks vary as much as the tanks do. You might install an underground tank to work in conjunction with a submersible pump. A pump-stand tank is sometimes a good choice for a jet pump. In-line tanks can also come in handy. Stand models are the type that you will use most often. While it is highly unlikely in a residential system, unless you are building a house on a farm, a multiple-tank setup might be called for. Your installer certainly should know how to make a good installation, and the diagrams I've provided you with can help you to understand the installation procedures.

### A Relief Valve

Make sure a relief valve is installed to protect the pressure tank from excessive pressure. Omitting this fitting could result in disaster. It is a code requirement, so someone should catch it if you and your installer miss it, but don't take this chance. Ratings on pressure tanks and relief valves can vary, but most tanks are rated at 100 PSI, and the relief valves used with them are rated at 75 PSI. If a relief valve isn't installed, a pressure tank can blow up, causing personal and property damage. The same thing can happen if the relief valve is rated higher than the tank's working pressure. This is a serious issue, so take an active part in making sure the right relief valve is installed properly.

### Keep It Dry

A pressure tank will last longer when you keep the exterior of it dry. In some cases, this may require installing a tank on blocks or a platform. If the basement, cellar, or other area where a tank is being installed tends to get wet during some seasons, elevate the tank to protect it. This is very simple to do at the time of installation, but it becomes a bit complicated after the fact. Spending a few bucks for some blocks can save your customers money down the road. Happy customers are what successful businesses are all about, so don't forget them when you are putting out the specs on a job.

**Chapter**

# 17

# Alternative Water Sources

An alternative water sources, for our purposes, is any water source that is not a dug, bored, or drilled well. Municipal water sources are excluded from the alternative category. Springs, lakes, cisterns, driven wells, and similar sources of water are the types that I will be calling alternatives. In some cases, the water will not have to be safe for drinking. There are times when rural builders have to provide water sources for homes and related structures being built, and some of the water is not required to be potable.

Let's say for example, you are building an exclusive home for an owner who enjoys horses. The homeowner might want water available for the stables, and this water doesn't necessarily have to be safe for human consumption. It might be practical to put the horse and barn water on a separate system. You might even suggest a solar-powered system. If the barn chores will require a lot of water, it could put a strain on the residential well. Rather than deplete the home's drinking water, an alternative source might be desired for the stable's needs. This would be a case where two water sources are used. They might both produce potable water, but it would not necessarily be critical that they do.

Even if you don't have a need for nonpotable water, an alternative water source may be worth considering. Many homes get their water from springs and driven wells. While these two sources of water are not normally considered to be dependable for a high quantity of water, a lot of houses do get by with them.

Irrigation systems for lawns and gardens can use nonpotable water. Like the horse example, irrigation is a prime candidate for a separate water source. The amount of water used for irrigation can be substantial, and it is usually needed most during dry seasons. Does it make sense to avoid using potable well water during a dry spell if you can? Of course it does, and an alternative water source can be the solution. There are, in fact, many reasons why a second water source might be wanted. In some cases, streams, ponds, cisterns, and similar water sources are ideal solutions to water problems.

An alternative water system can produce potable water. Springs are an excellent example. Have you noticed all the bottled spring water being sold in your local grocery store? Many people feel that the only good drinking water is spring water. If it's good enough to serve in fancy restaurants, it's good enough to drink at home. However, some special precautions should be exercised to avoid contamination of a spring. We'll talk more about this later in the chapter.

Driven wells can produce enough water to serve a full-time residence. Due to their design, driven wells can't be considered a dependable source of water when a lot is needed, but they are very inexpensive to install. Limitations on their use exist, but you may find that a driven well is just what you need.

Have you ever worked on old homes where a large cistern was installed in the cellar? They are quite common in many of the older homes in Maine. I never saw one in Virginia, but I've seen several during my plumbing and remodeling work in Maine. Cisterns saw a lot of use in the old days. There is no reason why they can't still be useful today.

As we move through this chapter, you are going to learn a lot about various types of water sources. Without a doubt, conventional wells will be your most likely source of water when building a house in the country. But alternative water sources have their place, and this is the place to learn about them.

## Driven Wells

Driven wells are very inexpensive to install. If ground conditions are suitable, a driven well is easy to make. No fancy truck or drilling rig is needed. A ladder, a sledgehammer, and a few specialized parts, along with a strong back and arms, are all that's needed.

What is a driven well? It is a pipe driven into the ground. Some people refer to these wells as driven points or just as points. Many of these wells exist in parts of Maine. They are especially popular along the coast, where summer cottages are built on sandy soil.

Personally, I wouldn't choose a driven well to supply a house with water when the property was going to be used as a full-time residence. I do, however, know of homes where a point provides the only source of potable water for entire families. Driven points have a very small reserve capability. If the water source for the well is strong, it can produce a lot of water. But if the water source is slow, it's easy to run this type of well dry. Let me give you some specifics on driven wells.

The components of a driven well are simple. The first piece is the drive point, which also serves as a filter. Pipe used to form the well makes up the second component, and the last piece of equipment needed is a pump. The size of the well pipe usually doesn't exceed 2 inches in diameter, and 1 inch diameters are common. Under ideal conditions, this type of well can be driven to a depth of about 50 feet.

## The Point

The point you select may be made of reinforced steel or it may have a bronze tip. The pointed end allows it to be driven into soft ground. At the opposite end of the point, there are pipe threads that allow additional sections of pipe to be added as the point is driven. These threads are protected by a drive cap during the driving phase of your work.

Well points are available with different types of screens. These screens act as a filter, and it's necessary for you to know the type of ground that the point will get its water from. For example, you would use a screen with a wide mesh if you were pulling water from coarse gravel. A fine-mesh screen would be used if water is being obtained from sand. The openings in the filter must be matched to the ground conditions to keep sand and similar particles from entering the well.

## The Well Pipe

The well pipe you choose to use may be standard galvanized water pipe or special piping designed just for driven wells. Go with the specialized piping. Galvanized pipe tends to rust. This happens along the interior sides of the pipe as well as at the threads. Since threads have thin walls, it's not unusual for this to be the first point of deterioration. I don't recommend the use of galvanized pipe for several reasons, and rust is one of them.

Another problem with galvanized pipe is that it doesn't stand up well to heavy pounding. To get the pipe short enough to drive effectively, you must cut and thread it. Making your own drive sections like this can certainly be done, but the work is time-consuming and hard on the arms, unless you have a power threader. Since galvanized pipe does not drive well, the depth of a well made with it is often limited to about one-half the depth of a well constructed with specially designed drive pipe.

To drive a pipe with power, it must not be too tall. A 5-foot section of pipe usually works well. A platform will be needed for the first few whacks with the hammer, but then a person can move down to the ground as the pipe is driven in deeper. The special drive pipe that I'm talking about is often called riser section. It runs about 5 feet in length, which makes it ideal for the purpose.

### Drive Caps

Drive caps are needed to protect pipe threads during the driving process. If you don't use a drive cap, the end of the pipe will roll out as it is hit. This, naturally, makes it impossible to thread couplings onto the pipe. Drive caps are available with either male or female threads. They are quite simple but very necessary.

### Power Driving

Power driving can accomplish a well depth of about 50 feet. Standard driving techniques, with a sledgehammer, will probably net only 30 feet of depth, which can be more than enough. A number of setups exist for using mechanical driving power. Impact drivers are also available, and they can make driving a well easier. If you were to set yourself up as a professional installer, mechanical means of driving would be the most sensible route to take. However, the majority of people whom I've known to drive wells have done it the old-fashioned way, with a sledgehammer and human sweat.

### Getting It Out

Getting a drive pipe into the ground can be difficult, but getting it out can be even more troublesome. When you drive a well point, you never know when some buried obstruction will stop your best efforts. When this happens, you must remove the riser sections and point to start somewhere else. You won't have much success doing this by hand if the point has reached a significant depth.

Pipe clamps are one solution to the problem of pipe removal. Putting the clamps around the pipe can allow you to drive the pipe out of the hole with your hammer. Lifting jacks, such as the type used to jack up automobiles, can also be used in conjunction with pipe clamps to raise sections of pipe. Blocks of wood should be placed on the ground to keep the jacks from sinking in. The jacks will exert upward pressure on the pipe clamp, thereby lifting the pipe. You will have to reset the jacks and clamps frequently, but this method will certainly pull the pipe out of the ground. Other methods of removal, such as the use of winches, can also be employed to remove pipe.

### Driving the Well Point

The actual driving of a well point should be started slowly. Many people use post-hole diggers or an auger to get a hole started, up to a depth of about 2 feet. Once this hole is in place, the well point is set in the hole. I should add that some type of pipe sealant compound must be applied to the males threads of all joints to make them watertight. It's essential that the well point and first section of well pipe be driven into the ground as close to a perfect vertical position as possible. A level can be used to line the drive point and pipe up in a vertical position. During this stage of the operation, don't bang the drive cap—tap it. Use light strokes so that you can control the position of the pipe. You must make sure that this first section goes into the ground straight.

As the first section gets into the ground, you must remove the drive cap, dope up the male threads on the riser, and add a new riser to the rig. Replace the drive cap and resume driving. As the first section disappears from sight, you can increase the power of your pounding. However, don't hit the pipe if it's vibrating. This could cause damage to underground sections.

The sequence of events in driving a well point is routine. You drive a section to a point near the top of the ground. Then you remove the drive cap, dope the threads, and add a new section. Replace the drive cap and keep driving. This goes on until you can go no further or until you have an abundant water supply.

### Towards the End

Towards the end, you must decide how the well will terminate. Will you install a pitless adapter? This is a good idea for a well that will have its water service running below the frost line. Will you install a simple pitcher pump on top of the pipe? This can be a suitable solution for seasonal use. Your well can be terminated above or below grade. Elbow fittings can allow you to turn the well pipe horizontally if you desire. Plan your termination in advance.

### Suitable Soils

Suitable soils must exist for driving a well. It is not possible, for example, to drive a point through bedrock. Soft, moist clay is a good soil for driving a well in. So is coarse sand and gravel. Fine sand and hard clay can be driven through, but the work will be difficult. Not all ground conditions will allow a well point to be driven, even with mechanical assistance. If you can do some test boring before sinking a point, it will certainly prove helpful.

### Quantity

There is no way to predict the quantity of water that will be produced by a well point. It may be 50 gallons of water, and it may be 250 gallons. The recovery rate is equally impossible to predict. Until you sink the point and test the well, there is no way to know what you will get. However, the combination of a driven well and a large pressure tank can produce adequate water for a full-time residence that is occupied by a family.

### Quality

The quality of water produced from a driven well can be compared to that of water coming from a dug or bored well. Since these types of wells normally connect with water at similar depths, the type of water being produced should be similar. All water should be tested periodically to make sure it's safe to drink if human consumption is the purpose of the water.

### Cisterns

Cisterns can be considered holding vessels for water. They might be constructed above ground or below grade. Brick, stone, or concrete can be used to build a cistern. Such a water holding area can be small enough to water a family vegetable garden or large enough to provide water for a family during several months of dry times.

Cisterns typically rely on surface water to fill them. For this reason, the water is not normally considered acceptable as drinking water. However, if filtering and treatment systems are used in conjunction with a cistern, the quality of the water can be raised to a potable level. Additionally, a cistern can be filled from a potable water source such as a well to create a large reserve of potable water. How long this water can be kept potable varies with conditions, but this is one way to make the most of a well during times of plentiful water.

In older homes, it was not uncommon for large cisterns to be built beneath the first floor. Some of these collection tanks resemble modern-day, aboveground swimming pools in size. In fact, a modern, aboveground pool can serve as a very effective cistern. Let me give you a quick example of this.

I purchased an aboveground pool for my young daughter to learn to swim in. The pool is about 42 inches deep, and it has a diameter of 15 feet. According to the paperwork provided with the pool, its water capacity is 4,400 gallons. This pool cost less than $300. Setting the pool up took only a few hours. If this holding tank was being used as a cistern, it would be quite effective, very efficient, and low in cost. Depending upon your cistern needs, don't overlook the possibility of aboveground pools as a material choice.

Many of the cisterns I've seen have been built in the cellars of homes. Some of these collection vessels were made to have water pumped into them. Many of them collected rainwater from the roof of the house they were built under. A lot of water runs off the roof area of a house during a rainstorm. If this water is diverted to a cistern, it doesn't take long to build up several hundred gallons of reserve water. It's entirely possible for a cistern to collect and contain enough water to provide a family with water for up to six months. In areas where rainfall comes in spurts, a cistern can be a lifesaver, so to speak.

Water from a cistern can be pumped in a manner similar to that used with shallow wells. This water can serve domestic needs when it's potable. Irrigation and farm animals are two reasons for using a cistern. Roof water can be collected easily with the use of gutters and piping. A cistern might be filled from a stream during wet times so that water is available in dry times when the stream has dried up.

Cisterns are rarely considered a potable water source for a new house. This type of water collection is not comparable to a drilled well. However, there are many circumstances that might warrant the use of a cistern.

## Ponds and Lakes

The use of existing ponds and lakes can do a lot for a property when high volumes of water are needed. One such situation might be a home where the owner operates a commercial greenhouse. A lot of water is used when growing plants commercially. If a lake or pond is handy, the expanse of reserve water works well for keeping plants healthy. The same can be said for irrigation purposes or livestock.

Ponds and lakes are not normally suitable as a source of potable water unless the water is purified. This is certainly possible, but the expense and trouble often outweigh the cost of drilling a well. For the most part, ponds, lakes, streams, and rivers should be ignored as potable water sources.

## Springs

Springs have long been a source of drinking water. However, the water is not always of a quality considered to be safe for drinking. If a spring is to be used as a primary source of potable water, some special provisions should be made. For instance, the spring should be protected from surface water.

If a spring is to be used for potable water, a watertight container of some type should be placed around the spring. Well casing works fine. Springs located on the sides of hills should have diversion ditches installed above them. This helps to keep surface water from running into the spring. Slotted pipe and gravel can be installed in trenches to intercept and divert surface water.

Fencing should be installed around a spring to protect it from animal entry. The uphill side of the spring should be fenced for a distance that is sufficient to prevent animal activity from contaminating the spring. Even with these protections, a spring should be tested regularly to assure its potability.

Springs, when available, are an inexpensive water source. They can provide large quantities of water, and the quality can be quite good. Since springs are not abundant, they cannot be counted on as a water source. But if one is available, it can provide numerous benefits to the property owner.

I remember visiting a spring on nearly a daily basis during the summers of my childhood. Going down to the spring was not a necessity for me. My parents and grandparents had municipal water supplies in their homes. I used to go to the spring as a pleasure. The spring was used as a water source by some people in the neighborhood. Looking back on this spring, I remember it having a metal lining. It seems to have been a 55-gallon drum. This probably wouldn't be considered a good lining, but as I recall, that's what was used. To the best of my knowledge, none of the people drinking from this spring ever became sick from the water. This doesn't mean that I would recommend using or installing such a water source.

There is a spring near where I live that is visited daily by dozens of people. At times, people are lined up one after another for a nearly unbelievable distance to get access to the spring water. Some of these people have a great number of half-gallon plastic jugs with them to collect the spring water. Unlike the spring I used to play around, this one has a pipe that extends out of the side of a hill. Water runs from it constantly. I've never drunk the water, but it appears a great number of people do. As often as I drive by this spring, I can recall only one or two occasions when the water wasn't running.

A gentleman I know gets all of his domestic water from a spring. He has done this for a long time. Last winter, his water-supply pipe froze. The spring didn't, but his water service did. When you consider that a spring in Maine can handle the winter temperatures without freezing, it's certainly possible for a spring to provide year-round water to a home. Still, you can't count on a spring having enough constant flow to be a primary water source.

## Many Needs

Alternative water sources can fill many needs. They can provide water for all domestic uses. Livestock can be watered with alternative sources. Gardens and lawns can be watered with nonpotable water. Fire protection can be boosted with the availability of a pond or cistern. Water-source heat pumps can gather their needs from alternative water sources and wells. There are so many potential uses for water that alternative water sources should always be considered.

As a builder, you should stay abreast of the many options available to you when providing water to the homes you build. Conventional wells will continue to provide most of the water used in rural locations, but other natural resources and man-made provisions can enhance the living conditions of people in the country. With a little research and some creative thinking, you can offer your customers options for water that other builders may not be aware of. This can help you win more bids and make more money.

# Chapter 18

# Troubleshooting Well Systems

Common problems with well installations can make life difficult for builders. When a person has purchased a new house from a builder and their well fails, the person is likely to call the builder. Some builders have no idea of how to troubleshoot well and pump problems, but they are the first people to be contacted. Homeowners who have wells that are not functioning properly are usually distressed. If ignored, the people can quickly become grumpy.

Problems with wells and well-pump systems are not unusual. These problems don't typically plague homeowners, but they do exist, sometimes with far more regularity than builders would like. Is there anything you can do to eliminate these problems? I doubt it. But, you can reduce the occurrence of such problems with quality workmanship and supervision. However, I believe you must be prepared to deal with a variety of problems associated with wells and their pumping systems.

Assuming that you use subcontractors to create your wells and to install your pump systems, you can call those people to help solve your problems. Having this option doesn't release you from responsibility, but it does make your job easier. Still, homeowners with well-related problems will be looking to you for support and help. The more you know about troubleshooting wells and septic systems, the more valuable you become to your customers. This is important.

As I'm sure you have gathered, this chapter is dedicated to dealing with problems. I wouldn't expect you to throw a toolbox in the back of your truck and rush out to fix well problems. If you have enough background, you might be able to give customers helpful advice by telephone.

Doing this might get their systems back in action right away. Your subcontractors will appreciate not having to make callbacks, and your customers will be pleased to have their water problems solved so quickly.

The information we will cover in this chapter is somewhat technical at times. Other aspects of the text are not so technical. I intend to break the chapter down into three primary categories. We will talk about water-source problems, problems with pumps and related equipment, and water quality. For example, if someone calls you and complains about a nasty smell in their water, you will be able to refer to this chapter for advice on what might be causing the odor. Just for the record, sulfur would be the most likely cause of this problem. Let's start with basic well problems.

### Basic Well Problems

Basic well problems are not common. In this category, we are talking about trouble with wells themselves rather than with pumps. Water quality is not a part of our present discussion. Since well problems are often dependent on the type of well being used, we will investigate the problems arising with specific well types.

### Driven Wells

Driven wells don't usually have a large holding capacity. It is not uncommon for flow rates or recovery rates to be low. Both of these factors can contribute to a home running out of water. If a customer calls with a complaint of having no water when a driven well is in use, you may be faced with a pump problem or a well problem. This is true of all types of wells. Driven wells are the most likely type of well to be run dry. This is a simple problem to troubleshoot.

If you suspect that a driven well is out of water, you can gain access to the well and drop a weighted line into the well to determine if any water is standing in the well. If there is little or no water present, your problem might be with the point filter or the water source. If the well point has become clogged, water will not be able to enter the well pipe. Assuming that there is insufficient water, you can take one of two actions.

You could pull the well point and inspect it. This will not be an easy task. If your area is experiencing an extremely dry spell, you may have to pull the well point to identify the problem. However, if area conditions don't point to a drop in local water tables, you might ask the customer to avoid using any water for a few hours and then try the pump again.

Given several hours for recovery, the well may produce a new supply of water. This doesn't rule out a partially clogged point, but it tends to indicate a low flow rate. To be sure of what is going on, the point will have to be pulled and inspected.

Due to the nature of a driven point, your options for finding out if water is present in the water table are limited. If water can't pass through the filter of a point, water won't enter the well. This is not the case with other types of wells. Unfortunately, some of the money saved by installing a well point can be lost through later problems, such as having to pull the point for inspection and possible replacement.

### Sand

Sand in a water distribution system that is served by a well point is an indication that the screen filter on the well point used has openings that are too large. This type of problem can be addressed by adding an in-line sediment filter, but the true solution lies in replacing the well point with one with a finer screen filter.

### Other Contaminants

Other contaminants can enter a water distribution system through a well point. These entries into the water system can be filtered out with water treatment conditioning. Replacing a well point may help to solve this type of problem. Essentially, some type of conditioning equipment will probably be needed to eliminate very small contaminants.

### Shallow Wells

Shallow wells are less likely to pose problems than driven wells are. However, these wells can provide builders with head-scratching trouble. Some shallow wells cave in over time. It doesn't always take a long time for this to happen. A new well can experience problems with cave-ins long before warranty periods are over. This is not a common problem, but it is one that can occur.

Shallow wells do sometimes run out of water. Given some time, these wells normally recover a water supply. If a shallow well runs dry, there is very little that can be done. You must either wait for water to return or create a new well.

Problems are sometimes caused by sand or other sediment pumping out of a shallow well. This is usually a result of the foot valve or drop pipe hanging too low in the well. If a well has worked well for a few months and then begins producing sand or other particles, it can be an indication that the well is caving in. Sometimes a foot valve will become clogged under these conditions. The simple act of shaking the drop pipe can clear a foot valve of debris and allow a pump to return to normal operation.

Checking a shallow well to see if water is in reserve is easy. A weighted line can be dropped into a well to establish water depth. Assuming that water is present in sufficient quantity, you can rule out a dry well. But you cannot rule out the fact that the drop pipe or foot valve in the pump may be installed above the water level. This can be checked by pulling the drop pipe out of the well and measuring it. The length of a drop pipe can then be compared to the depth at which water is located. If you have water in the well and your drop pipe is submerged in it, you can rule the well out as your problem.

### Drilled Wells

Drilled wells very rarely run out of water. It is, however, possible that a drilled well could be run dry. A weighted line will allow you to test for existing water. Pulling the drop pipe and comparing its length to the depth at which water is contacted will determine whether a lack of water in a house is the well's fault or some fault of a pumping system.

In all of my years as a plumber, I've never know a drilled well to run out of water without some type of outside interference. By outside interference, I mean some form of man-made trouble, such as blasting with explosives somewhere in the general area. Let me give you an example of this from my recent past.

A friend of mine has enjoyed a drilled well for decades. In all of these years the well had never given its owner any problem until this past summer. Roadwork was being done within a mile or so of my friend's house in the early summer. Part of the work involved the blasting of bedrock. Shortly after this blasting took place, my friend's well quit producing water. Why? My guess, and it's only a guess, is that the blasting caused a change in the underground water path. It may be that the blasting shifted the rock formations and diverted the water that was at one time serving the well of my friend. I've seen similar situations occur at other times. It's impossible for me to say with certainty that blasting ruined the well, but it's my opinion that it did.

## Troubleshooting Jet Pumps

You are already aware that there are differences between jet pumps and submersible pumps. Knowing this, it only makes sense that there will be differences in the types of problems encountered with the different types of pumps. Let's start our troubleshooting session with jet pumps.

### Will Not Run

A pump that will not run can be suffering from one of many failures. The first task is to check the fuse or circuit breaker. If the fuse is blown, replace it. When the circuit breaker has tripped, reset it. This is something you could ask your customer to check.

When the fuse or circuit breaker is not at fault, check for broken or loose wiring connections. Bad connections account for a lot of pump failures. It is possible the pump won't run due to a motor overload protection device. If the protection contacts are open, the pump will not function. This is usually a temporary condition that corrects itself.

If the pump is attempting to operate at the wrong voltage, it may not run. Test the voltage with a voltammeter. The power must be on when this test is conducted. With the leads attached to the meter and the meter set in the proper voltage range, touch the black lead to the white wire and the red lead to the black wire in the disconnect box near the pump. Test both the incoming and outgoing wiring.

Your next step in the testing process should be at the pressure switch. The black lead should be placed on the black wire and the red lead should be put on the white wire for this test. There should be a plate on the pump that identifies the proper working voltage. Your test should reveal voltage that is within 10% of the recommended rating.

An additional problem that you may encounter is a pump that is mechanically bound. You can check this by removing the end cap and turning the motor shaft by hand. It should rotate freely.

A bad pressure switch can cause a pump to stop running. With the cover removed from the pressure switch, you will see two springs, one tall and one short. These springs are depressed and held in place by individual nuts. The short spring is preset at the factory and should not need adjustment. This adjustment controls the cutout sequence for the pump. If you turn the nut down, the cutout pressure will be increased. Loosening the nut will lower the cutout pressure.

The long spring can be adjusted to change the cut-in and cutout pressure for the pump. If you want to set a higher cut-in pressure, turn the nut tighter to depress the spring further. To reduce the cut-in pressure, you should loosen the nut to allow more height on the spring. If the pressure switch fails to respond to the adjustments, it should be replaced.

It is also possible that the tubing or fittings on the pressure switch are plugged. Take the tubing and fittings apart and inspect them. Remove any obstructions and reinstall them.

The last possibility for the pump failure is a bad motor. You will use an ohmmeter to check the motor, and the power to the pump should be turned off. Start checking the motor by disconnecting the motor leads. We will call these leads L1 and L2. The instructions you are about to receive are for Goulds pumps with motors rated at 230 volts. When you are conducting the test on different types of pumps, you should refer to the manufacturer's recommendations.

Set the ohmmeter to RX100 and adjust the meter to zero. Put one of the meter's leads on a ground screw. The other lead should systematically be touched to all terminals on the terminal board, switch, capacitor, and protector. If the needle on your ohmmeter doesn't move as these tests are made, the ground check of the motor is okay.

The next check to be conducted is for winding continuity. Set the ohmmeter to RX1 and adjust it to zero. You will need a thick piece of paper for this test; it should be placed between the motor switch points and the discharge capacitor.

You should read the resistance between L1 and A to see that it is the same as the resistance between A and yellow. The reading between yellow to red should be the same as L1 to the same red terminal.

The next test is for the contact points of the switch. Set the ohmmeter to RX1 and adjust it to zero. Remove the leads from the switch and attach the meter leads to each side of the switch; you should see a reading of zero. If you flip the governor weight to the run position, the reading on your meter should be infinity.

Now let's check the overload protector. Set your meter to RX1 and adjust it to zero. With the overload leads disconnected, check the resistance between terminals one and two and then between two and three. If a reading of more than one occurs, replace the overload protector.

The capacitor can also be tested with an ohmmeter. Set the meter to RX1000 and adjust it to zero. With the leads disconnected from the capacitor, attach the meter leads to each terminal. When you do this, you should see the meter's needle go to the right and drift slowly to the left. To confirm your reading, switch positions with the meter leads and see if you get the same results. A reading that moves toward zero or a needle that doesn't move at all indicates a bad capacitor.

I realize the instructions I've just given you may seem quite complicated. In a way, they are. Pump work can be very complex. I recommend that you leave major troubleshooting to the person who installed your problem pump. If you are not familiar with controls, electrical meters, and working around electrical wires, you should not attempt many of the procedures I am describing. The depth of knowledge I'm providing may be deeper than you ever expect to use, but it will be here for you if you need it.

**Runs But Gives No Water**

When a pump runs but gives no water, you have seven possible problems to check out. Let's take a look at each troubleshooting phase in its logical order.

The first consideration should be that of the pump's prime. If the pump or the pump's pipes is not completely primed, water will not be delivered. For a shallow-well pump you should remove the priming plug and fill the pump completely with water. You may want to disconnect the well pipe at the pump and make sure it is holding water. You could spend considerable time pouring water into a priming hole only to find out the pipe was not holding the water.

For deep-well jet pumps, you must check the pressure-control valves. The setting must match the horsepower and jet assembly used, so refer to the manufacturer's recommendations.

Turning the adjustment screw to the left will reduce pressure and turning it to the right will increase pressure. When the pressure-control valve is set too high, the air-volume control cannot work. If the pressure setting is too low, the pump may shut itself off.

If the foot valve or the end of the suction pipe has become obstructed or is suspended above the water level, the pump cannot produce water. Sometimes shaking the suction pipe will clear the foot valve and get the pump back into normal operation. If you are working with a two-pipe system, you will have to pull the pipes and do a visual inspection. However, if the pump you are working on is a one-pipe pump, you can use a vacuum gauge to determine if the suction pipe is blocked.

If you install a vacuum gauge in the shallow-well adapter on the pump, you can take a suction reading. When the pump is running, the gauge will not register any vacuum if the end of the pipe is not below the water level or if there is a leak in the suction pipe.

An extremely high vacuum reading, such as 22 inches or more, indicates that the end of the pipe or the foot valve is blocked or buried in mud. It can also indicates the suction lift exceeds the capabilities of the pump.

If the pump is running without delivering water, the most likely cause is a leak on the suction side of the pump. You can pressurize the system and inspect it for these leaks.

The air-volume control can be at fault with a pump that runs dry. If you disconnect the tubing and plug the hole in the pump, you can tell if the air-volume control has a punctured diaphragm. If plugging the pump corrects the problem, you must replace the air-volume control.

Sometimes the jet assembly will become plugged up. When this happens with a shallow-well pump, you can insert a wire through the 1/2" plug in the shallow-well adapter to clear the obstruction. With a deep-well jet pump, you must pull the piping out of the well and clean the jet assembly.

An incorrect nozzle or diffuser combination can result in a pump that runs but that produces no water. Check the ratings in the manufacturer's literature to be sure the existing equipment is the proper equipment.

The foot valve or an in-line check valve could be stuck in the closed position. This type of situation requires a physical inspection and the probable replacement of the faulty part.

### Cycles Too Often

When a pump cycles on and off too often it can wear itself out prematurely. This type of problem can have several causes. For example, any leaks in the piping or pressure tank would cause frequent cycling of the pump.

The pressure switch may be responsible for a pump that cuts on and off too often. If the cut-in setting on the pressure gauge is set too high, the pump will work harder than it should.

If the pressure tank becomes waterlogged (filled with too much water and not enough air), the pump will cycle frequently. If the tank is waterlogged, it will have to be recharged with air. This would also lead you to suspect that the air-volume control is defective.

An insufficient vacuum could cause the pump to run too often. If the vacuum does not hold at 3 inches for 15 seconds, it might be the problem.

The last thing to consider is the suction lift. It's possible that the pump is getting too much water and creating a flooded suction. This can be remedied by installing and partially closing a valve in the suction pipe.

### Won't Develop Pressure

Sometimes a pump will produce water but will not build the desired pressure in the holding tank. Leaks in the piping or pressure tank can cause this condition to occur.

If the jet or the screen on the foot valve is partially obstructed, the same problem may result.

A defective air-volume control may prevent the pump from building suitable pressure. You can test for this by removing the air-volume control and plugging the hole where it was removed. If this solves the problem, you know the air-volume control is bad.

A worn impeller hub or guide vane bore could result in a pump that will not build enough pressure. The proper clearance should be .012 on a side or .025 diametrically.

With a shallow-well system, the problem could be being caused by the suction lift being too high. You can test for this with a vacuum gauge. The vacuum should not exceed 22 inches at sea level. Deep-well jet pumps require you to check the rating tables to establish their maximum jet depth. You should also check the pressure-control valve to see that it is set properly.

### Switch Fails

If the pressure switch fails to cut out when the pump has developed sufficient pressure, you should check the settings. Adjust the nut on the short spring and see if the switch responds; if it doesn't, replace the switch.

Another cause for this type of problem could be debris in the tubing or fittings between the switch and the pump. Disconnect the tubing and fittings and inspect them for obstructions.

We have now covered the troubleshooting steps for jet pumps, so we will move on to submersible pumps.

### Troubleshooting Submersible Pumps

There are some major differences between troubleshooting submersible pumps and troubleshooting jet pumps. One of the most obvious differences is that jet pumps are installed outside of wells and submersible pumps are installed below the water level of wells.

There are times when a submersible pump must be pulled out of a well, and this can be quite a chore. Even with today's lightweight well pipe, the strength and endurance needed to pull a submersible pump up from a deep well are considerable. Plumbers that work with submersible pumps regularly often have a pump-puller to make removing the pumps easier.

When a submersible pump is pulled, you must allow for the length of the well pipe when planning the direction to pull from and where the pipe and pump will lie once removed from the well. It is not unusual to have between 100 and 200 feet of well pipe to deal with, and some wells are even deeper.

It is important, when pulling a pump or lowering one back into a well, that the electrical wiring does not rub against the well casing. If the insulation on the wiring is cut, the pump will not work properly. Let's look now at some specific troubleshooting situations.

**Won't Start**

Pumps that won't start may be the victims of a blown fuse or tripped circuit breaker. If these conditions check out okay, turn your attention to the voltage.

In the following scenarios we will be dealing with Goulds pumps and Q-D-type control boxes.

To check the voltage, remove the cover of the control box to break all motor connections. Be advised: wires L1 and L2 are still connected to electrical power. These are the wires running to the control box from the power source.

Press the red lead from your voltmeter to the white wire and the black lead to the black wire. Keep in mind that any major electrical appliance that might be running at the same time, like a clothes dryer, should be turned on while you are conducting your voltage test.

Once you have a voltage reading, compare it to the manufacturer's recommended ratings. For example, with a Goulds pump that is rated for 115 volts, the measured volts should range from 105 to 125. A pump with a rating of 208 volts should measure a range from 188 to 228 volts. A pump rated at 230 volts should measure between 210 and 250 volts.

If the voltage checks out okay, check the points on the pressure switch. If the switch is defective, replace it.

The third likely cause of this condition is a loose electrical connection in the control box, the cable, or the motor. Troubleshooting for this condition requires extensive work with your meters.

To begin the electrical troubleshooting, we will look for electrical shorts by measuring the insulation resistance. You will use an ohmmeter for this test, and the power to the wires you are testing should be turned off.

Set the ohmmeter scale to RX100K and adjust it to zero. You will be testing the wires coming out of the well from the pump at the well head. Put one of the ohmmeter's leads to any one of the pump wires and place the other ohmmeter lead on the well casing or a metal pipe. As you test the wires for resistance, you will need to know what the various readings mean, so let's examine this issue.

You will be dealing with normal ohm values and megohm values. Insulation resistance will not vary with ratings. Regardless of the motor, horsepower, voltage, or phase rating, the insulation resistance will remain the same.

A new motor that has not been installed should have an ohm value of 20,000,000 or more and a megohm value of 20. A motor that has been used but is capable of being reinstalled should produce an ohm reading of 10,000,000 or more and a megohm reading of 10.

Once a motor is installed in the well, which will be the case in most troubleshooting, the readings will be different. A new motor installed with its drop cable should give an ohm reading of 2,000,000 or more and a megohm value of 2.

An installed motor in a well that is in good condition will present an ohm reading of between 500,000 and 2,000,000. Its megohm value will be between 0.5 and 2.

A motor that gives a reading in ohms of between 20,000 and 500,000 and a megohm reading of between 0.02 and 0.5 may have damaged leads or may have been hit by lightning; however, don't pull the pump yet.

You should pull the pump when the ohm reading ranges from 10,000 to 20,000 and the megohm value drops to between 0.01 and 0.02. These readings indicate a damaged motor or cables. While a motor in this condition may run, it probably won't run for long.

When a motor has failed completely or the insulation on the cables has been destroyed, the ohm reading will be less than 10,000 and the megohm value will be between 0 and 0.01.

With this phase of the electrical troubleshooting done, we are ready to check the winding resistance. You will have to refer to the chart on page 176 for correct resistance values, and you will have to make adjustments if you are reading the resistance through the drop cables. I'll explain more about this in a moment.

If the ohm value is normal during your test, the motor windings are not grounded and the cable insulation is intact. When the ohm readings are below normal, you will have discovered that either the insulation on the cables is damaged or the motor windings are grounded.

To measure winding resistance with the pump still installed in the well, you will have to allow for the size and length of the drop cable. Assuming you are working with copper wire, you can use the following figures to obtain the resistance of cable for each 100 feet in length and ohms per pair of leads:

Cable Size: 14    Resistance: .5150

Cable Size: 12    Resistance: .3238

Cable Size: 10    Resistance: .2036

Cable Size: 8     Resistance: .1281

Cable Size: 6     Resistance: .08056

Cable Size: 4     Resistance: .0506

Cable Size: 2     Resistance: .0318

If aluminum wire is being tested, the readings will be higher. Divide the ohm readings above by 0.61 to determine the actual resistance of aluminum wiring.

If you pull the pump and check the resistance for the motor only (without testing the drop cables), you will use different ratings. You should refer to a chart supplied by the manufacturer of the motor for the proper ratings.

When all the ohm readings are normal, the motor windings are fine. If any of the ohm values are below normal, the motor is shorted. An ohm value that is higher than normal indicates that the winding or cable is open or that there is a poor cable joint or connection. Should you encounter some ohm values higher than normal while others are lower than normal, you have found a situation where the motor leads are mixed up and need to be attached in their proper order.

If you want to check an electrical cable or a cable splice, you will need to disconnect the cable and have a container of water to submerge the cable in; a bathtub will work.

Start by submerging the entire cable, except for the two ends, in water. Set your ohmmeter to RX100K and adjust it to zero. Put one of the meter leads on a cable wire and the other to a ground. Test each wire in the cable with this same procedure.

If at any time the meter's needle goes to zero, remove the splice connection from the water and watch the needle. A needle that falls back to give no reading indicates that the leak is in the splice.

Once the splice is ruled out, you have to test sections of the cable in a similar manner. In other words, once you have activity on the meter, you should slowly remove sections of the cable until the meter settles back into a no-reading position. When this happens, you have found the section that is defective. At this point, the leak can be covered with waterproof electrical tape and reinstalled, or you can replace the cable.

## Will Not Run

A pump that will not run can require extensive troubleshooting. Start with the obvious and make sure the fuse is not blown and the circuit breaker is not tripped. Also check to see that the fuse is of the proper size.

Incorrect voltage can cause a pump to fail. You can check the voltage as described in the electrical troubleshooting section above.

Loose connections, damaged cable insulation, and bad splices, as discussed above, can prevent a pump from running.

The control box can have a lot to do with whether or not a pump will run. If the wrong control box has been installed or if the box is located in an area where temperatures rise to over 122 degrees F., the pump may not run.

When a pump will not run, you should check the control box carefully. We will be working with a quick-disconnect type box. Start by checking the capacitor with an ohmmeter. First, discharge the capacitor before testing. You can do this by putting the metal end of a screwdriver between the capacitor's clips. Set the meter to RX1000 and connect the leads to the black and orange wires leading out of the capacitor case. You should see the needle start towards zero and then swing back to infinity. Should you have to recheck the capacitor, reverse the ohmmeter leads.

The next check involves the relay coil. If the box has a potential relay (three terminals), set your meter on RX1000 and connect the leads to the red and yellow wires. The reading should be between 700 and 1800 ohms for 115-volt boxes. A 230-volt box should read between 4500 and 7000 ohms.

If the box has a current relay coil (four terminals), set the meter on RX1 and connect the leads to black wires at terminals one and three. The reading should be less than one ohm.

In order to check the contact points, you will set your meter on RX1 and connect to the orange and red wires in a three-terminal box. The reading should be zero. For a four-terminal box, you will set the meter at RX1000 and connect to the orange and red wires. The reading should be near infinity.

Now you are ready to check the overload protector with your ohmmeter. Set the meter at RX1, and connect the leads to the black wire and the blue wire. The reading should be at a maximum of 0.5.

If you are checking the overload protector for a control box designed for 1 horsepower or more, you will set your meter at RX1 and connect the leads to terminal number one and to terminal number three on each overload protector. The maximum reading should not exceed 0.5 ohms.

A defective pressure switch or an obstruction in the tubing and fittings for the pressure switch could cause the pump not to run.

As a final option, the pump may have to be pulled and checked to see if it is bound. There should be a high amperage reading if this is the case.

### Doesn't Produce Water

When a submersible pump runs but doesn't produce water, there are several things that could be wrong. The first thing to determine is if the pump is submerged in water. If you find that the pump is submerged, you must begin your regular troubleshooting.

Loose connections or wires connected incorrectly in the control box could be at fault. The problem could be related to the voltage. A leak in the piping system could easily cause the pump to run without producing adequate water.

A check valve could be stuck in the closed position. If the pump was just installed, the check valve may be installed backwards. Other options include a worn pump or motor, a clogged suction screen or impeller, and a broken pump shaft or coupling. You will have to pull the pump if any of these options are suspected.

### Tank Pressure

If you don't have enough tank pressure, check the setting on the pressure switch. If that's okay, check the voltage. Next, check for leaks in the piping system, and as a last resort, check the pump for excessive wear.

### Frequent Cycling

Frequent cycling is often caused by a waterlogged tank, as was described in the section on jet pumps. Of course, an improper setting on the pressure switch can cause a pump to cut on too often, and leaks in the piping can be responsible for the trouble. You may find that the problem is caused by a check valve that has stuck in an open position.

Occasionally the pressure tank will be sized improperly and cause problems. The tank should allow a minimum of 1 minute of running time for each cycle.

## Water Quality

Water quality in private water supplies can vary greatly from one building lot to another. It can even fluctuate within a single water source. A well that tests fine one month may test differently six months later. Frequent testing is the only way to assure a good quality of water. Many times some type of treatment is either desirable or needed before water can be used for domestic purposes.

Many people have heard of hard water and soft water. While these names may be familiar, a lot of people don't know the differences between the two types of water. Some water supplies have hazardous concentrations of contaminants, but most wells are affected more often by mineral contents that, while not necessarily harmful to a person's health, can create other problems.

Is it a builder's responsibility to provide customers with water that is free of mineral content? It shouldn't be. Water conditioning equipment can get very expensive. Spending $1,500 or more for such equipment would not raise any eyebrows in the plumbing community. If you want to make sure that you don't become entangled in a lot of litigation over water quality, have your lawyer create a disclaimer clause for your contracts.

There are four prime substances that can affect the quality of water. Physical characteristics are the first. By physical characteristics, we are talking about such factors as color, turbidity, taste, odor, and so forth. Chemical differences create a second category. This can involve such topics as hard and soft water. Biological contents can modify both the physical and chemical characteristics of water. This third factor in determining water quality can often render water unsafe for drinking. The fourth consideration is radiological factors, such as radon. To understand these four groups better, let's look at them individually.

### Physical Characteristics

Water quality can be assessed by its physical characteristics. Taste and odor rank high as concerns. Water that doesn't taste good or that smells bad is undesirable. It may perfectly safe to drink, but its physical qualities make it unpleasant to live with.

What causes problems with taste and odors? Foreign matter is at fault. It may come in the form of organic compounds, inorganic salts, or dissolved gases. Sulfur content is one of the best known causes of odor in water.

Water that doesn't look good can be difficult to drink enjoyably. Color is a physical characteristic that can taint someone's opinion of water quality. Colored water rarely indicates a health concern, but it is a sure way of drawing some customer complaints. Dissolved organic mater from decaying vegetation and certain inorganic matter typically gives water a distinct color.

Do you know what turbid water is? It's simply water that is cloudy in appearance. Turbidity is caused by suspended materials in the water. Such materials might be clay, silt, very small organic material, plankton, and inorganic materials. Turbidity doesn't usually pose a health risk, but it makes water distasteful to look at in a drinking glass.

Technically, temperature falls into the category of physical characteristics. This, however, is not a problem that many people complain about. Deep wells produce water at consistent temperatures. Shallow wells are more likely to have water temperatures that fluctuate. In either case, temperature is rarely a concern with wells.

Foam is something that you will not normally encounter in well water. However, foamability is a physical characteristic of water, and it can be an indicator that serious water problems exist. Water that foams is usually being affected by some concentration of detergents. While the foam itself may not be dangerous, the fact that detergents are reaching a water source should raise some alarm. If detergents can invade the water, more dangerous elements may also be present.

## Chemical Characteristics

Chemical characteristics are often monitored in well water. A multitude of chemical solutions may be present in a well. As long as quantities are within acceptable, safe guidelines, the presence of chemicals does not automatically require action. However, concentrations of some chemicals can prove harmful. Following is a list of some chemicals that might be present in the next well you install:

| Arsenic | Barium | Cadmium |
|---------|--------|---------|
| Chromium | Cyanides | Fluoride |
| Lead | Selenium | Silver |

## Chlorides

Chlorides in solution are often present in well water. If an excessive quantity of chlorides is present, it may indicate pollution of the water source.

## Copper

Copper can be found in some wells as a natural element. Aside from giving water a poor taste, small amounts of copper are not usually considered harmful.

## Fluorides

Why pay a dentist to give your children fluoride treatments when fluorides may be present in your drinking water? Natural fluorides can be found in some well water. Too much fluoride in drinking water is not good for teeth, so quantities should be measured and assessed by experts.

## Iron

Iron is a common substance found in well water. If water has a high iron content, it will be difficult to avoid brown stains on freshly washed laundry. Plumbing fixtures can be stained by water containing too much iron. The taste of water containing iron can be objectionable.

## Lead

Lead is one contaminant you don't want to show up on a well test. It's possible for dangerous levels of lead to exist in a water source, but most water containing lead derives the detrimental element from plumbing pipes. Modern plumbing codes have provisions to guard against lead being introduced to potable water from piping, but older homes don't share this safeguard.

## Manganese

Manganese, like iron, is very common in well water. Staining of laundry and plumbing fixtures is one reason to limit the amount of manganese in domestic water. It is not unusual for manganese to affect the taste of water adversely. Additionally, excessive consumption of manganese can cause health problems.

### Nitrates

Nitrates show up most often in shallow wells. They can cause what is known as blue-baby disease in infants who ingest water containing it. Shallow wells that are located near livestock are susceptible to nitrate invasion.

### Pesticides

As we all know, pesticides and well water don't mix; at least they shouldn't. Shallow wells located in areas where pesticides are used should be checked often to confirm the suitability of the well's water for domestic use. Wells can also be contaminated during ground treatments for termite control around houses.

### Sodium

Sodium can show up in well water. For average, healthy people, this is not a problem. However, individuals who are forced to maintain low-sodium diets can be affected by the sodium content in well water.

### Sulfates

Sulfates in well water can act as a natural laxative. You can imagine why this condition would not be desirable in most homes.

### Zinc

Zinc doesn't normally draw attention to itself as a health risk, and it is not a common substance in well water. But, it can sometimes be present. Taste is normally the only objection to a content of zinc in well water.

## Hard Water

Hard water is a common problem among well users. This type of water doesn't work well with soap and detergents. If you heat a pot of water on a stove and find a coating of white dustlike substance left in the pan, it is a strong indication that hard water is present. This same basic coating can attack plumbing pipes and storage tanks, creating a number of plumbing problems. Hard water can even cause flush holes in the rims of toilets to clog up and make the toilets flush slowly or poorly.

## Acidic Water

Acidic water is not uncommon in wells. When water has a high acid content, it can eat holes through the copper tubing used in plumbing systems. Plumbing fixtures can be damaged from acidic water. People with sensitive stomachs can suffer from a high acid content. The acidity of water is measured on a pH scale. This scale runs from zero to 14. A reading of seven indicates neutral water. Any reading below seven is moving into an acidic range. Numbers above seven are inching into an alkaline status.

## Biological Factors

Biological factors can be a big concern when it comes to well water. To call water potable or suitable for domestic use, it must be free of disease-producing organisms. What are some of the organisms? Bacteria, especially of the coliform group, is one of them. Protozoa, virus, and helminths (worms) are others.

Biological problems can be avoided in many ways. One way is to use a water source that doesn't support much plant or animal life, such as a well. Springs, ponds, lakes, and streams are more likely to produce biological problems.

It is also necessary to protect a potable water source from contamination. The casing around a well does this. Light should not be able to shine on a water source, and a well cap or cover meets this requirement. Temperature can also play a part in bacterial growth, but the temperature of most wells is not something to be concerned about. If biological activity is a problem, various treatments to the water can solve the problem.

## Radiological Factors

Radiological factors are not a big threat to most well users, but some risk does exist. Special testing can be done to determine if radioactive materials are a significant health risk in any given well. This test should, in my opinion, but conducted by experienced professionals. The same goes for biological testing.

## Solving Water-Quality Problems

There are enough ways of solving water-quality problems that a small book could be written on the subject. Rather than give you a full tour of all aspects of water treatment, I will concentrate on the methods most often used in average homes. There is some form of treatment available for nearly any problem you encounter.

### Bacteria

Bacteria in well water is serious. The most common method of dealing with this problem doesn't require any fancy treatment equipment. A quantity of chlorine bleach is usually all that is needed. Bleach is added to the contaminated well water and allowed to settle for awhile. After a prescribed time, the well is drained or run until no trace of the bleach is evident. This normally clears up biological activity. Sophisticated treatment systems do exist for nasty water, but the odds of needing it are remote.

### Acid Neutralizers

Acid neutralizers are available at a reasonable cost to control high acid contents in domestic water. These units are fairly small, easy to install, not difficult to maintain, and they don't cost a small fortune to purchase.

### Iron

Iron and manganese can be controlled with iron-removal filters. Like acid neutralizers, these units are not extremely expensive, and they can be installed in a relatively small area. If both a water softener and iron-removal system are needed, the iron-removal system should treat water before it reaches a water softener. Otherwise, the iron or manganese may foul the mineral bed in the water softener.

## Water Softeners

Water softeners can treat hard water and bring it back to a satisfactory condition. The use of such a treatment system can prolong the life of plumbing equipment, while providing users with more desirable water quality.

## Activated Carbon Filters

Activated carbon filters provide a solution for water that has a foul taste or odor. Many of these filters are simple, in-line units that are inexpensive and easy to install. When conditions are severe, a more extensive type of activated carbon filter may be required. This type of filter can remove the ill effects of sulfur water (hydrogen sulfide).

## Turbidity

Turbidity can be controlled with simple, in-line filters. If the water being treated contains high amounts of particles, the cartridges in these filters will have to be changed frequently. Left unchanged, they can collect so many particles that water pressure is reduced greatly.

Understanding troubleshooting steps and knowing what symptoms point to specific problems are essential for anyone in a service-related business. Take the time to refine your troubleshooting skills, and you will find that your work is both more enjoyable and more profitable.

# Chapter 19

# Site Limitations for Septic Systems

There are site limitations for private sewage-disposal systems. Not all pieces of land are suitable. Land developers and builders need to be able to spot land that is likely to give them trouble. Experience does much to help in this area. After several years of buying and developing land, a person begins to know good soil by sight.

There are little signs that can give you hints about the quality of land. For example, bulges and occasional glimpses of rock on the land's surface could mean that bedrock is close to the surface. This will certainly interfere with a private sewage-disposal system.

When I was developing land and building homes in Virginia, septic systems were a common part of my job. In all the jobs I did with septic systems, I never encountered bedrock. Most of the soil conditions were very good for leach fields. This has not been the case since I moved to Maine.

Land in Maine, where I work, is rocky. Most of the rock is underground but not by much. Bedrock, or ledge, as it's called in Maine, can be within inches of the topsoil. It took me awhile to get used to this fact. It affects the installation of septic systems and foundations. Quoting a price for a job without knowing that ledge is present can be a financial disaster. Blasting out bedrock to get a full foundation in can destroy any profits a builder hopes to make if the work is not planned for.

Rock is not the only risk encountered in installing septic systems. Some ground just won't perk. If this is the case, the land may be deemed unbuildable. Buying land for development and then discovering that there are no acceptable septic sites can really ruin your day. Experienced land buyers have clauses in their purchase agreements to protect them from this type of risk. A typical land agreement will contain a contingency clause that gives the buyer a chance to have the soil tested before an absolute commitment is made to buy the land. If the tests prove favorable, the deal goes through. When soil studies turn up problems, the contract might be voided or some compensation might be made in the sales price.

The size of a building lot can affect its ability to be approved for a septic system. Many houses that require a septic system for sewage disposal also require a water well for drinking water. For obvious reasons, septic systems must not be installed too closely to a water well. The minimum distance between the two is normally 100 feet or more. I have seen some exceptions to this rule, but not many.

Since moving to Maine, I learned a lot about building practice that I never encountered in Virginia. For one thing, I can't imagine anyone getting a building permit and putting in a foundation before the well and septic locations have been chosen and approved. Yet I know of a case where this happened recently. Let me tell you about it, so the same problem won't come up with your building business.

A man in Maine bought a piece of land. He hired a builder, whom I know, to build his house. The builder had his site contractor clear the lot and install the septic system. It was passed without any problem by the local code officer. Then the builder had the foundation put in—not just footings but full basement walls. It was at this time that a problem was discovered.

The house needed a well. But because of the placement of the septic system, there was no place left on the lot where a water well could be installed in compliance with code requirements. Now what do you do? It is not that the lot was too small for both a well and a septic system. The trouble arose from where the septic system was placed on the lot. Basically, the builder created an unbuildable lot from a piece of land that had been buildable. I would not have wanted to be in that builder's shoes.

In an effort to salvage his situation, the builder went to adjoining landowners and asked them to give an easement so that a well could be drilled on their property for the other person's house. It took some doing, but I believe an agreement of some type was worked out with the neighbors. Had the neighbors refused to cooperate, I don't know what would have happened. I suspect the builder would have lost a costly lawsuit.

The builder we have just been discussing made some very serious errors in judgment. He was just lucky that the circumstances didn't turn too ugly. Any time that you must rely on a private septic system for a house that you're building, you're at risk. And, it's not always just houses. Some commercial buildings are dependent upon private waste-disposal systems. To protect yourself, you have to know what to look out for.

Let me tell you about a builder in Virginia who made a big mistake when bidding on a job. This builder went out to a building lot with the lot owner. After walking over the property, the builder accepted a set of plans and went back to his office to work up a price for a new house.

After completing a quote, the builder was awarded a contract to build the lot owner's new house. It was not until the permit process started that the builder found out that the municipal sewer, which served many houses in the area, didn't extend down to the building lot on which he was scheduled to build. There was a public water supply, but no sewer. This left the builder with two options. He could either pay to have the street cut and the sewer extended, or he could pay for a septic system.

When the builder priced his work, he was planning on paying a tap fee to connect to a city sewer. As I remember, the tap fee was around $3,000. At any rate, the only money he had budgeted was the amount figured for the tap fee. Paying to have the sewer extended would cost much more. Installing a septic system would be difficult, due to the size and shape of the lot. A deal was worked out with the lot owner and the sewer was extended. I don't know what type of financial adjustments were made between the builder and his customer, but I expect the builder made considerably less money on the job that what he had hoped to.

### As They Appear

Things are not always as they appear. In the case of our last example, the builder assumed that public water and sewer were available to the building lot. Fortunately, water was. Unfortunately, a sewer connection was not readily available. This situation is not as rare as you might think.

I've run into a number of building lots where municipal water hookups were available but a private waste-disposal system was needed. With my contingency clauses and inspections, I've never been put in a bind from this type of thing. But a buyer or builder who doesn't research what options are available could get in big trouble very quickly.

Some people assume that since building lots on either side of a particular lot are able to use septic systems, the lot in question should be suitable for such a system. This is not always the case. I've seen land where out of five acres there might be only one or two sites suitable for a standard septic system. This situation is not uncommon, so watch out for it. Make sure that you have an approved septic location established before you make any firm commitments to buy or build.

## Grade

The grade of a building lot can have a lot to do with the type of septic system that must be installed. If the grade will not allow for a gravity system, the price that you pay will go up considerably. Pump stations can be used, but they are not inexpensive. If you're building houses on spec, you might have trouble selling one that relies on a pump station. People don't like the idea of having to replace pumps at some time in the future. And many people are afraid the pumps will fail, leaving them without sanitary conditions until they can be replaced.

The naked eye is a natural wonder, but it cannot always detect the slope of a piece of land accurately. Looks can be deceiving. Unless the land you are looking at leaves no room for doubt about its elevation, check the land with a transit. You can't afford to bid jobs with regular septic systems and then wind up having to install pump systems.

## A Safe Way

There is a safe way to work with land where a septic system is needed. You can go by reports that are provided to you by experts. Even this is not foolproof, but it's as good as it gets. If you have a soil engineer design a septic system for your job, you can be pretty sure that the system will work out close to the way it has been planned. Since every jurisdiction I know of requires a septic design before a building permit will be issued, it makes sense to go ahead and get the design early.

I've dealt with two types of situations when getting septic designs. In Virginia, the designs were drawn by a county official. The cost, if there was one, was very minimal. Let me tell you how this process worked.

A representative of the county in which the land was located would come out and bore test pits with an auger. This was done by hand, so no clearing was needed for heavy equipment. A perk test was performed by the official. A few days later, a finished design was provided. It didn't take long or cost much to have a detailed plan created. The plan was official, so it could be counted on for permit acquisition.

Things are done differently where I work in Maine. Here, soils studies are done by independent engineers. The cost for these studies is normally around $250. As a builder, I pick an engineer of my preference to test the soils and draw a septic design. I can then use that design to obtain a septic permit from the code-enforcement office. This is a more expensive process than what I was used to in Virginia, but in the scheme of things, it is still a bargain. I'd much rather pay $250 to discover a problem before I walk into it than to spend thousands of dollars trying to fix a mistake I've made.

If you are buying land, make sure that your purchase agreement provides a contingency that will allow you to have the land approved for a septic system before you are committed to going through with the sale. It is wise to specify in your contingency what type or types of septic designs you will accept. Some types of systems cost much more than others. If your contract merely states that the land must be suitable for a septic system, you will have to complete the sale regardless of what the cost of a system is, so long as one can be installed.

When you are bidding jobs, make clear in your quote what your limits are in regards to site conditions. Specify the type of septic system your quote is based on, and provide language that will protect you if some other type of system is required. Don't take any chances when dealing with septic systems, because the money you lose can be significant.

## Standard Systems

Standard septic systems are built with a gravity flow. The drain field, or leach field, as it is often called, is made up of perforated pipe and crushed stone. This is the least expensive type of septic system to install. Unless the soil will not perk well, you will normally be able to use a standard system.

## Chamber Systems

Chamber systems are much more expensive than pipe-and-gravel systems. Chamber systems are used when soil doesn't perk well. If the soil will perk but not well, a chamber system may be your only choice. How do these systems work? Basically, the chambers hold effluent from a septic tank until the ground can absorb it. Unlike a perforated pipe, which would release effluent quickly, the chamber controls the flow of effluent at a rate acceptable to the soil conditions. Where a pipe-and-gravel system might flood an area with effluent, a chamber system can distribute the liquid more slowly and under controlled circumstances.

The cost of a chamber system can easily be twice that of a pipe-and-gravel system. It's not unusual in my area for a chamber system to cost up to $10,000. This is a lot of money for a septic system. And, if you happen to bid a job based on a gravel system at say $4,000, the extra $6,000 spent on chambers will deflate your profit very quickly.

## Pump Stations

Pump stations are another big expense in some septic systems. The cost of the pump station, the pump and its control, and the additional labor required to install such a system can add thousands of dollars to the cost of a standard septic system. If you were unlucky enough to get stuck with a pumped chamber system when you had planned on a gravity gravel system, you could lose much of your building profit all at once.

## Trees

Trees are another factor you must consider when doing a site inspection for a septic system. Tree roots and drain fields don't mix very well. If there are trees in the area of the septic system, they must be removed. Even trees that are not directly in the septic site should be removed from the edges of the area. How far should open space exist between a septic system and trees? It depends on the types of trees that are growing in the area. Some trees have roots that reach out much further than others. You can get a good idea of how far a tree's roots extend by looking at the branches on the tree. The spread of the branches will often be similar to the spread of the roots. If there is any doubt, ask the professional who draws your septic design to advise you on which trees can be left standing.

## Burying A Septic Tank

Burying a septic tank requires a fairly deep hole. Even if you are using a low-profile tank, the depth requirement will be several feet. If you are working in an area where bedrock is present, like I do, you must be cautious. You could run into a situation where rock will prevent you from burying a septic tank. This was almost the case on my personal home.

When I built my most recent home, the land I chose consisted of a lot of bedrock. When digging footings for the house, my site contractor hit solid rock in less than two feet of excavation. I knew we would hit rock at a shallow depth, so this didn't alarm me. It did, however, make it difficult to install the water line from my well at a depth where it would not freeze. In fact, I had to use an in-line heat tape to protect the water service. The rock also created some concern as to how we would bury my septic tank.

When I bought my land, I probed it with a metal rod to determine how far below the surface I would encounter ledge. My many probe sites showed that 18 inches was the most I could count on digging in many areas. Rock in some areas was deeper. With the use of my probe, I was able to plot a location for my foundation, my well, and my septic system.

My site work turned out pretty much as I had expected it to. Since I had probed the earth, there were no big surprises. By using the natural slope of the land and some fill, I was able to bury a low-profile septic tank without any real trouble. However, if the land had not consisted of a natural slope, I might not have been able to get my septic tank buried without blasting the bedrock. It might have been necessary for me to build some type of mound system to overcome the problems associated with rock. This would have been very expensive. My preliminary site evaluation proved very helpful in setting a budget and staying within it.

A stiff, small-diameter steel rod with a sharp point is very helpful in probing septic locations. The rod can be used before a system is installed to establish any underground obstacles. But make sure that the area you are probing does not contain any underground utilities. Jamming your probe rod through buried wires can be a shocking experience.

After a septic system is installed, a probe rod can be used to locate the septic tank if you ever have to find it again. Someone should pinpoint the tank location and record it for future reference. Then tank's owner will need access to the tank openings to have the system pumped out periodically. In any event, a good probe rod comes in handy.

### Underground Water

Underground water could present problems for the installation of a septic system. This type of problem should be detected when test pits are dug for perk tests. However, it is possible that the path of the water would evade detection until full-scale excavation was started. For this reason, you should have some type of language in your contracts with customers to indemnify you against underground obstacles such as water.

I've never had a problem with water when installing a septic system, but I have had it get in the way while remodeling a basement. My helper and I broke up a concrete floor in a basement once and found a fast-flowing stream just below it. The water was so abundant that it had washed out the crushed stone used under the concrete. To install the basement bath that we had broken the floor up for, we had to use cast-iron pipe. There was so much water that it was floating our plastic pipe, and it was impossible to make glue joints in the water. If this much water can run under someone's basement floor, it could certainly pose a problem for a septic system.

## Driveways and Parking Areas

When you assess a lot for a septic system, consider the placement of driveways and parking areas. Even though a septic system will be below ground, it is not wise to drive vehicles over it. The weight and movement can damage the drain field to a point where replacement is required. You certainly don't need this type of warranty work, so make sure that all vehicular traffic will avoid the septic system.

## Erosion

Erosion can be a problem with some building lots and land. If you install a septic field on the side of a hill, you must make sure that the soil covering the field will remain in place. This can be done by planting grass or some other ground cover. But, when you check out a piece of land, you need to take the erosion factor into consideration. The cost of preventing a washout over the septic system could add a significant amount of expense to your job.

## Setbacks

Setbacks are something else that you should check on before committing to a septic design. Many localities require all improvements made on a piece of land be kept a minimum distance from the property lines. A typical setback for a side property line is 15 feet, but this is not always the case. Where I live, there are no setbacks. But, I've seen setback requirements that were more than 15 feet. This can become a very big factor in the installation of a septic system. Let me give you an example.

Let's say that you are buying a piece of land to develop into building lots. You've had your perk tests done and your septic designs drawn. On paper there doesn't seem to be any problem. All of the wells and septic systems fit on their individual lots according to the plot plan. But, when you go to get your permits, you find that the zoning requirements establish side setbacks of 20 feet. Based on these requirements, some of your septic systems encroach on the setback zone. This, of course, makes them unacceptable. You might be able to obtain a variance that would allow the drain fields to run into the setback zone, but you might not. If you can't bend the rules, you can't install the septic systems. Do your homework, and make sure that what you have sketched on paper can actually be installed.

## Read The Fine Print

When you are looking at a septic design, read the fine print. A design will give you all the specifications needed to bid a job. You should read the design carefully to determine exactly what types of materials are being called for. It is a very good idea to take the design with you when you do an on-site inspection. A long tape measure should also be considered standard equipment in your field inspections.

Once you arrive at the building site, use the septic design to lay out the system. Once you have the septic site staked out, you can see more clearly what obstacles, such as trees, might be in your way. When you stake the field out, you don't normally have to be precise. The idea here is to see approximately what you will be dealing with in terms of a work area. How difficult will it be to get equipment to the site? This is something you might not think about, but you should.

A few years ago I oversaw the construction of a house where the leach field was high up on the top of a hill. A pump station was installed behind the house to pump effluent up to the field. Due to the topography, getting crushed stone to the site was very difficult. Trucks couldn't back up to the work area and dump their loads. Instead, the gravel had to be ferried up the hill in the bucket of a backhoe. This certainly didn't make job impossible, but it did make much more expensive. As a bidding contractor, you have to keep your eyes open for these types of problems.

As you refer to the septic design, you can determine how long the drain lines must be. There will also be details on how deep the trenches must be. You can use your probe rod to take random tests for underground obstructions.

Will the work area be large enough to accommodate piles of dirt and gravel as you are working? If you have to work the field in sections, due to limited space, this can increase the amount of time needed to construct the system. Whenever there is an increase in time, there is usually an increase in cost.

If you will be required to remove trees, what will you do with the stumps? Getting rid of tree stumps has become a problem in the area where I work. It used to be common practice to bury the stumps, but this is no longer the case. More often than not, the stumps are hauled to some location where they can be ground up. The hauling and disposal fees for stumps can add a lot to the cost of a septic system. Keep this in mind if you are responsible for clearing the work area and disposing of unwanted debris.

Pay attention to all details contained in the septic design. Everything that you need to know about a system to price it should be on the design. Once you combine the design with a physical inspection, you should be in a good position to work up an accurate price.

## Site Visits

Site visits should be considered a mandatory step when bidding jobs that require private waste-disposal systems. You can tell a lot from a septic design, but until you walk the land where the system will be installed, you can't formulate a safe bid.

Many contractors try to work off averages when they figure jobs. For example, they will figure so much per square foot for framing costs and so much per square foot for siding work. They may use a per-fixture price to estimate their plumbing costs. Using averages can work. I've estimated jobs based on some square-foot or fixture factor, but you can never be sure of these types of prices. This is even more true when a septic system is involved.

Unless you do a full-blown takeoff and work-up on a job, you can't be sure of your prices. Even when you do go to the trouble of figuring every little cost, there is some risk of error. Considering how much money could be at stake with a septic system, it would be very dangerous to simply plug a generic figure into a bid proposal.

If you are not set up to install septic systems yourself, you should get quotes from contractors who are. This should be done before you offer prices to customers. Builders and general contractors are not always experienced enough in each trade and field to produce accurate estimates on their own. This is why wise builders seek firm prices from subcontractors before presenting any price to their customers.

Assuming that you will be acting in the role of a general contractor, you should take your septic contractor out to the job site for a first-hand look at the lay of the land. Unless you require your subcontractor to visit the work area, you can't be sure that the price you are given will be accurate. Most septic contractors wouldn't offer you a price without a field inspection. Ones who are willing to give you phone quotes are playing the odds. As long as you know they will stand behind their prices, you can go along with them. But only you can decide if you are willing to take the risk of doing without a physical inspection of the work site. Personally, I wouldn't trust any quote that was not based on an inspection of the working conditions.

# Chapter 20

# Septic Designs and Soil Studies

As a builder, it is not your job to do your own soil studies. These will be done by soil engineers or county officials. Drawing septic designs will not be a part of your job description either. But, you do have to know how to interpret them. Code-related issues will be incumbent on the contractor you hire to install your septic systems. However, if you don't have a cursory knowledge of code issues, you may find yourself feeling very foolish. So, what are we going to do about this? Well, I'm going to prepare you with enough background so that you can hold your own with any builder when it comes to talking turkey about septic systems.

## Septic Designs

I think we should start our discussion with septic designs. Chronologically, soil studies would come first, but it may help you to understand the soil studies if you first have knowledge of design criteria.

As we run through the information required on a septic design, you will see that there are differences in what is required for a new system and a replacement system. As a builder, you will most often be dealing with new systems. However, you may become involved in the utilization of a building lot that once supported a house and a septic system. If, for example, the home was destroyed by fire, the new house you are contracted to build

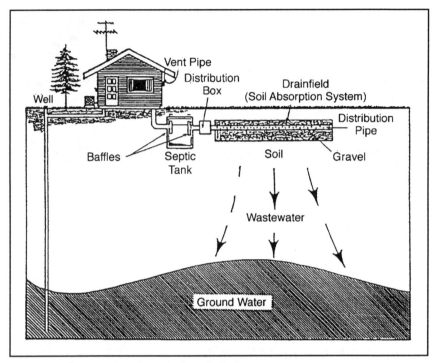

**Figure 20.1** Cross-section of a typical septic design

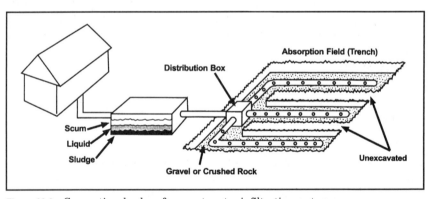

**Figure 20.2** Conventional subsurface wastewater infiltration system

might require a replacement septic system. For this reason, I will go over both areas of the design form.

The top section of the design form calls for information pertaining to the job location and the property owner's name and address. This information is located at the top left corner of the form. At the right corner, you will find an area where a septic permit must be attached once the design has been approved.

As we move into the main part of the design, there are many individual boxes that request pertinent information. For example, there is a box where you must indicate if the system will be a new system, a replacement system, an expanded system, or an experimental system.

The next question box is one to be filled out by a code official. It asks if the system complies with rules, if it is connected to a sanitary sewer, if the system design is recorded and attached, and if system is installed.

If you are installing a replacement system, you must provide information on when the failing system was installed. You must also indicate if the failing system is of a bed, trench, or chamber type. If it is of some other type, you must describe it.

Another question asks for the size of the property. In my case, this was figured in acres, but some lot sizes would be given in actual dimensions. In addition to the lot size, it is necessary to identify the type of zoning in which the property is located.

The next big box of questions deals with what the application requires. The first question asks if a rule variance is required. Is a new system variance needed? Will a replacement variance be needed? Subquestions ask if a variance will require approval from a local plumbing inspector or from both a state and a local plumbing inspector. The final question in this box asks if a minimum-lot-size variance is needed.

Moving along, the next box deals with the type of building being served. Will the building be a single-family home? You must indicate if the dwelling is of a modular or mobile design. If the building will be used as a multi-family dwelling, you must report it. Buildings used for purposes other than those already discussed must be described.

As you shift to the right side of the design form, there is a box that asks if the system is engineered, non-engineered, or primitive (referring to the use of an alternative toilet). The next step is to determine the individually installed components of the system, such as a treatment tank, a holding tank, an alternative toilet, a non-engineered disposal area, or a separate laundry system. Just below this question box is another that requires you to describe the type of water supply being used for the property.

Boxes along the bottom of the first page of this design report start with questions about the type of treatment tank to be used. Is the tank aerobic or septic? What is its capacity? Will the tank be of a standard style, or will it be a low-profile unit?

The next box asks about soil characteristics, such as the soil profile and condition. Another question asks for the depth limiting factor. In my case, due to bedrock, the depth limiting factor at the septic area was 36 inches.

A third box in the bottom section asks about water conservation. Will there be any special procedures used to conserve water? If a house will be equipped with low-volume toilets, there is a box to check off. When separate facilities will be used for laundry waste, there is a place to indicate it. Alternative toilets have a box to be checked.

When you enter the box pertaining to the size of a design, you are given five options: small, medium, medium-large, large, and extra large.

The next box deals with septic systems where pumping will be required. A box must be checked to state whether a pump will be needed or not. There are actually three boxes to check, one if a pump is not required, another if a pump may be required, and a third for when a pump is definitely required.

The next-to-last box on the first page asks about the disposal area's type and size. Is the system to be made with a bed layout? If so, what will the size of the bed be in square feet? Will chambers be needed? Are you planning a trench system? If so, how many linear feet will be required? For other types of systems, you must specify your plans.

The last box on the first page asks for the criteria used for design flow. Options include the number of bedrooms, seating capacity, employees, and water records, depending upon the type of building the system will serve. In the case of a residence, the number of bedrooms is most often used as a guide to determining the number of gallons that will be introduced into the system on any given day.

The very bottom of the first page provides a statement regarding an on-site inspection. This section must be dated and completed by the individual who has designed the system. Then, you move onto page two.

The top half of page two of this septic design form has grid boxes. This grid system is supplied so that the design professional can draw a site plan to scale. Part of the site plan will show the location of the building being served, its well (if one is to be installed), and the septic system. Other information may be included in the drawing, such as roads, rivers, ponds, property boundaries, and so forth.

The bottom section of the second page deals with soil descriptions and classifications as they were determined at observation holes. The first question asks if the hole was a test pit or created by boring. Subjects covered for the soil include: texture, consistency, color, and mottling.

The box provided for soil data is ruled to allow for depths ranging from zero to 50 inches. In the case of my tests, the soil texture was sandy loam to a depth of 36 inches, where bedrock was encountered. The consistency was friable throughout. Color ran from brown in the first 6 inches to reddish in the 6- to 16-inch depths to light brown in the final depths.

Additional information in the bottom section states the soil profile, which in my case was a two. The soil classification was condition "A."

Septic Designs and Soil Studies 203

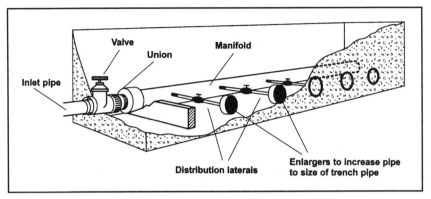

Figure 20.3  Pressure manifold detail

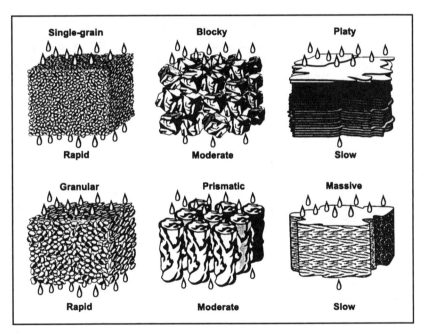

Figure 20.4  Types of soil structure

My slope was 10 to 15 percent, and my limiting factor was 36 inches, due to bedrock.

Page three of my septic design consists mostly of a detailed drawing of the septic layout. It mandates a low-profile, 1,000-gallon septic tank and no pump. All of the details of the septic system, including the tank, the distribution box, and the bed, are drawn to scale.

Directly below the drawing of my septic system are some fill-in-the-blank spaces. The first one indicates the depth of fill required on the upslope. In my case, this was 12 inches. For depth of fill on the downslope, I was required to have between 30 and 60 inches.

Construction elevations are also given in this section of the report. My reference elevation was set at zero and was marked with a nail and red flag on a tree. The bottom of my disposal area was set at 72 inches below the benchmark (this was the nail and red flag). For the top of my distribution lines, a calculation was made for 60 inches below the benchmark.

Two cross-section drawings were attached to my design. These drawings showed all the details of the installation. For example, the drawing started at the bedrock and showed the original soil surface. It then showed a 12-inch layer of crushed stone. It indicated a layer of hay and 4-inch perforated pipe. This was to be covered by a 12-inch layer of sandy-type fill. Further details showed the rest of the fill needed to accommodate the slope of my system.

Anyone with a reasonable understanding of construction terms could look at my septic design and see exactly what was involved in a satisfactory installation. Even if you are a builder who is not familiar with septic systems, reviewing a septic design will bring you up to speed quickly.

**Design Criteria**

Design criteria for septic systems can vary from one jurisdiction to the next. You should always consult local authorities to determine what requirements are in effect within your region. However, I can give you a broad-brush understanding of how criteria are often set.

**Trench and Bed Systems**

Trench and bed systems are both types of drain fields that do not require the use of chambers. Bed systems are the more common of the two. The design criteria for either of these systems are different from what would be used to lay out a chamber system.

The landscape position that is normally considered suitable for a trench system should not have a slope of more than 25 percent. A slope greater than this can impair the use of equipment needed to install a system. Bed systems may be limited to a slope of no more than 5 percent. Keep in mind that the numbers I'm giving you are only suggestions. They do not necessarily represent the requirements in your area.

| Method | Typical applications |
|---|---|
| **Gravity flow** | |
| 4-inch perforated pipe | Single or looped trenches at the same elevation; beds. |
| Distribution box | Multiple independent trenches on flat or sloping sites. |
| Serial relief line | Multiple serially connected trenches on a sloping site. |
| Drop box | Multiple independent trenches on a sloping site. |
| **Dosed distribution** | |
| 4-inch perforated pipe (with or without a distribution box) | Single (or multiple) trenches, looped trenches at the same elevation, and beds. |
| Pressure manifold | Multiple independent trenches on sloping sites. |
| Rigid pipe pressure network | Multiple independent trenches at the same elevation (a preferred method for larger SWISs) |
| Dripline pressure network | Multiple independent trenches on flat or sloping sites (a preferred method for larger SWISs) |

**Figure 20.5** Distribution methods and applications

| Contingency option | Description | Comments |
|---|---|---|
| Reserve area | Unencumbered area of suitable soils set aside for a future replacement system. | Does not provide immediate relief from performance problems because the replacement system must be constructed. The replacement system should be constructed such that use can be alternated with use of the original system. |
| Multiple cells | Two or more infiltration cells with a total hydraulic capacity of 100% to 200% of the required area that are alternated into service. | Provide immediate relief from performance problems by providing stand-by capacity. Rotating cells in and out of service on an annual or other regular schedule helps to maintain system capacity. Alternating valves are commercially available to implement this option. The risk from performance problems is reduced because the malfunction of a single cell involves a smaller proportion of the daily flow. |
| Water conservation | Water-conserving actions taken to reduce the hydraulic load to the system, which may alleviate the problem. | A temporary solution that may necessitate a significant lifestyle change by the residents, which creates a disincentive for continued implementation. The organic loading will remain the same unless specific water uses or waste inputs are eliminated from the building or the wastewaters are removed from the site. |
| Pump and haul | Conversion of the septic tank to a holding tank that must be periodically pumped. The raw waste must be hauled to a suitable treatment and/or disposal site. | Holding tanks are a temporary or permanent solution that can be effective but costly, creating a disincentive for long-term use. |

Figure 20.6  Malfunction contingencies

| | 1. Low Risk / Safest Situation | 2. Medium Risk / Potential Hazard | 3. High Risk / Unsafe Situation | Your Risk |
|---|---|---|---|---|
| Capacity of system | Tank designed to handle more than enough wastewater, based on the size of the home and occupancy. | No excess capacity. | Undersized system. Bedrooms or water-using appliances added without expending the wastewater system. | |
| Separation distances | Drainfield at least 100 feet from any well or sensitive area. | Drainfield between 50 and 100 feet from a well or surface water. | Drainfield less than 50 feet from a well or surface water. | |
| Age of system or holding tank (year installed) | Less than 5 years old. | Between 6 and 20 years old. | System more than 20 years old. | |

Figure 20.7  Assessing risks with septic system placement

These systems can be installed on land that is level and well-drained and on the crests of slopes. Convex slopes are considered the best location. Areas where these systems should not be installed include depressions and the bases of slopes where suitable surface drainage is not available.

In terms of texture, a sandy or loamy soil is best suited to trench and bed systems. Gravelly and cobbled soils are not as desirable. Clay soil is the least desirable.

When it comes to soil structure, a strong, granular, blocky, or prismatic structure is best. Platy or unstructured massive soils are the least desirable.

| Treatment objective | Treatment process | Treatment methods |
|---|---|---|
| Suspended solids removal | Sedimentation | Septic tank<br>Free water surface constructed wetland<br>Vegetated submerged bed |
| | Filtration | Septic tank effluent screens<br>Packed-bed media filters (incl. dosed systems)<br>Granular (sand, gravel, glass, bottom ash)<br>Peat, textile<br>Mechanical disk filters<br>Soil infiltration |
| Soluble carbonaceous BOD and ammonium removal | Aerobic, suspended-growth reactors | Extended aeration<br>Fixed-film activated sludge<br>Sequencing batch reactors (SBRs) |
| | Fixed-film aerobic bioreactor | Soil infiltration<br>Packed-bed media filters (incl. dosed systems)<br>Granular (sand, gravel, glass)<br>Peat, textile, foam<br>Trickling filter<br>Fixed-film activated sludge<br>Rotating biological contactors |
| | Lagoons | Facultative and aerobic lagoons<br>Free water surface constructed wetlands |
| Nitrogen transformation | Biological<br>Nitrification (N)<br>Denitrification (D) | Activated sludge (N)<br>Sequencing batch reactors (N)<br>Fixed film bio-reactor (N)<br>Recirculating media filter (N, D)<br>Fixed-film activated sludge (N)<br>Anaerobic upflow filter (N)<br>Anaerobic submerged media reactor (D)<br>Submerged vegetated bed (D)<br>Free-water surface constructed wetland (N, D) |
| | Ion exchange | Cation exchange (ammonium removal)<br>Anion exchange (nitrate removal) |
| Phosphorus removal | Physical/Chemical | Infiltration by soil and other media<br>Chemical flocculation and settling<br>Iron-rich packed-bed media filter |
| | Biological | Sequencing batch reactors |
| Pathogen removal (bacteria, viruses, parasites) | Filtration/Predation/Inactivation | Soil infiltration<br>Packed-bed media filters<br>Granular (sand, gravel, glass bottom ash)<br>Peat, textile |
| | Disinfection | Hypochlorite feed<br>Ultraviolet light |
| Grease removal | Flotation | Grease trap<br>Septic tank |
| | Adsorption | Mechanical skimmer |
| | Aerobic biological treatment (incidental removal will occur; overloading is possible) | Aerobic biological systems |

Figure 20.8  Treatment processes and methods

When you are looking at the color of soil with the intention of installing a trench or bed system, look for bright, uniform colors. This indicates a well-drained, well-aerated soil. Ground that has a dull, gray, or mottled appearance is usually a sign of seasonal saturation. This makes soil unsuitable for a trench or bed system.

If you find soil that is layered with distinct textural or structural differences, be careful. This may indicate that water movement will be hindered, and this is not good.

Ideally, there should be between 2 and 4 feet of unsaturated soil between the bottom of a drain system and the top of a seasonal high-water table or bedrock.

Check with your local authorities to confirm the information I've given you here. You can also ask them to give you acceptable ratings for percolation tests. If you want to do your own perk tests for reference purposes, you simple dig a hole and pour water in it. Then you watch and wait to see how long it takes the soil to absorb the water. You should measure and note the number of inches of water in the hole when you begin the test and the number of minutes it takes for the water to be absorbed. This type of test will have much to do with the type of drain field that must be installed. I will tell you more about perk tests in a moment.

### Mound and Chamber Systems

Mound and chamber systems can have different design criteria from those discussed for bed and trench systems. I chose bed and trench systems to use as our example since they are the types of systems most often used. Your local code office or county extension office can provide you with more detailed criteria for all types of systems.

### Soil Types

What soil types are suitable for an absorption-based septic system? There are a great many types of soils that can accommodate a standard septic system. Naturally, some are better than others. Let's take a few moments to discuss briefly what you should look for in terms of soil types.

### The Best

What is the best type of soil to have when you want to install a normal septic system? There is not necessarily one particular type of soil that is best. However, there are several types of soil that would fall into a category of very desirable. Gravels and gravel-sand mixtures are some of the best soils to work with. Sandy soil is also very good. Soil that is made up of silty gravel or a combination of gravel, sand, and silt can be considered good to work with. Even silty sand and sand-silt combinations rate a good report card. In all of these soil types, it is best to avoid what is known as fines.

**Pretty Good**

Just as there are a number of good soils, there are a good many soil types that are pretty good to install a septic system in. Gravel that has clay mixed with it is fairly good to deal with, and so is a gravel-sand-clay mixture. The same can be said for sand-clay mixtures. Moving down the list of acceptable soil types, you can find inorganic silts, fine sands, silty or clayey fine sands.

**Not So Good**

Inorganic clay, fat clay, and inorganic silt are not so good when it comes to drainage values. This is also true of micaceous or diatomaceous fine sandy or silty soils. These types of soils can be used in conjunction with absorption-based septic systems, but the systems will have to be designed to make up for the poor drainage characteristics of the soils.

**Just Won't Do**

Some types of soils just won't do when it comes to installing a standard septic system. Of these types of soils, organic silts and clays are unacceptable. So are peat and other soils that have a high organic rating.

I've read some quite sophisticated books on the subject of soils. Believe me when I tell you that an entire book could be written on the subject of soils alone. Operating on the assumption that you are a builder and don't wish to become a soil scientist, I will spare you all the technical information that is available on various types of soils. As a builder, you will have to rely on someone else to design your septic systems, so I can see no need to go into great detail on the finer points of soil types.

**Chapter**

# 21

# Gravel-and-Pipe Septic Systems

Gravel-and-pipe septic systems are about the least expensive option available. Just because these systems are less expensive than other types of systems doesn't mean that they are of a lower quality or don't perform as well. The reason pipe-and-gravel systems cost less is simple: they don't require as much material or time to build. Why is this? It's because the soil in which the systems are installed is of a better drainage quality than soils where more complex systems, such as chamber systems, would be required.

If you build houses on spec, you should be particularly alert to the type of septic system your houses will require. When people buy a new house, they don't normally care whether the septic system is of a pipe-and-gravel type or a chamber type, so long as the system meets code requirements and is designed and installed to give years of worry-free service. If you have two houses sitting side by side, one with a chamber system and one with a pipe-and-gravel system, no one will know the difference until you tell them. There will be, however, one big difference that potential buyers will notice, and that's the price. A chamber system can easily cost twice what a pipe-and-gravel system costs. To make the same profit on both spec houses, you will have to price one thousands of dollars higher than the other. This is where you and your customers will notice the biggest difference.

When I was a builder in Virginia, I never had to install a chamber system. All of the septic systems installed through my building company were pipe-and-gravel systems. After moving to Maine, I saw a big

## 212  Chapter Twenty-one

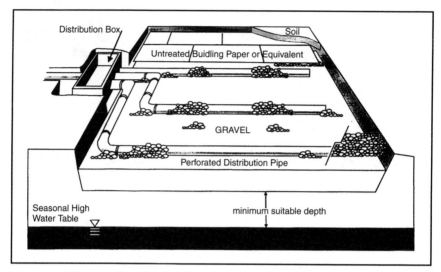

**Figure 21.1**  Cross-section of typical disposal bed

swing in the odds. As a builder in Maine, I see a lot more chamber systems than I do pipe-and-gravel systems.

How much difference is there in the cost between a pipe-and-gravel system and a chamber system? Prices can vary. I'm sure costs in Maine are not comparable to costs in Nebraska or California. From my personal experience, an average pipe-and-gravel system in Maine costs me, at builder's rates, a little over $4,000. A chamber system for the same type of house will come in somewhere around $8,500, although I've

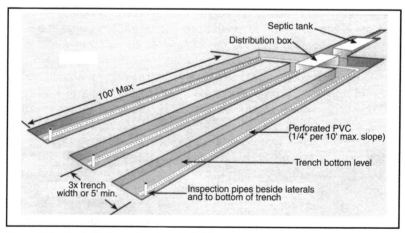

**Figure 21.2**  Typical absorption trench field

seen them at more than $10,000. This is a huge difference in cost. And the only reason for this increase is the suitability of the soil in which the system is being installed.

Maine has a lot of wetlands and a lot of bedrock. Both of these are factors forcing the use of chamber systems. I was lucky when I built my house, since I was able to use a pipe-and-gravel system. Was this only luck? Not exactly. Knowing that a chamber system would cost at least twice as much as a simple septic system, I considered this factor in my search for land. You can use the same approach when you are looking for lots and land to develop spec houses on.

Consider what I'm about to tell you. I was involved in a fairly large land-development project some years back. The building lots were in a rural area and consisted of 20-acre building sites. Each lot was required to have its own septic system. The land chosen for this project perked well, and we were able to get approval for pipe-and-gravel systems. With this particular project, we were dealing with ten lots. If those lots had required more expensive septic systems, we might have lost $40,000 or more in sales potential. So you see, the difference between a pipe-and-gravel system and a complex system can have a lot of impact on your profit picture.

## The Components

Let's talk about the basic components of a pipe-and-gravel septic system. Starting near the foundation of a building, there is a sewer. The sewer pipe should be solid, not perforated. I know this seems obvious, but I did find a house about three years ago where the person who installed the sewer used perforated drain-field pipe. It was quite a mess. Most jobs today involve the use of schedule-40 plastic pipe for the sewer. Cast-iron pipe can be used, but plastic is the most common and is certainly acceptable.

The sewer pipe runs to the septic tank. There are many types of materials that septic tanks can be made of, but most of them are constructed of concrete. It is possible to build a septic tank on site, but every contractor I've ever known has bought precast tanks. An average-size tank holds about 1,000 gallons. The connection between the sewer and the septic tank should be watertight.

The discharge pipe from the septic tank should be solid, just like the sewer pipe. This pipe runs from the septic tank to a distribution box, which is also normally made of concrete. Once the discharge pipe reaches the distribution box, the type of materials used changes.

### 214 Chapter Twenty-one

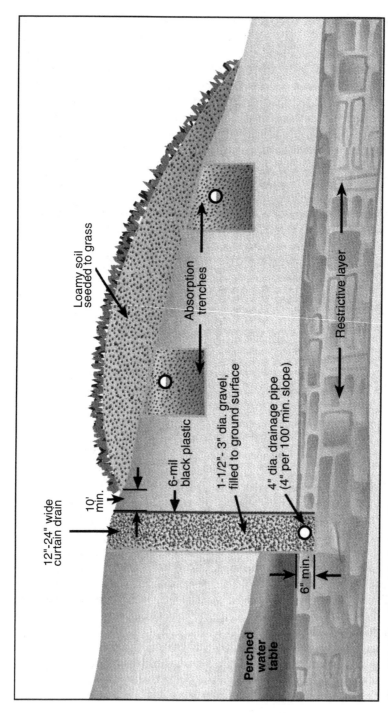

**Figure 21.3** Use of a curtain drain to divert water around an absorption field.

Gravel-and-Pipe Septic Systems 215

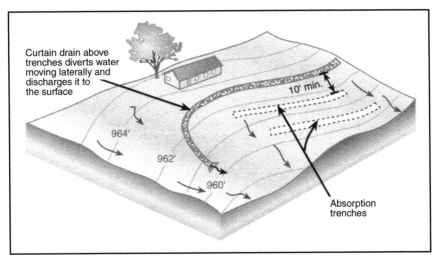

**Figure 21.4** Sample of the use of a curtain drain.

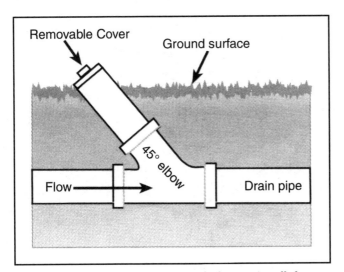

**Figure 21.5** Typical PVC drain pipe with cleanout installed.

The drain field is constructed according to an approved septic design. In basic terms, the excavated area for the septic bed is lined with crushed stone. Perforated plastic pipe is installed in rows. The distance between the drainpipes and the number of drain pipes are controlled by the septic design. All of the drain-field pipes connect to the distribution box. The septic field is then covered with material specified in the septic design.

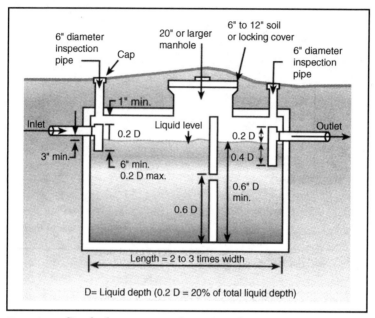

**Figure 21.6** Standard components of a septic tank.

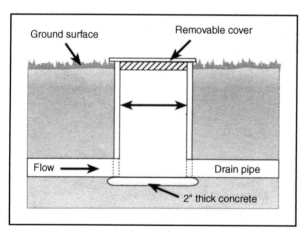

**Figure 21.7** A manhole cleanout for a drain pipe.

As you can see, the list of materials is not a long one. Some schedule-40 plastic pipe, a septic tank, a distribution box, some crushed stone, and some perforated plastic pipe are the main ingredients. This is the primary reason why the cost of a pipe-and-gravel system is so low when compared to other types of systems.

## Types Of Septic Tanks

There are many types of septic tanks in use today. Precast concrete tanks are by far the most common. However, they are not the only type of septic tank available. For this reason, let's discuss some of the material options that are available.

### Precast Concrete

Precast concrete, as I've already said, is the most popular type of septic tank. When this type of tank is installed properly and is not abused, it can last almost indefinitely. However, heavy vehicular traffic running over the tank can damage it, so this should be avoided.

### Metal

Metal septic tanks were once common. There are still a great number of them in use, but new installations rarely involve metal tanks. The reason is simple: metal tends to rust, and that's not good for a septic tank. Some metal tanks are said to have given 20 years of good service. This may be true, but there are no guarantees that a metal tank will last even 10 years. In all my years of being a contractor, I've never seen a metal septic tank installed. I've dug up old ones, but I've never seen a new one go in the ground.

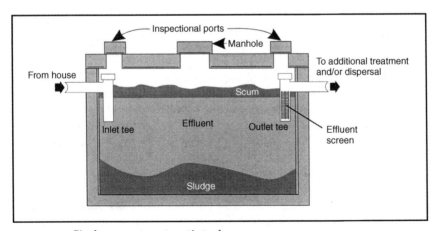

**Figure 21.8** Single-compartment septic tank.

### Fiberglass

I don't have any personal experience with fiberglass septic tanks, but I can see some advantages to them. Their light weight is one nice benefit for anyone installing the tank. Durability is another strong point in their favor. However, I'm not sure how the tanks perform under the stress of being buried. I assume that their performance is good, but again, I have no first-hand experience with them.

### Wood

Wood seems like a strange material to use for the construction of a septic tank, but it does happen. The wood of choice, as I understand it, is redwood. I guess if you can make hot tubs and spas, you can make a septic tank. However, I don't think I would be eager to warranty a septic tank made of wood.

### Brick and Block

Brick and block have also been used to form septic tanks. When these methods are employed, some type of parging and waterproofing must be done on the interior of the vessel. Personally, I would not feel very comfortable with this type of setup. This is, again, material that I have never worked with, so I can't give you much in the way of case histories.

### Installing a Simple Septic System

Installing a simple septic system is pretty easy if you have the right tools, equipment, and knowledge. As a homebuilder or general contractor, you may have access to all the tools and equipment needed to make an installation. To illustrate what's involved, let's run through a typical installation.

The first step in the installation of a septic system is the septic design. This will give you all the details needed to make an acceptable installation. The next step is a permit to do the installation. Your local jurisdiction might require a special license to install septic systems, so check on this if you are planning to do your own installations without the aid of an outside contractor.

Excavation is where your installation will begin. A backhoe is usually all that is needed for the digging. There will be a benchmark set

# Gravel-and-Pipe Septic Systems 219

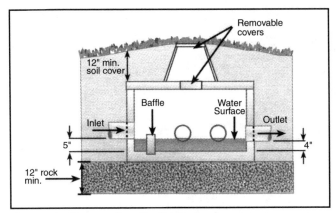

**Figure 21.9** Cross-section of a distribution box.

somewhere, probably in association with a tree, which you will use for elevation measurements. These measurements should be precise, so a transit is needed. Once the septic system is laid out on the ground, you are ready to dig. A lot of contractors use ordinary flour to mark dig locations.

Unless there are extenuating circumstances, a standard backhoe will be capable of digging the sewer trench, the leach bed, and the hole for the septic tank. It is helpful to have your own dump truck, but you can get by with a contract hauler to do the hauling for you. Dirt and stone can be moved around with the front bucket on the backhoe. The main thing to remember when installing a septic system is to follow the septic design to the letter. It is also wise to check your work periodically to make sure that everything is in keeping with the requirements of the design.

After all excavation is done, you have to install the septic components. The leach field is often done first. With today's modern materials, this work doesn't involve complicated plumbing practices or equipment. A carpenter's saw can be used to cut plastic pipe. Joints are made with a solvent-weld (glue). As you install the drainpipes, you must refer to the septic design and follow all requirements. Make sure that the septic system is far enough away from the house and the well to meet local requirements.

After the lines of the leach field are installed, they are connected to the distribution box. Again, this work is detailed on the septic design. Once the distribution box and field are set up, you are ready to set the septic tank. Keep in mind that this sequence is only a suggestion. Some contractors might prefer to set their septic tanks first. It doesn't matter where you start, so long as the job ends up properly installed.

## 220  Chapter Twenty-one

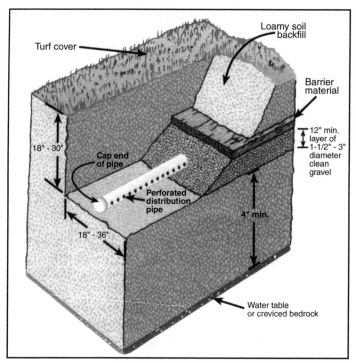

**Figure 21.10**  View of typical trench construction.

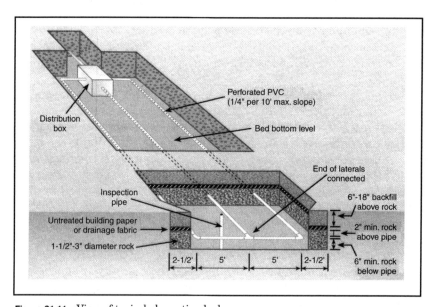

**Figure 21.11**  View of typical absorption bed.

Gravel-and-Pipe Septic Systems 221

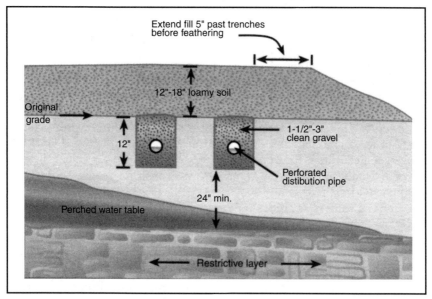

**Figure 21.12** Shallow placement of absorption trenches.

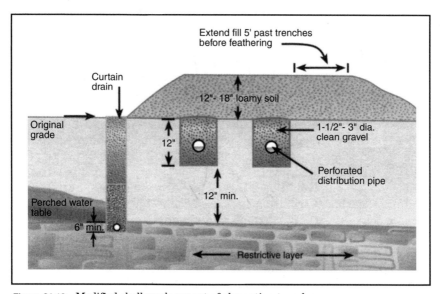

**Figure 21.13** Modified shallow placement of absorption trenches.

When setting the septic tank, you must make sure that it is being positioned on solid ground. This is rarely a problem. If the ground is weak, there will probably be some guidance provided in the septic design. It might be necessary to install a layer of stone under the tank, but this is not typical. You may have to use a tamper to compact the soil, but again, this is not a normal procedure. Usually, a hole is dug and the tank is set into place.

Concrete septic tanks are heavy. A couple of workers can't just horse them into a hole. You will probably use the front bucket of your backhoe to manipulate the tank. In doing this, make sure that the chain used to lift the tank is strong enough to hold the load. Dropping a concrete tank can damage it, not to mention what it could do to a worker's foot trapped under it. Don't take any chances where safety is concerned. Don't put yourself or others in a position to get hurt if the chains should slip or snap.

After the tank is set in its hole, you will probably have to position it with the help of the backhoe bucket and some backfilling. Get the tank to sit in place where you want it, and then connect it to the distribution box. As a reminder, the pipe used for the run from the tank to the distribution box should be solid, not perforated.

When you connect a pipe to the septic tank, you will not be using a glue joint. Instead, the pipe will push through a precast opening and extend into the tank for several inches. The annular space between the pipe and the opening of the hole in the tank will be filled with a cement mixture. You need to make this connection watertight. Otherwise, groundwater could run into the tank, and sewage could seep out into the ground. Neither of these possibilities is desirable.

An elbow fitting is normally attached to the end of the drainpipe that protrudes into the tank. A short piece of pipe is extended from the elbow into the liquid level of the tank. As the liquid level rises in the tank, it also rises up the short length of pipe and is drained out of the tank into the distribution box.

The run of drainpipe from the septic tank to the distribution pipe should be installed with a consistent grade or fall. In other words, the pipe should slope downward toward the distribution box. A standard fall is 1/4 inch of fall for every foot the pipe travels. For example, a pipe that is 8 feet long would be 2 inches higher at the septic tank than it would be at the distribution box. This same type of grading is used when a sewer is extending from a building to a septic tank.

Licensed plumbers typically connect building sewers to main sewers. However, most plumbers limit their rough-in work to a point not to exceed 5 feet past the foundation. If this is the case, you would have to run the sewer from a septic tank to within 5 feet of a foundation in order to have a plumber make the connection within the budget of rough-in work. When a plumber is asked to run sewer piping for more than the 5 feet we have discussed, you can expect some extra charge.

Most septic installers do run a sewer from the tank to a point near the foundation of the building being served. The pipe used for this can be thin-wall sewer pipe. Many installers prefer to use schedule-40 plastic pipe. This is the same type of pipe used for drains and vents inside a building. You should consult your local code requirements to determine what options are available to you. Personally, I prefer schedule-40 plastic, even if thin-wall pipe is an approved material. Long-turn fittings should be used instead of their short-turn cousins. I recommend installing a clean-out just outside the building's foundation.

The sewer that extends between a building and a septic tank must be installed with a consistent grade. The pipe must be supported by solid ground. It is not acceptable to stick blocks of wood or rocks under a pipe to support it, and loose fill dirt is not acceptable. The bed of the trench must be solid and graded, so keep this in mind as you dig. If your digging produces a trench in which there are gaps under a sewer pipe, the trench must be filled in to avoid these gaps. This can be done with crushed stone or dirt that is tamped into place.

The connection of the sewer pipe is parged with cement, just as the pipe to the distribution box is. Once the sewer pipe is extended into the septic tank, a tee fitting is installed on it. Sewage coming into the tank will hit the back of the tee fitting and be directed downward.

Don't cover any of your work up until it has been inspected and approved. Every jurisdiction that I've ever worked in required an on-site inspection of septic installations before they could be buried. Failure to get such an inspection prior to backfilling your work could be very expensive, so make sure the inspection has been made and is approved.

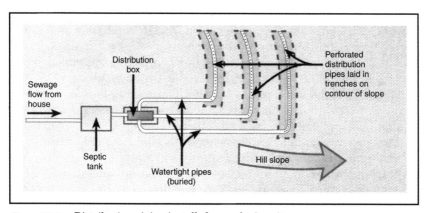

**Figure 21.14** Distribution piping installed on a sloping site.

## 224 Chapter Twenty-one

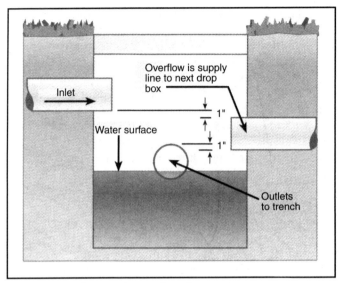

**Figure 21.15** Side view of a stepdown box

### Do It Yourself

When your next septic job comes along, should you do it yourself? There are several things to consider before you can answer this question. The first one is easy. Can you install your own septic systems legally? Check with your local code-enforcement office to see if any special qualifications are needed to install septic systems. You may find that a license of some type is required to obtain septic permits. It may be the case that anyone can install a septic system so long as the installation complies with an approved design. Before you spend any time debating over whether or not to install your own systems, get the answer to this question. If you are not allowed to make your own installations, there is nothing else to think about.

Let's assume that you can install your own septic systems. Do you want to? If you don't want to get involved with the hands-on work, you can simply sub the work out to another contractor. This is what I've always done. However, there are some advantages to doing your own installations, and you may wish to consider them.

A builder who installs septic systems can eliminate one subcontractor. Anytime you can reduce the number of people whom you must count on in the production of a new home, you are one step closer to success. Depending upon other people will often get you in trouble. At least it has always been a sore spot in my business endeavors. I've

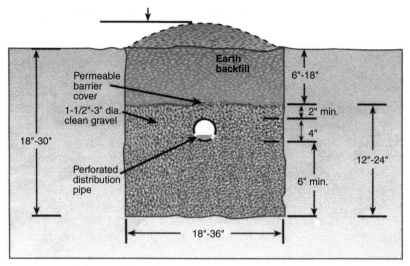

**Figure 21.16** Side view of an absorption trench.

found that work that I have complete control over goes much better than jobs where I must rely on others.

**Money**

Money is almost always a good motivator for expanding your services. If you could make a few extra thousand dollars on a job by installing a septic system yourself, this might be all the reason you need to get into doing septic systems. There is money to be made by installing your own septic systems. The amount of money varies from job to job, but is always enough to make the effort worth considering.

Builders who work primarily with subcontractors may not see the monetary gains of installing septic systems as advantageous. If you don't have your own employees and equipment, it will definitely be easier to just sub the work out to a septic installer. You could rent equipment and use your own payroll labor, if you have any, and this should result in some positive cash flow. However, you might find that your time is better spent selling new jobs than it is overseeing the installation of a septic system. This is the position that I've often found myself in.

I've never owned heavy equipment. Piece workers and subcontractors have always made up most of my work force. While I believe in having some employees available, my building business centers around subcontractors much more than it does employees. For these reasons, it has never seemed cost effective for me to get into doing my own septic installations.

My situation is a little unusual in that I'm a licensed builder and a licensed master plumber. The fact that I maintain both a building and a plumbing company makes me an ideal candidate to create a septic crew. Most of my work is presently done in areas where public sewers are not available. Even with all of this going for me, I've still never jumped into the septic business. Why? Well, I'm really not sure. It does seem that I should, but I've stayed busy doing what I do.

One reason why I've probably not made a transition into septic work is the need for equipment and operators. My volume of septic work would not support the purchase and upkeep of a backhoe, trailer, and dump truck. Neither would it keep an operator and driver busy. Come to think of it, this is probably why I haven't pursued the opportunity. If I had the equipment and personnel, I'm quite sure I would do my own septic installations.

If I were looking for an expansion option for my existing businesses, septic work might be a consideration. However, site contractors often have a lock on septic work. Since these contractors bid clearing, driveway installations, grading, excavation, and so forth, they are a natural source for septic installations. This, too, is probably another reason why I haven't made financial commitments to get into the septic business.

**Work Is Work**

When times are tough, work is work. Installing your own septic systems can put some extra money in your pockets. It can also help to keep your crews busy. This is a plausible reason to consider doing your own installations, even if you have to rent the equipment to work with. If you have a choice of sending your crews home for lack of work, having them push inventory around for something to do, or installing a septic system, you should come out ahead to have them putting in septic systems.

**Control and Quality**

Control and quality are both good reasons for doing your own septic installations. If your crews are doing the job, you have more control over the work than you would if a subcontractor was doing it. It also stands to reason that you would have better control over the quality of the work being done. Outside of money, this may be the best reason to do your own septic work.

## The Technical Side

The technical side of installing septic systems is not difficult to understand. If you can read blueprints, you should be able to interpret a septic design. Once this goal is accomplished, there is very little to stop you from installing septic systems properly. Field experience helps, just as it does in any type of work, but installing a septic system is not a job that requires a lot of hands-on experience to accomplish. I don't wish to minimize the skill required for septic work, but compared to other types of trade-related work, it is pretty easy to understand.

## Bad Ground

Bad ground can force you away from a simple pipe-and-gravel septic system. If the percolation rate is not sufficient, you will have to look to some other type of system, such as a chamber system. We will talk about these expensive systems in the next chapter.

Some soils are not absorbent enough to accommodate a typical pipe-and-gravel septic system. You must be sure of what your septic options are before you bid a job. If you typically plug in a generic figure to represent the price of a septic system for a job without reviewing the septic requirements for a particular job, you are setting yourself up for big trouble. A day will come when you are forced to use a chamber or pump system, and your typical price will be way too low.

# Chapter 22

# Chamber-Type Septic Systems

If you are dealing with a job where a typical pipe-and-gravel septic system can't be used, you are faced with chamber-type systems and other special-use systems. Any of these systems will typically cost more to install than a simple pipe-and-gravel system. The specific type of system used will depend greatly on existing soil conditions. Choosing a special-use system is rarely left to the discretion of a builder. Engineers and county officials are normally the ones to dictate what type of special system will be used.

There are four common types of special-use septic systems. Chamber systems are the most common. Trench systems are usually the most inexpensive type of special-use system available. A mound system might be prescribed, and there are times when a pump system is required. Your installation cost with any of these systems is likely to be higher than what you might expect.

Before moving north, I'd never encountered a chamber system. The soil in Virginia tended to be well suited to septic systems, so gravel-and-pipe systems were normally installed. Pump systems were required occasionally, but chamber systems were never required on any of my jobs in Virginia. While they were very unusual in Virginia, they account for a large number of systems in Maine.

My personal experience with septic systems is limited to the East Coast. I don't have first-hand experience with other parts of the country. However, a little research on your part will reveal the types of systems that are the most common in your area. Since septic systems require septic designs and permits to be installed, it is fairly easy to research what types of systems are being used in your region.

## Chamber Systems

Chamber septic systems are used most often when the perk rate on ground is low. Soil with a rapid absorption rate can support a standard pipe-and-gravel septic system. Clay and other types of soil may not. When bedrock is close to the ground surface, as is the case in much of Maine, chambers are often used.

What is a chamber system? A chamber system is installed very much like a pipe-and-gravel system except for the use of chambers. The chambers might be made of concrete or plastic. Concrete chambers are naturally more expensive to install. Plastic chambers are shipped in halves and put together in the field. Since plastic is a very durable and relatively cheap material, plastic chambers are more popular than concrete chambers.

When a chamber system is called for, there are typically many chambers involved. These chambers are installed in the leach field between sections of pipe. As effluent is released from a septic tank, it is sent into the chambers. The chambers collect and hold the effluent for a period of time. The liquid is gradually released into the leach field and absorbed by the earth. The primary role of the chambers is to retard the distribution rate of the effluent.

Building a chamber system allows you to take advantage of land that would not be buildable with a standard pipe-and-gravel system. In these cases chamber systems are good. However, when you look at the

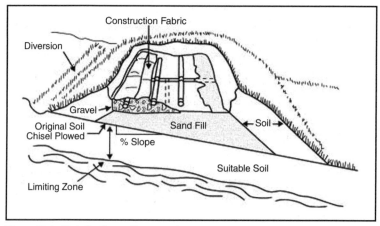

**Figure 22.1** Detail of a septic mound.

price tag of a chamber system, you may need a few moments to catch your breath. I've seen a number of quotes for these systems that pushed the $10,000 mark. Ten grand is a lot of money for a septic system. But, if you don't have any choice, what are you going to do?

A chamber system is simple enough in its design. Liquid leaves a septic tank and enters the first chamber. As more liquid is released from the septic tank, it is transferred into additional chambers that are further away. This process continues, with the chambers releasing a predetermined amount of liquid into the soil as time goes on. The process allows more time for bacterial action to attack raw sewage, and it controls the flow of liquid into the ground.

If a perforated pipe system was used in ground where a chamber system is recommended, the result could be a flooded leach field. This might create health risks. It would most likely produce unpleasant odors, and it might even shorten the life of the septic field.

Chambers are installed between sections of pipe within the drain field. The chambers are then covered with soil. The finished system is not visible above ground. All of the action takes place below grade. The only real downside to a chamber system is the cost.

## Trench Systems

Trench systems are the least expensive version of special septic systems. They are comparable in many ways to a standard pipe-and-gravel bed system. The main difference between a trench system and a bed system is that the drain lines in a trench system are separated by a physical barrier. Bed systems consist of drainpipes situated in a rock bed. All of the pipes are in one large bed. Trench fields depend on separation to work properly. To expand on this, let me give you some technical information.

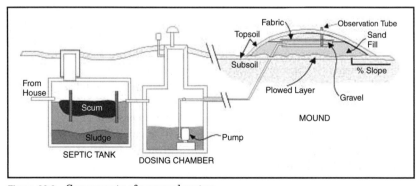

**Figure 22.2** Components of a mound system.

A typical installation is set into trenches that are between 1 to 5 feet deep. The width of the trench tends to run from 1 to 3 feet. Perforated pipe is placed in these trenches on a 6-inch bed of crushed stone. A second layer of stone is placed on top of the drainpipe. This rock is covered with a barrier of some type to protect it from the backfilling process. The type of barrier used will be specified in the septic design.

When a trench system is used, both the sides of the trench and the bottom of the excavation are outlets for liquid. Only one pipe is placed in each trench. These two factors differentiate a trench system from a standard bed system. Bed systems have all of the drainpipes in one large excavation. In a bed system, the bottom of the bed is the only significant infiltrative surface. Since trench systems use both the bottoms and sides of trenches as infiltrative surfaces, more absorption is potentially possible.

Neither bed or trench systems should be used in soils where the percolation rate is either very fast or very slow. For example, if the soil will accept 1 inch of liquid per minute, it is too fast for a standard absorption system. This can be overcome by lining the infiltrative surface with a thick layer (about 2 feet or more) of sandy loam soil. Conversely, land that drains at a rate of 1 inch an hour is too slow for a bed or trench system. This is a situation where a chamber system might be recommended as an alternative.

### More Land Area

Because of their design, trench systems require more land area than bed systems do. This can be a problem on small building lots. It can also add to the expense of clearing land for a septic field. However, trench systems are normally considered to be better than bed systems. There are many reasons for this.

Trench systems are said to offer up to five times more side area for infiltration to take place. This is based on a trench system with a bottom area identical to a bed system. The difference is in the depth and separation of the trenches. Experts like trench systems because digging equipment can straddle the trench locations during excavation. This reduces damage to the bottom soil and improves performance. In a bed system, equipment must operate within the bed, compacting soil and reducing efficiency.

If you are faced with hilly land to work with, a trench system is ideal. The trenches can be dug to follow the contour of the land. This gives you maximum utilization of the sloping ground. Infiltrative surfaces are maintained while excessive excavation is eliminated.

The advantages of a trench system are numerous. For example, trenches can be run between trees. This reduces clearing costs and allows trees to remain for shade and aesthetic purposes. However, roots may still be a consideration. Most people agree that a trench system performs better than a bed system. When you combine performance with the many other advantages of a trench system, you may want to consider trenching for your next septic system. It costs more to dig individual trenches than it does to create a group bed, but the benefits may outweigh the costs.

## Mound Systems

Mound systems, as you might suspect, are septic systems that are constructed in mounds that rise above the natural topography. This is done to compensate for high water tables and soils with slow absorption rates. Due to the amount of fill material to create a mound, the cost is naturally higher than it would be for a bed system.

Coarse gravel is normally used to build a septic mound. The stone is piled on top of the existing ground. However, topsoil is removed before the stone is installed. When a mound is built, it contains suitable fill material, an absorption area, a distribution network, a cap, and topsoil. Due to the raised height, a mound system depends on either pumping or siphon action to work properly. Essentially, effluent is either pumped or siphoned into the distribution network.

As the effluent is passing through the coarse gravel and infiltrating the fill material, treatment of the wastewater occurs. This continues as the liquid passes through the unsaturated zone of the natural soil.

The purpose of the cap is to retard frost action, deflect precipitation, and retain moisture that will stimulate the growth of groundcover. Without adequate groundcover, erosion can be a problem. There are a

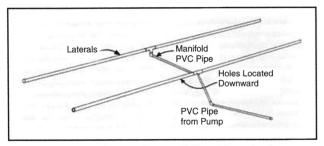

**Figure 22.3** Piping diagram for manifold and laterals of a wastewater distribution network.

multitude of choices available as acceptable groundcovers. Grass is the most common choice.

Mounds should be used only in areas that drain well. The topography can be level or slightly sloping. The amount of slope allowable depends on the perk rate. For example, soil that perks at a rate of 1 inch every 60 minutes or less should not have a slope of more than 6 percent if a mound system is to be installed. If the soil absorbs water from a perk test faster than 1 inch per hour, the slope could be increased to 12 percent. These numbers are only examples. A professional who designs a mound system will set the true criteria for slope values.

Ideally, about 2 feet of unsaturated soil should exist between the original soil surface and the seasonally saturated topsoil. There should be 3 to 5 feet of depth to the impermeable barrier. An overall range of perk rates could go as high as 1 inch in 2 hours, but this, of course, is subject to local approval. Perk tests for this type of system are best when done at a depth of about 20 inches. However, they can be performed at shallow depths of only 12 inches. Again, you must consult and follow local requirements.

The design and construction of mound systems can get quite complicated. This is not a problem to you as a builder, because experts will provide the design criteria. It will then be up to you or your septic installer to follow instructions and see to it that the mound is built as specified.

## Pump Systems

Pump systems are expensive. They are needed when a septic system is installed at an elevation higher than the sewer being served. This situation is not as uncommon as you might think, especially when dealing with houses that have plumbing in their basements. However, houses with basements are not the only structures where pump systems might be needed.

If you are bidding a job where a pump system will be needed, you must be careful not to overlook the pump and related elements that are to be installed. Additional labor and material costs for pump systems can be very substantial.

## Holding Tanks

Holding tanks are sometimes used as a last resort for building sites that have very poor perk rates. The use of holding tanks may or may not be acceptable in your area. Don't buy land on the assumption that you will be able to install holding tanks as an alternative to an absorption septic system. Even if holding tanks are allowed in your area, the use of such a waste-disposal system is likely to have a negative impact on the value of any building you build.

Some land simply isn't suitable for any reasonable type of absorption septic system. This typically labels the land as unbuildable. Under these conditions, holding tanks might be a solution. Basically, a holding tank is just a large vessel that collects wastewater and sewage. The container holds the waste until a certain level is reached in the tank. At this point, the tank must be emptied. This is done with a truck setup to pump out septic tanks. Depending on the size of the holding tank and the volume of use, periodic pumping intervals vary.

Having holding tanks pumped out gets expensive after awhile. The process is also something of an inconvenience. Some people are willing to put up with the hassle and expense, and others aren't. If you are building on spec, you should avoid sites where holding tanks will be required, unless you have some very good plan of attack for selling what you build. Average homebuyers will not normally like the idea of living with a holding tank in place of a septic system.

I can think of very few occasions when holding tanks have been used for full-time residences. This is a septic alternative that is better suited to seasonal cottages and camps. However, you might find some occasion when a holding tank can help you out of a tight spot. Remember, check with your local code office to see if holding tanks may be used before you count on them as a solution to your septic-system problems.

### Other Types of Systems

Engineers may be able to offer other types of systems to meet your special needs. It may be that a combination of systems will be pieced together to accomplish a difficult goal. If you run across a piece of land that doesn't perk well, avoid it if you can. Unless you are able to buy the land very inexpensively as an offset to the poor perk rate, the cost of a special septic system could ruin the profitability of building on the land. You will have to weigh each individual set of circumstances to determine what is worth pursuing. When you are asked to custom-build something for a customer on land that exhibits difficult septic characteristics, have the customer consult engineers or other professionals for a detailed septic design. As long as you are working from an approved septic design, you should be able to keep yourself out of trouble.

Chapter

# 23

# Wastewater Pump Systems

Pump stations are sometimes the only answer for difficult septic situations. We have talked about avoiding pump systems whenever possible, but sometimes a suitable alternative is just not available. It is during these times that a pump system must be installed. It's not that pump stations are very difficult to put in; it is the cost that makes them hard on a builder. The routine maintenance and potential for failure, along with the cost, make it tough for the property owner. But if a pump system is the only way to have a septic system, it's the only way.

If you build in rural areas long enough, you will eventually come across a piece of land where a pump is necessary to drain the plumbing fixtures in the house you are about to build. I've only installed one whole-house pump station during my building career. But I've installed countless pump stations for plumbing fixtures located in basements, garages, and similar locations. It is not always necessary to pump the discharge of all fixtures. Sometimes it is only one fixture that needs to be pumped. It may be an entire fixture group that has to be pumped, such as a full bathroom. If you can avoid installing a whole-house pump station, do it.

There are three types of drainage pumping that we will discuss in this chapter. The first is a single-fixture pump. Our second topic will be basin pumps that are located within a house. The third type of pumping situation to be discussed will be a whole-house pump station. Any of these methods of pumping might solve your problems. Some are cheaper and easier to install than others. All of them add some cost to a job, so you must be aware of this during the bidding phase.

## Single-Fixture Pumps

Single-fixture pumps are inexpensive and easy to install. They don't have any real effect on the plumbing other than the fixtures they are installed on. This type of pump is often used on laundry tubs. In fact, a lot of plumbers call them laundry-tray pumps. The pumps can, however, be installed on any type of residential sink or lavatory. They are not suitable for bathtubs, showers, toilets, or washing machines. (There is one occasion when these pumps are suitable for use with washers. If the indirect waste of the washing machine dumps into a laundry tub, the pump can be used to empty the contents of the laundry sink. Even though the pump is limited in its flow rate, the holding capacity of a deep laundry sink is enough to allow the pump to keep up with the volume of water discharged by an automatic clothes washer.)

A single-fixture pump is small. It installs directly under the bottom of the fixture it is serving. Electricity is needed to make the pump run. An experienced plumber can add a single-fixture pump to a new installation in just a matter of minutes. You must take into consideration, however, the expense of the pump and the electrical circuit that will be needed. All of these factors tend to push the price of your job up.

If you have only one sink that must be installed below the gravity level of other plumbing, a single-fixture pump is a cheap, easy way out. These pumps are often used on basement bar sinks. The discharge pipe from a single-fixture pump normally has only a 3/4-inch-diameter drain. This makes piping the drain in concealed locations easier.

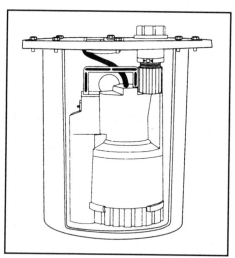

**Figure 23.1** Single-fixture pump: *(Courtesy A.Y. McDonald Mfg. Co.)*

There are some disadvantages to small, single-fixture pumps. Depending upon the type of fixture being served, the pumps may become clogged with sediment and debris. This is often the case when they are used with laundry tubs. When a clothes washer empties into a laundry tub, it often deposits lint in the sink. This lint, if it gets into the drainage system, can plug up the strainer in a small pump. One way to avoid this is by using a strainer to protect the drain opening of the laundry tub. But the strainer could clog and cause the sink to overflow. You could be fighting a losing battle either way.

Single-fixture pumps can be adjusted to cut on as soon as water is detected in a drain, or they can be set to wait until a specific amount of water pressure builds up. The pumps last longer if they don't have to cut on and off frequently and for short durations. Therefore, a setting that allows water to collect in a plumbing fixture before being pumped out is advantageous. It is also possible to use the pumps in a manual fashion, where someone turns the pump on and off as needed. This, of course, would not be suitable in the case of draining an automatic clothes washer, but it can work well for other applications.

The strainers in single-fixture pumps are sensitive. Dirt, sand, lint, and other items are all capable of blocking the pump filter, causing a backup of water in a fixture. Most pump failures that I've witnessed have occurred when the pumps were used on laundry trays with nothing more than cross-bar protection over the drain. Pumps installed on lavatories and bar sinks are less likely to fail.

While single-fixture pumps are inexpensive and effective, they are not always the best choice. I've just explained to you how the small pumps can become clogged and cause back-ups in fixtures. This type of failure could result in flooding. For example, if you were to have an automatic clothes washer dumping into a laundry tray where a single-fixture pump was installed, there could be more than enough water discharged to overflow the flood rim of the laundry tray if the pump were to fail.

How often do single-fixture pumps fail? Provided that they are protected from contaminants and are not abused, they perform very well. My plumbing company has responded to a lot of calls for failed pumps on single fixtures, but in most cases the failure was related to operator error. By this I mean that people allowed objects to go down a drain that were never intended to pass through a pump of such diminutive size.

I would not hesitate to install a single-fixture pump in my own home, but I'm always a little nervous when I install them for other people. Since I don't know how my customers will treat their pumps, I have no way of knowing if I will be getting frantic calls for help from people who are faced with a house that is flooding.

When I install a single-fixture pump professionally, I have a little instruction sheet that I give my customers. It warns them of the risks associated with careless use of the pump. Before I leave a job, I ask the customers to read and sign the form. A copy is given to them, and I keep a copy for my records. This provides written documentation that I instructed the individuals in the proper use of their pumps. If a problem arises from abuse or negligence on the part of a homeowner, I have my little piece of paper to help protect me.

Most plumbers don't go to the extremes that I do in getting paperwork signed for a pump installation. In all my years of plumbing and building, I've only had to produce my form as a defense on one occasion. This was after two free service calls to clear blockages in a strainer of a pump. On both occasions, the pump strainer was full of lint and sand. After the second trip I produced the paperwork and instructed the homeowner that if I were required to respond again and found lint or sand in the strainer, a charge for the service call would be made. The customer has never called me back with a pump problem since.

Is there an alternative to single-fixture pumps when only one fixture needs to be pumped? Yes, there is. A more costly but better pumping arrangement is available for all types of plumbing fixtures. Depending on the type and number of fixtures being pumped, requirements vary. Let me explain with individual examples.

### Gray-Water Sumps

Gray-water sumps and pumps can be used to collect and pump water from all plumbing fixtures that don't receive or discharge human waste. In a typical residence this would include all fixtures except toilets. Many commercial buildings use gray-water sumps to handle wastewater from sinks, and a lot of gray-water sumps are installed in houses for one reason or another.

What is a gray-water sump? The sump itself is not much more than a covered bucket. It is simply a watertight container that is used to collect wastewater from a plumbing fixture. The sump can be installed by sitting it on the floor under a fixture, or it can be buried in a concrete floor. Burying the sump or suspending it beneath a floor would be necessary if it were to serve a bathtub or shower.

A drain is run from a plumbing fixture to the sump, where a watertight connection is made. Another pipe is connected to the discharge outlet at the sump and run to a drainpipe that empties by the force of gravity. The pipe used for this type of pump often has a diameter of 1 inches, although, a larger pipe could be used.

The cover of the sump is often fitted with a gasket that comes into contact with the rim of the sump. A sump cover is normally held in place with machine screws. An air vent should extend from the top of the sump to either open airspace or to a connection with another plumbing vent that does terminate into open air. The diameter of the vent pipe will normally be 2 inches.

Gray-water sumps normally have a holding capacity of about 5 gallons. The sumps that I use are made of polyethylene, and are therefore corrosion resistant. When equipped with a gas-tight cover and vent, a gray-water sump does not emit odors. A key part of a sump system is the pump used. There are many options available for pump purchases.

When I'm required to install a gray-water sump system, I buy a kit. The kit includes the sump and a pump. The pump is a 1/3 horsepower unit that requires 115-volt electrical service. The amp draw is 2.3, and the pump is shipped with a 10-foot plug-in cord. When an existing outlet is within reach, additional electric work may not be needed.

Let me give you an idea of how much water a small pump like I use is capable of moving. Pumping water 10 feet high, one of these pumps can produce up to 2,250 gallons per hour. That's a lot of water. The pump cuts on and off automatically with the use of a float switch. A check valve should be installed in the discharge line to prevent water that has been pumped into the vertical discharge line from returning to the sump.

A gray-water sump doesn't take up a lot of room. The sumps I use stand 15 inches tall and have a width of 15 inches. They weigh about 16 pounds. The supplier of my systems calls them sink-tray systems. Call them what you want, they are an economical alternative to a whole-house pump station. Due the size and design of gray-water sumps, they are not affected by small particles of debris in the same way that a single-fixture pump is. While more expensive than a single-fixture pump, a gray-water sump system is more dependable and can handle a lot more water.

## Black-Water Sumps

Black-water sumps work on a principle similar to that of gray-water sumps. However, they may receive the discharge of toilets and other fixtures of a similar nature. The pumps used in a black-water sump are of a different type than what are used in a gray-water set-up. Black-water sumps are normally installed below a finished floor level. They are quite often buried in concrete floors. It is possible to install such a sump in a crawlspace by simply creating a stable base for it on the ground.

Black-water basins or sumps are available in different sizes. A typical residential sump may be 30 inches deep with an 18-inch diameter at the lid. Basin packages can be purchased that include all parts necessary to set up the basin.

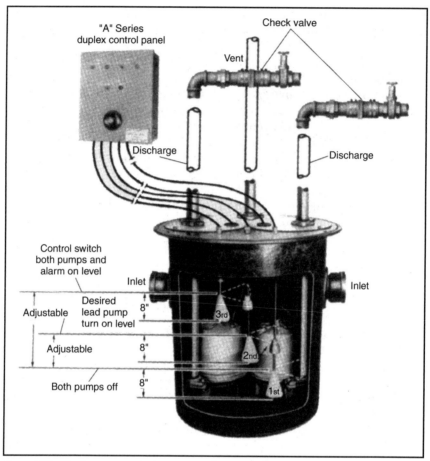

**Figure 23.2** Duplex sewage ejector system: *Courtesy of Goulds Pumps, Inc.*

A 2-inch vent pipe should extend from the basin cover to open air outside a building. A check valve is used in the vertical discharge line. A 4-inch inlet opening is molded into the basin to accept the waste of all types of residential plumbing fixtures. Some type of float system is used to activate the pump that is housed in the basin. The exact type of float system used depends on the type of pump being used.

Pumps used for a typical in-house black-water sump are known as effluent pumps and sewage ejectors. The discharge pipe from the pump normally has a diameter of 2 inches, although a 3-inch discharge flange is available. Even if the waste of a toilet is being pumped, a 2-inch discharge pipe is sufficient.

The cost of a complete black-water sump system will run into several hundred dollars, but this is much less than what a whole-house pump station would cost. If you have a basement bathroom that

Wastewater Pump Systems 243

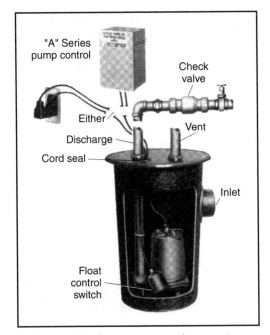

**Figure 23.3** Single-pump sewage ejector system: *Courtesy of Goulds Pumps, Inc.*

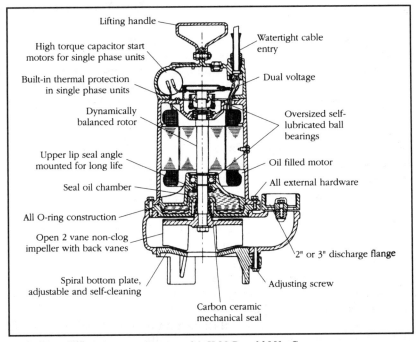

**Figure 23.4** Effluent pump: *Courtesy of A. Y. McDonald Mfg. Co.*

requires pumping, this type of system is ideal. I've installed dozens if not hundreds of them, and I've never been called back for a failure. During my plumbing career I have responded to failures in similar systems. The most common problem is a float that has become wedged against the side of the sump. This won't happen if the system is installed properly.

If you have a house where most of the plumbing can be drained by gravity and only a few fixtures need to be pumped, such as in the case of a basement bathroom, an in-house black-water sump is the way to go.

## Whole-House Pump Stations

Whole-house pump stations are about as bad as it gets for a builder. These systems involve a lot of material, labor, and costs. They also pose the potential threat of disabling all the plumbing if the pump fails. As bad as they are, pump stations are sometimes a blessing. If you have land that cannot be built on under any other circumstances, pump stations look pretty darn good.

Since all the plumbing in a house with a pump station is dependent on a pump's operation, you must take some special precautions when installing such a system. Code requirements are often very stringent on this issue, although they vary from place to place. We will talk more about code issues a little later.

It is standard procedure to install a whole-house pump station in a location outside of the home being served. A sewer pipe runs from the house or building to the storage sump. The size and capacity of this sump are determined by local code requirements and anticipated use. A sump can be made of many types of material, such as concrete or fiberglass.

Fiberglass sumps are not uncommon. One fiberglass sump that I know of is 3 feet deep and has a 30-inch diameter. This particular sump is designed for interior use with a grinder pump. It has a 4-inch inlet and an outlet for a 2-inch vent. The basin kit comes complete with a control panel, a hand-off automatic switch, a terminal strip, and an audible alarm. There are also three encapsulated mercury switches that provide positions for on, off, and alarm. This is only one example of a whole-house pumping system; there are many others available to choose from.

The potential designs for a whole-house pump system are too numerous to detail. I can, however, explain to you how one such system in a house I built for a customer was installed.

The house that required a full-scale pump station sat low on its building lot. Even with a full basement, half of which was exposed above the ground due to bedrock, a pump station was needed. A septic field was located behind and above the house at a considerable distance from the foundation. Chambers had to be installed in the leach field. The basic septic system was a gravity-fed system, but the sewage from the house had to be pumped up to the septic tank. Once waste was delivered to the tank, the septic disposal process could take its normal course.

To accomplish our goal, a black-water sump was installed underground outside of the house. I think it was about 15 or 20 feet from the foundation in the rear of the house. Plumbing fixtures in the home drained by gravity to the holding tank. A grinder pump, which was float-operated and equipped with an alarm system, pumped sewage up the hill to the septic tank. Since a fairly large holding tank was used, the pump didn't have to cycle each time a toilet was flushed. The holding tank was large enough to allow occupants of the house to use minimal plumbing fixtures for a short time even if electrical power or a pump failed.

If for some reason the grinder pump failed, lights and an audible alarm would come on in the house. This would alert the homeowners of a problem in an attempt to prevent flooding the holding tank, also known as the pump station. This system worked well when it was installed, and it is still functioning just fine, to the best of my knowledge, after several years of use.

Requirements for pump stations vary from place to place. Even plumbing codes disagree on some points of installation procedures. For example, one major plumbing code requires the use of two pumps in certain pump stations. This provides a backup pump in the case of a failure with the main pump. Another major plumbing code doesn't require a second pump. For reasons like this one, you must investigate the code requirements in your area to avoid costly mistakes.

Regardless of how they are installed, pump stations add to the cost of a project. This is something all builders should be aware of. If an alarm system is required, and it is in many cases, the cost will be higher than if one is not mandated. Before you bid a job with a pump station, get all of your facts straight.

# Chapter 24

# Landscaping Septic-System Areas

Landscaping septic-system areas is often needed to prevent erosion, but these areas have limitations on the types of landscaping that can be done. Obviously, it would not be wise to plant willow trees over a distribution area. The roots of this type of tree would get into the system and wreak havoc.

There are two prime considerations when landscaping a septic area. The first is appearance, and the second is plant choice. Picking the right plants is very important. Other considerations include creating a landscaping plan that will not require a lot of hands-on attention. Traffic, even foot traffic, should be avoided on septic mounds. If the landscaping plan requires frequent maintenance, the foot traffic in the area can be detrimental to the system.

## Appearance

The appearance of a septic area can be enhanced greatly with good landscaping. Some distribution fields are flat and require no more effort than a lawn to fit into the landscape. But mound systems, for example, can be quite noticeable if they are merely planted with grass. A staggering of plants that grow to various heights is a good way to draw attention away from the shape of distribution mounds. Since appearance is a personal matter, we are not going to spend a lot of time on the subject. You and your customers will know the look you are after. However, you might not know what your limitations are, so let's talk about the types of plants that should and should not be used.

## General Information

There is some general information to keep in mind when planting over a septic system. Don't do excessive tilling of soil over the distribution area when planting. Too much tilling can result in erosion of a septic mound. It's a good idea to wear rugged gloves when working with soil over a septic system. Coming into direct contact with septic soil could create a potential health risk. Avoid skin contact with the soil if the system has been in operation.

The types of plants used on a mound or distribution field should be of a type that does not require or seek out a lot of water. You don't want the roots getting into the distribution system. Trees and shrubs should not be planted directly over mounds or distribution pipes. In fact, no woody plant should be planted on a mound or over a distribution field. Herbaceous (nonwoody) plants are best for installations on mounds and over septic fields.

Trees and shrubs can be used as a part of the landscaping plan, but should be too close to the distribution area. A rule-of-thumb is to keep trees and shrubs at least 20 feet away from the edge of the mound or field. If the trees being planted are known for seeking water, they should be planted at least 50 feet away. Some examples of these types of trees are poplar, maple, and elm.

The use of irrigation and fertilization on mounds is not recommended. Erosion is always a concern with mounds, so irrigation should not be a part of your landscaping plan. Root barriers can be installed to reduce the risk of roots reaching down into a septic system. A common form of root barrier is a geotextile membrane that is impregnated with a long-lasting herbicide that will kill plant roots. If animal burrowing or tunneling is discovered on a mound, you must take steps quickly to eliminate the activity. This type of destruction can lead to serious erosion and a costly repair to the septic system. A final piece of general advice is to avoid planting edible plants, such as vegetables and herbs, on a mound or drain field.

## Plant Selection

Plant selection is a critical part of successful landscaping over septic mounds and fields. Geographical location has a lot to do with the types of plants that are suitable for landscaping. A job in Virginia is likely to have very different plants than a job in Maine would have. The climate drives much of the decision on what plants will thrive. In the simplest of terms, grasses and flowers are the two best choices for landscaping septic mounds and fields. Deciding on what specific types of flowers and grasses to use will be dependent on climatic conditions. For this reason, I am not in a position to tell you precisely what types of plants are best in your region. To do so would require far more room than we have here to discuss the topic. But, I can give you some pointers.

## Grass

Grass is usually the first thing planted on a septic mound or distribution field. A fine fescue grass can create dense cover that doesn't need a lot of mowing. This is good, since foot traffic over the septic system should be minimal. Fescue grass is a traditional lawn grass. It can tolerate dry soil and shady sites. This makes it ideal for a septic application. Remember to obtain grass that does not require frequent cutting or watering.

## Flowers

Flowers are an excellent choice for landscaping septic areas. Annual flowers can be used to make a seasonal splash of color. Perennial flowers, such as daylilies, can be grown in the septic area. Erosion can be a concern with flowers. The flowers should either be planted close together or mulched well to prevent erosion. Wildflower assortments do well in septic situations. Violets, butterfly weed, prairie clover, wild geranium, and similar flowers all work well if they are hardy enough to grow in your region.

## Other Options

In addition to grasses and flowers, there are other options. Trees and shrubs need to be kept away from the primary septic drainage area, but they can be used at a distance to add to the appearance of a mound or distribution field. What else can you use to dress up a mound? How about some colorful gravel? Have you considered attractive lawn ornaments? Would a birdbath in an area where foot traffic is acceptable look good? Pine-bark mulch can be used to accent landscaping. Short fencing, the type that is used around small flowerbeds, can be used to add both color and shape to a mound.

Creativity is the key when it comes to landscaping. Septic systems do demand that certain limitations be placed on the materials used for landscaping, but there are a lot of options available. Talk to your local nurseries to determine what plants will grow well in your area and not invade the septic distribution system. With a creative eye and some help from local plant professionals, you can create a fantastic look on what might otherwise be an eyesore.

Chapter

# 25

# Septic-System Maintenance

How long will an average septic system last? The life of a septic system can vary considerably. Local environmental conditions and general maintenance can affect the longevity of a soil absorption system. In theory, a septic system that is designed and installed properly can function indefinitely. However, because many private sewage-disposal systems are not maintained properly, a typical system may last only 20 years. It is common for responsible contractors to assess and reserve a second area of suitable soil in the event that a repair area is needed.

Soil absorption systems work best when the amount of solids, greases, and fats that will settle are kept to a minimum. It is impossible to block all these items from entering a septic system, but much can be done to reduce the amount of unwanted material. For example, let's say that you have had bacon for breakfast. You placed a paper towel on a plate and drained the bacon. This usually results in a grease-soaked paper towel. Grease and fat are present. Neither of these is good for a septic system. Many people will discard the paper towel and wash the plate. The film layer of fat and grease on the plate goes down the drain and into the septic system. This is bad for the drainage piping in the home and for the septic system. Take the plate to the trashcan and wipe it clean with a disposable towel. Once the plate is free of most fat and grease, you can wash it and minimize the amount of unwanted material entering your septic system.

## Garbage Disposals

There has been ongoing debate over the use of garbage disposals when there is a septic system waiting to accept the waste from the appliance. Data can be found to support the use of garbage disposals, but there is plenty of advice that indicates that using a garbage disposal with a septic system is not wise.

As a master plumber, I don't think garbage disposers should be used in conjunction with septic systems. We know that fat and grease should not be allowed to enter the drainage system. If a garbage disposer is employed and used with meats, it's inevitable that grease and fat will get into the drainage system. Bones might be processed through a disposal. Anything that gets past the baffles in a septic tank can get into the distribution system and clog the perforated pipe and soil. Some jurisdictions prohibit the use of garbage disposals in conjunction with private sewage systems. I concur with this line of thinking.

## Unwanted Objects

In addition to fat and grease, there are other unwanted objects that should not be disposed of through a septic system. Some of these items include:

- Paper towels

- Cigarettes

- Condoms

- Cat litter

- Feminine hygiene products

- Plastics

- Disposable diapers

## Temperature

Temperature is an important factor in the successful operation of a septic system. Septic systems essentially digest waste. This process is dependent on temperature. A septic tank has to be somewhat warm to work properly. In cold climates, septic tanks must be buried deeper than they would be in warmer climates. In extreme cases, septic tanks may have to be insulated to perform properly.

## Pathogens

Nitrogen, phosphorus, organic matter, bacteria, and viral pathogens can all be present due to a septic system. These elements can be harmful to human health and the environment. To prevent serious risks from these elements, a septic system must be installed and maintained properly.

## Failure Symptoms

There are a number of potential symptoms that may indicate a failing septic system. Some of these signs of trouble include the following:

- Strong odors

- Ponding in the septic area

- Wastewater backups in a building

- A measurable decline in water quality

## Beneficial Bacteria

Septic systems need beneficial bacteria in the tank and drain field to work properly. Substances such as poisons, solvents, paints, motor oil, pesticides, fertilizers, disinfectants, caustic drain openers, and other household chemicals can interfere with the beneficial bacteria in a septic system. If a home is equipped with a water softener, the salt recharge solution should not be allowed to enter the septic tank.

### Additives

How about those additives that you see advertised to save a septic system? Do they work? I can't say definitively that they don't, but there is a lot of data from reputable sources that indicate that the benefits of these additives are not worth the risk. Many professional reports indicate that the use of additive cleaners will produce little benefit and may cause serious problems. The destruction of useful microbes, which some additives are suspected of being responsible for, can result in the increased discharge of pollutants. There is also a strong belief that the chemicals in the additives can contaminate receiving waters.

### Excess Water

Excess water can have a negative effect on a septic system. Don't divert storm water from gutters and sump pumps to a septic tank. Make sure that the landscaping around a septic system is done in a way to prevent runoff water from collecting in the septic area.

### Cleaning

How often should a typical septic tank be cleaned? The answer to this question depends on the size of the tank and the amount of usage the tank receives. A rule-of-thumb for cleaning an average system is to have the tank pumped every 3 to 5 years. Tanks should be inspected once every 2 years to confirm that baffles are operating properly, that no leaking is present, and that the levels of sludge and scum are acceptable. When the sludge layer thickness exceeds 25 percent of the working liquid capacity of a tank, the tank should be pumped out. If the bottom of the scum layer is within 3 inches of the bottom of the outlet baffle, the tank should be pumped.

With proper installation and maintenance, a residential septic system should last for 20 to 30 years. But this requires common sense and routine maintenance. Disregarding the needs of a septic system can shorten its useful life by many years. Maintenance is much less expensive than replacement, so take care of those systems.

Chapter
# 26

# Troubleshooting Septic Systems

This chapter is filled with routine remedies for sorrowful septic systems. If you install new septic systems for long enough, you are bound to run into some problems after the job is done on some of the systems you are responsible for. Like any other type of new installation, there is a risk that something will go wrong while the installation is still under warranty. This makes a homeowner's problem a builder's problem.

If a builder, such as yourself, subs septic work out to an independent contractor, it does not make the independent contractor solely responsible for problems with a septic system. Since the builder is the general contractor, a customer for whom work is done has a right to look for help from the builder. In other words, even if you don't put a septic system in yourself, you are still responsible for it.

As a general contractor who has subcontracted a septic system out, you are fortunate enough to have someone to call to solve your septic problems. If you install septic systems with an in-house crew, you or your employees will have to assume full responsibility. Either way, you need to be able to talk intelligently to your customers when problems arise.

There are several types of potential failures to consider. Many of the possibilities are avoidable if a builder informs customers of the do's and don'ts when a house is sold. Do you know what recommendations to make to your customers in regard to the use of their septic systems? Many contractors don't. Even the ones who know what to say often never take the time to inform their customers. To me, this is stupid. If

I can avoid warranty work by informing my customers, you can bet I'll spend a few minutes giving them the facts.

We are about to discuss many potential septic problems. But before we do, let me ask you a question. Pretend I'm a homebuyer whom you are building a house for. Will I have any problems with my septic tank if you install a garbage disposer under my kitchen sink? Well, will I? Do you know what to tell me? If you don't, you should.

Garbage disposers are subject to conflicting reports from experts. Some plumbing codes prohibit the use of garbage disposers in houses where septic systems are used for waste disposal. Other plumbing codes do permit such installations. This would be the first thing you would need to know in order to answer my installation question. If it's against code to install a disposer, you have a quick answer to give me.

Assuming that the code will allow the use of a disposer, your answer should be more complex. For example, I might need a large septic tank to accommodate the use of a disposer. There is some risk that chunks of food from the disposer might clog up my drain field. If the field becomes clogged, it has to be dug up and repaired or replaced. It is not an inexpensive undertaking.

How you answer my installation question could have some serious effects on your company. For example, if you have me—the homeowner—sign a notice of what precautions to take and the potential risk I'm assuming by requiring you to install a garbage disposer, you're pretty well off the hook. If you don't advise me of the potential risks and simply tell me that it's no problem to install a disposer, I might be able to build a lawsuit against you when my septic system fails. Keep this type of scenario in mind as we discuss the various potential problems with septic systems. The more you can disclose to a customer in writing, the better off you are. It is, of course, always best to have customers sign a copy of the disclosure for you to keep in your files.

**An Overflowing Toilet**

Some homeowners associate an overflowing toilet with a problem in their septic system. It is possible that the septic system is responsible for the toilet backup, but in a house that is still under warranty, this is not very likely. More likely, it is a stoppage either in the toilet trap or in the drainpipe. Knowing this can help you decide whom to send out on the call.

If you get a call from a customer who has a toilet flooding the bathroom, there is a quick, simple test you can have the homeowner perform to tell you more about the problem. Will the kitchen sink drain? Will other toilets in the house drain? If other fixtures drain just fine, the problem is not with the septic tank.

There are some special instructions that you should give your customers prior to asking them to test other fixtures. First, it is best if they use fixtures that are not in the same bathroom with the plugged-up toilet. Lavatories and bathing units often share the same main drain that a toilet uses. Testing a lavatory that is near a stopped-up toilet can tell you if the toilet is the only fixture affected. It can, in fact, narrow the likelihood of the problem down to the toilet's trap. But if the stoppage is some way down the drainpipe, it's conceivable that the entire bathroom group will be affected. It is also likely that, if the septic tank is the problem, water will back up in a bathtub.

When an entire plumbing system is unable to drain, water will rise to the lowest fixture, which is usually a bathtub or shower. So, if there is no backup in a bathing unit, there probably isn't a problem with a septic tank. But backups in bathing units can happen even when the major part of a plumbing system is working fine. A stoppage in a main drain could cause the liquids to back up into a bathing unit.

To determine if there is a total backup, have homeowners fill their kitchen sinks and then release all the water at once. Get them to do this several times. A volume of water may be needed to expose a problem. Simply running the faucet for a short while might not reveal a problem with the kitchen drain. If the kitchen sink drains successfully after several attempts, it's highly unlikely that there is a problem with the septic tank. This would mean that you should call your plumber, not your septic installer.

## Whole-House Backups

Whole-house backups (where none of the plumbing fixtures drain) indicate either a problem in the building drain, the sewer, or the septic system. There is no way to know where the problem is until some investigative work is done. This is not a good job to assign to homeowners. Your plumber is the most logical subcontractor to call when this type of problem exists. It's possible that the problem is associated with the septic tank, but your plumber will be able to pinpoint the location where trouble is occurring.

For all the plumbing in a house to back up, there must be some obstruction at a point in the drainage or septic system beyond where the last plumbing drain enters the system. Plumbing codes require clean-out plugs along drainage pipes. There should be a clean-out either just inside the foundation wall of a home or just outside the wall. This clean-out location and the access panel of a septic tank are the two places to begin a search for the problem.

If the access cover of the septic system is not buried too deeply, I would start there. But, if extensive digging would be required to expose the cover, I would start with the clean-out at the foundation, hopefully on the outside of the house. Your plumber should be able to remove the clean-out plug and snake the drain. This will normally clear the stoppage, but you may not know what caused the problem. Habitual stoppages point to a problem in the drainage piping or septic tank.

Removing the inspection cover from the inlet area of a septic tank can show you a lot. For example, you may see that the inlet pipe doesn't have a tee fitting on it and has been jammed into a tank baffle. This could obviously account for some stoppages. Cutting the pipe off and installing the diversion fitting will solve this problem.

Sometimes pipes sink in the ground after they are buried. Pipes sometimes become damaged when a trench is backfilled. If a pipe is broken or depressed during backfilling, there can be drainage problems. When a pipe sinks in uncompacted earth, the grade of the pipe is altered, and stoppages become more likely. You might be able to see some of these problems from the access hole over the inlet opening of a septic tank.

Once you remove the inspection cover of a septic tank, look at the inlet pipe. It should come into the tank with a slight downward pitch. If the pipe is pointing upward, it indicates improper grading and a probable cause for stoppages. If the inlet pipe is either not present or partially pulled out of the tank, there's a very good chance that you have found the cause of your backup. If a pipe is hit with a heavy load of dirt during backfilling, it can be broken off or pulled out of position. This won't happen if the pipe is supported properly before backfilling, but someone may have cheated a little during the installation.

In the case of a new septic system, a total backup is most likely to be the result of some failure in the piping system between the house and the septic tank. If your problem is occurring during very cold weather, it is possible that the drainpipe has retained water in a low spot and that the water has since frozen. I've seen this happen several times in Maine with older homes (not ones that I've built).

Running a plumber's snake from the house to the septic tank will tell you if the problem is in the piping, assuming that the snake used is a pretty big one. Little snakes might slip past a blockage that is capable of causing a backup. An electric drain cleaner with a full-size head is the best tool to use.

### The Problem is in the Tank

There are times, even with new systems, when the problem causing a whole-house backup is in the septic tank. These occasions are

Table 26.1  Troubleshooting tips.

| Observation | Condition and Cause |
|---|---|
| Liquid level is approximately 2 inches below the inlet and even with the outlet bottom. There is no apparent wastewater flow in the tank. | Tank is installed properly and at rest with no indication of backup based on liquid level. |
| Liquid level is below the inlet and elevated less than 2 inches above the bottom of the outlet. Free flow of wastewater from inlet to outlet is apparent. | Tank is installed properly and is currently in use with no indication of backup based on liquid level. |
| Regardless of observed wastewater flowage in septic tank, liquid level is at or above inlet bottom or elevated by 2 inches or more above the outlet bottom. | Tank is probably installed properly, but elevated wastewater levels indicate probable backup in the system down-flow of the the tank. The inspector should perform a flow trial. |
| Regardless of observed wastewater flowage in the septic tank, the liquid level is at or below the outlet and the inlet is submerged. | Tank is installed up gradient or installed backwards (i.e., with the inlet in the outlet's position). Up-gradient tanks may appear to slope up towards the outlet end. Tanks installed backwards may have tees and baffles in reverse positions. Either condition should be corrected by a licensed installer. |
| Regardless of observed flowage in tank, liquid level is more than 2 inches below the inlet and the outlet appears and no more than 2 inches above the outlet bottom. | Tank is sloped down gradient. Depending on the severity of the slope, the tank may actually appear to slope downward toward the outlet. If the slope is minimal, no repair is necessary. Consider evaluation by a licensed installer. |
| Regardless of observed flowage in tank, liquid level is below inlet and outlet. | Tank may be leaking and may have structural problems. Pump the system and have a licensed installer make repairs as necessary. |

rare, but they do exist. When this is the case, the top of the septic tank must be uncovered. Some tanks, like the one at my house, are only a few inches beneath the surface. Other tanks can be buried several feet below the finished grade. If you built the house recently, you should know where the tank is located and how deeply it is buried.

Once a septic tank is in full operation, it works on a balance basis. The inlet opening of a septic tank is slightly higher than the outlet opening. When water enters a working septic tank, an equal amount of effluent leaves the tank. This maintains the needed balance. But if the outlet opening is blocked by an obstruction, water can't get out. This will cause a backup.

Strange things sometimes happen on construction sites, so don't rule out any possibilities. It may not seem logical that a relatively new septic tank could be full or clogged, but don't bet on it. I can give you all kinds of things to think about. Suppose your septic installer was using up old scraps of pipe for drops and short pieces, and one of the pieces had a plastic test cap glued into the end of it? This could certainly render the septic system inoperative once the liquid rose to a point where it would attempt to enter the outlet drain. Could this really happen? I've seen the same type of situation happen with interior plumbing, so it could happen with the piping at a septic tank.

What else could block the outlet of a new septic tank? Maybe a piece of scrap wood found its way into the tank during construction and is now blocking the outlet. If the wood floated in the tank and became aligned with the outlet drop, pressure could hold it in place and create a blockage. The point is that almost anything could block the outlet opening, so take a snake and see if it is clear.

If the outlet opening is free of obstructions and all drainage to the septic tank has been ruled out as a potential problem, you must look further down the line. Expose the distribution box and check it. Run a snake from the tank to the box. If it comes through without a hitch, the problem is somewhere in the leach field. In many cases, a leach-field problem will cause the distribution box to flood. If liquid rushes out of the distribution box, you should be alerted to a probable field problem.

## Problems With a Leach Field

Problems with a leach field are uncommon among new installations. Unless the field was poorly designed or installed improperly, there is very little reason why it should fail. However, extremely wet ground conditions due to heavy or constant rains could force a field to become saturated. If the field saturates with ground water, it cannot accept the effluent from a septic tank. This in turn causes backups in houses. When this is the case, the person who created the septic design should be looked to in terms of fault.

I've never built a house with a septic system where the field failed. But, as a plumbing contractor, I've responded to such calls, even with fairly new houses. The problem is usually a matter of poor workmanship. If you keep a watchful eye on your septic crew during an installation, you should not have to be awakened in the middle of the night by an irate customer with a failed leach field.

Assuming that you are unfortunate enough to experience a saturated drain field, your options are limited. You might wait a few days in hopes that the field will dry out, and it might. Beyond that option, you've got some digging to do. Extending the size of the drain field is the only true solution when ground-water saturation is a seasonal problem.

## Older Fields

Older fields often clog up and fail. This type of problem is not likely to happen while a field is under warranty. But you may be interested to know what types of situations can cause a field failure.

### Clogged With Solids

Some drain fields become clogged with solids. Financially, this is a devastating discovery. A clogged field has to be dug up and replaced. Much of the crushed stone might be salvageable, but the pipe, the excavation, and whatever new stone is needed can cost thousands of dollars. The reasons for a problem of this nature is either a poor design, bad workmanship, or abuse.

If the septic tank is too small, solids are likely to enter the drain field. An undersized tank could be the result of a poor septic design, or it could come about as a family grows and adds onto their home. A tank that is adequate for two people may not be able to keep up with usage when four people are involved. Unfortunately, finding out that a tank is too small often doesn't happen until the damage has already been done.

As a builder, you might take on jobs that involve the building of room additions. If you do, and if a septic system is involved, you should check to see that you are not putting too much burden on the septic system. Local officials may require upgrades to an existing system when bedrooms are added to a house.

Why would a small septic tank create problems with a drain field? Septic tanks accept solids and liquids. Ideally, only liquids should leave the septic tank and enter the leach field. Bacterial action occurs in a tank to break down solids. If a tank is too small, there is not adequate time for the breakdown of solids to occur. Increased loads on a small tank can force solids down into the drain field. After this happens over time, the solids plug up the drainage areas in the field. This is when digging and replacement are needed.

### Too Much Pitch

Can you have too much pitch on a drainpipe? Yes, you can. A pipe that is graded with too much pitch can cause several problems. In interior plumbing, a pipe with a fast pitch may allow water to race by without removing all the solids. A properly graded pipe floats the solids in the liquid as drainage occurs. If the water is allowed to rush out, leaving the solids behind, a stoppage will eventually occur.

In terms of a septic tank, a pipe with a fast grade can cause solids to be stirred up and sent down the outlet pipe. When a 4-inch wall of water dumps into a septic tank at a rapid rate, it can create quite a ripple effect. The force of the water might generate enough stir to float solids that should be sinking. If these solids find their way into a leach field, clogging is likely.

### Garbage Disposers

We talked a little bit about garbage disposers earlier. When a disposer is used in conjunction with a septic system, there are more solids than would be present without a disposer. This, where code allows, calls for a larger septic tank. Due to the increase in solids, a larger tank is needed for satisfactory operation and a reduction in the risk of a clogged field. I remind you again, some plumbing codes prohibit the use of garbage disposers where a septic system is present.

### Other Causes

Other causes for field failures can be related to collapsed piping. This is not common with today's modern materials, but it is a fact of life with some old drain fields. Heavy vehicular traffic over a field can compress it and cause the field to fail. This is true even of modern fields. Saturation of a drain field will cause it to fail. This could be the result of seasonal water tables or prolonged use of a field that is giving up the ghost.

Septic tanks should have the solids pumped out of them on a regular basis. For a normal residential system, pumping once every two years should be adequate. Septic professionals can measure sludge levels and determine if pumping is needed. Failure to pump a system routinely can result in a build-up of solids that may invade and clog a leach field.

### House Stinks

Have you ever had a customer call and say, "My house stinks"? This has never happened to me as a builder, but it has been a complaint that my plumbing company has dealt with. Houses sometimes develop unpleasant odors associated with a drainage or septic system. This can occur even in new houses. In fact, it happened in my mother-in-law's new house a couple of years ago.

Sewer gas is a natural element in drains and septic systems. The odor associated with this gas, which by the way is extremely flammable, is normally controlled with the use of vent pipes and water-filled fixture traps. If a fixture trap loses its water seal, gas can escape into a home. This is dangerous and unpleasant. A faulty wax ring under a toilet can lead to the same problem. So can blocked vents and leaking pipe joints.

If you have a customer who is complaining of sewer odors inside their house, call your plumber. The first thing an experienced plumber will check is the trap seals. In a fixture that is not used often the water in its trap will evaporate. When this happens, sewer gas has a direct path into a home. This is common with floor drains and seldom used fixtures.

In my mother-in-law's case, the problem was a dry trap at her downstairs shower. The shower was almost never used, and the trap seal evaporated. I ran the shower for a few minutes to fill the trap and solved her problem. She now runs water in the shower periodically to prevent a reoccurrence of the odor.

Floor drains are frequently a cause of interior odors. Since these drains see so little use, their seals evaporate. Unless the trap is fitted with a trap primer, water must be poured down the drain now and them. A primer is a little water line that maintains a trap seal automatically, but it is not common in residential applications.

When the source of the odor seems to be a toilet, the wax seal between the toilet bowl and its flange should be replaced. More complex problems can also occur in a plumbing system. For example, I once had a new house where the plumbing crew forgot to remove their test caps from the roof vents. Capped off, the vents couldn't work. This affected the drainage more than the odors. A poorly vented drain will drain very slowly. But a plugged vent, such as one that is filled with ice or that has a bird's nest in it, can start an odor problem. If leaking joints are suspected in a vent or drainage system, there are special tests that your plumber can perform. Colored smoke can be used to reveal leak locations. Another test uses a peppermint smell to pinpoint odor leaks. Your plumber should be aware of both of these tests.

## Outside Odors

Outside odors normally have to do with a leach field. But the problem could be with the plumbing vents on top of a house. Sewer gas escapes from plumbing vents. Under most conditions it goes unnoticed. But under the right weather conditions, such as heavy air with no breeze, the odor from a vent might be forced down to where you can smell it. If a vent is too close to an open window, the gas can come into a house. The plumbing code sets standards for vent placement and height. If these regulations are observed, a problem should not exist. However, a short vent pipe or a vent that is close to a window or other ventilating opening could cause some problems.

### Puddles and Odors

Puddles and odors are sometimes found in and near leach fields. If you have septic puddles, you're going to have septic odors. Pumping out the septic tank won't help here. Your problem is with the field itself. It is not too unusual for this type of problem to attack fairly new systems that were not installed properly.

One main reason for puddles in a new system is the grade on the distribution pipes in the leach field. These pipes should be installed relatively level. If they have a lot of pitch on them, effluent will run to the low end and build up. This causes the puddle and the odor. A sloppy installer who doesn't maintain an even, nearly nonexistent grade while installing pipes in a septic field can be the root cause of your problem. And, it's an expensive problem to solve.

Since effluent is running quickly to a low spot, most of the absorption field is not being utilized. It doesn't mean that the field is too small, although this could cause liquid to surface. In the case of overgraded pipes, the problem is that most of the leach field is not being used, and the volume being dumped in the low spot cannot be absorbed quickly enough. This will require excavating the field and correcting the pipe grade. Not a cheap situation. If, as a builder, you find the field pipes to be installed with too much grade, you should have some opportunity to force the septic installer to pay for corrections.

Saturation in part of a septic field will cause outside odors. The ground may drain naturally and solve the problem in time. This could be the case if odors occur only after heavy or prolonged rains. Outside odors shouldn't exist with septic systems. When they are present, especially for several days, you are probably looking at an expensive problem.

### Supervision

Supervision of the work being done as septic systems are installed is one of your best protections as a builder. Homeowners might not expect you to be an expert in the installation of septic systems, but you can bet that they will expect you to provide them with a good job. If this means crawling down into a septic bed and putting a level on the distribution pipes to check for excessive grading, do it. As the builder and general contractor of a house with a septic system, you cannot avoid responsibility for the system.

You can take the attitude that you shouldn't have to get personally involved with the installation of a septic system. Your position might be that you hire competent professionals to make your septic installations, and a code officer inspects the work, so why should you inspect? I don't think that this position will hold up in court. Even if you never go to court, you have your reputation as a builder to think of. If your customers are getting bad septic systems, you're going to get a bad reputation. Some supervision on your part can avoid this problem.

### When a Problem Occurs

When a problem occurs with a septic system that you are involved with, take fast action. Don't put your customers off. Most people are forgiving of mistakes as long as corrective action is taken quickly. Since you are not likely to know where the problem is coming from, a call to your plumber is usually a good first step. If you can't get your plumber quickly, go to the customer's house personally. An inspection done by you might not reveal the cause of a septic problem, but it will have a favorable impact on the customer. A visit of this type also buys you time to get your plumber or septic installer on the job.

A quick response is necessary when a customer's plumbing is failing. However, haste in making decisions as to what's causing the problem should never be considered. Take your time. Make your plumber investigate the situation thoroughly. Solve the problem right the first time. If you make some token gesture at fixing the problem and it doesn't work, you are going to look incompetent. It's better to spend enough time to fix a problem right the first time than it is to run back and forth trying one fix after another until you hit upon the right one.

### Don't Be Afraid to Ask For Help

Don't be afraid to ask for help when you are faced with a septic problem. Unless you installed the system yourself, you should have some professionals available to help you. Your plumber and your septic installer can both be of help, and they both have a stake in your problems. Call them. Ask them to come out and investigate the problem. If they like working for you, they'll come.

Even if you installed a system yourself, consider calling in some experts when you are up against tough troubleshooting problems. Septic installers should be willing to help you for a price. County officials are another potential source of help. The person who drew your septic design might be able to shed some light on the cause of your problem. Make some phone calls. Get advice, even if you have to pay for it. You owe it to your customers to solve their warranty problems. If you don't offer your help freely, you may find yourself being served with legal papers.

Fortunately, problems with new septic systems are rare. Unless someone made a mistake in the design or installation, a septic system should function without failure for many, many years. A good system could easily go 20 years or more without anything more than routine pumping. With some exceptions, a leach field should last indefinitely when installed with modern materials and proper workmanship.

Don't be afraid of septic systems. It's okay to respect them, but there is no reason to fear them. I've worked with septic systems for decades, and I've never had any significant problem with one. If you make it a rule not to build houses where septic systems are needed, you will be throwing away a lot of good work potential.

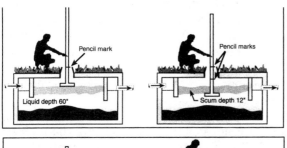

Figure 14a. Measuring the scum level in the septic tank. Left: Lower the bottom of the board to the top of the mat of scum. Mark the stick at the top of the tank opening. Right: Force the board all the way through the mat. Bring it back until you feel the bottom of the mat. Mark the stick again. The distance between these two marks is the scum depth.

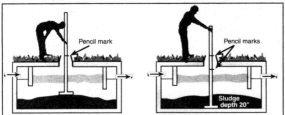

Figure 14b. Measuring the sludge depth in the septic tank. Left: Measure the depth of liquid and the depth of sludge. Let the stick drop until you feel resistance. Mark the stick. Right: Force the board all the way through the sludge to the bottom of the tank. Mark the stick again. The distance between these two marks is the sludge depth. If you do not know the maximum available liquid depth for your tank, measure from the bottom of the board to the wet line on the stick to find out.

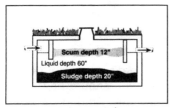

Figure 14c. Determine if the tank should be cleaned. Add the scum and sludge depths. If this figure is more than half the liquid depth, clean the tank. The maximum available liquid depth in this example is 60 inches. The scum depth is 12 inches and the sludge depth is 20 inches — a total of 32 inches. Since this is more than half of the maximum liquid depth, this tank should be cleaned.

**Figure 26.1** Making measurements to determine when to pump out a septic tank.

Troubleshooting Septic Systems 267

Table 26.2  Suggested frequency for pumping septic tanks.

| Tank size, (gallons) | Household size (number of people) | | | | | | | | |
|---|---|---|---|---|---|---|---|---|---|
| | 1 | 2 | 3 | 4 | 5 | 6 | 7 | 8 | 9 |
| | Pumping frequency (years) | | | | | | | | |
| 500 | 5.8 | 2.6 | 1.5 | 1.0 | 0.7 | 0.4 | 0.3 | 0.2 | 0.1 |
| 750 | 9.1 | 4.2 | 2.6 | 1.8 | 1.3 | 1.0 | 0.7 | 0.6 | 0.4 |
| 1,000 | 12.4 | 5.9 | 3.7 | 2.6 | 2.0 | 1.5 | 1.2 | 1.0 | 0.7 |
| 1,250 | 15.6 | 7.5 | 4.8 | 3.4 | 2.6 | 2.0 | 1.7 | 1.4 | 1.2 |
| 1,500 | 18.9 | 9.1 | 5.9 | 4.2 | 3.3 | 2.6 | 2.1 | 1.8 | 1.5 |
| 1,750 | 22.1 | 10.7 | 6.9 | 5.0 | 3.9 | 3.1 | 2.6 | 2.2 | 1.9 |
| 2,000 | 25.4 | 12.4 | 8.0 | 5.9 | 4.5 | 3.7 | 3.1 | 2.6 | 2.2 |
| 2,250 | 28.6 | 14.0 | 9.1 | 6.7 | 5.2 | 4.2 | 3.5 | 3.0 | 2.6 |
| 2,500 | 31.9 | 15.6 | 10.2 | 7.5 | 5.9 | 4.8 | 4.0 | 4.0 | 3.0 |

Note: Pumping frequency can increase by 50 percent if garbage disposal is used.

## Do

**Do** obtain necessary permits from the appropriate local agency before doing any construction or repairs.

**Do** use professional certified installers when needed.

**Do** keep your septic tank and distribution box accessible for pumping and adjustment. Install risers if necessary. The covers should be locked or of sufficient weight to prevent a child from lifting them.

**Do** have your septic system inspected annually and tank pumped out every 2 to 5 years by a professional contractor.

**Do** keep a detailed record of repairs, pumpings, inspections, permits issued and other maintenance activities.

**Do** conserve water to avoid overloading the system. Repair dripping faucets and leaking toilets, avoid long showers and run washing machines and dishwashers only when full. Use water-saving features in faucets, shower heads and toilets.

**Do** divert other sources of water, such as roof drains, house footing drains, sump pump outlets, and driveway and hillside runoff away from the septic system. Use curtain drains, surface diversions, downspout extensions, retaining walls, etc. to divert water.

**Do** take leftover hazardous household chemicals to an approved hazardous waste collection center for disposal. Use bleach, disinfectants and drain and toilet bowl cleaners sparingly and in accordance with product labels.

## Don't

**Don't** go down into a septic tank for any reason. Toxic gases in the tank can be explosive and can cause asphyxiation.

**Don't** allow anyone to drive or park over any part of the system.

**Don't** cover the absorption field with a hard surface, such as concrete or asphalt. Grass is the best cover for promoting proper functioning of the field. The grass will not only prevent erosion but will help remove excess water.

**Don't** plant a garden, trees or shrubbery over or near the absorption field area. Tillage may cut absorption trenches. The roots can clog and damage the drain lines.

**Don't** make or allow repairs to your septic system without obtaining the necessary permits.

**Don't** pour into drains any grease, cooking fats, chemical drain openers, paint, varnishes, solvents, fuels, waste oil, photographic solutions, pesticides or other organic chemicals. They can upset the bacterial action in the tank and pollute groundwater.

**Don't** use your toilet as a trash can. Keep out coffee grounds, bones, cigarette butts, disposable diapers, feminine hygiene products, paper towels, facial tissues and other materials that decompose very slowly.

**Don't** add enzyme or yeast additives to the septic tank in hopes of improving bacterial action. None have been proven beneficial and some actually cause damage to soil and vegetation and may pollute groundwater.

**Figure 26.2**  Maintenance suggestions for septic tanks.

## Chapter Twenty-six

**Table 26.3 Troubleshooting a septic system.**

| Problem | Possible causes | Suggested remedies |
|---|---|---|
| Sewage backs up into house and/or plumbing fixtures don't drain or are sluggish. | 1. Pipe may be clogged between house and septic tank. | a. Remove blockage with sewer routing tool. This usually is a job for a professional.<br>b. If roots have penetrated pipe joints, reseal joints after routing or replace pipe.<br>c. Avoid planting fast-growing trees near system. |
| | 2. Scum layer may be plugging the septic tank inlet. | a. Pump the septic tank.<br>b. Check inlet baffle or tee after pumping tank. |
| | 3. Tank outlet may be plugged by a collapsed baffle or by solids overflow. The pipe from the tank to the absorption field may be partially collapsed or plugged by tree roots. | a. Pump the septic tank.<br>b. Route out the line and replace defective baffles, tees or pipe sections.<br>c. Establish a regular pumping schedule. |
| | 4. Blockage that reoccurs in a new system is likely caused by improperly installed sewer line(s). | Reconstruct the sewer line(s) using the correct slope. |
| | 5. Blockage that reoccurs in a previously trouble-free system is probably caused by a broken sewer pipe connection. | Locate and replace broken pipe. |
| | 6. Excess water is entering the system. | a. Install interceptor drains to lower or divert the high water table.<br>b. Fix water leaks and use water conservation practices.<br>c. Replumb backwash from water softener system out of septic system.<br>d. Check septic tank for water tightness.<br>e. Check to see if dripline or downspout water is running onto absorption field.<br>f. Install surface diversion where surface run-on to absorption field is a problem.<br>g. Replumb sump pump out of septic system. |
| | 7. House sewer vent (soil stack) is plugged or frozen closed. | Clear plugged vents using a garden hose. Enlarge undersized vents. In winter, check for ice buildup on vents. |
| | 8. Lift-station pump has failed. | Repair or replace the pump. |
| | 9. Septic tank effluent filter is plugged. | Pump septic tank, then remove and clean filter. To be performed by certified professional. |
| Sewage is surfacing in yard. | 1. Absorption field is too small. | a. Consult county health department sanitarian or Missouri Department of Health for proper size of field based on soil conditions and household sewage input.<br>b. If undersized, enlarge the existing field or build a new one. |

*(Continued)*

**Table 26.3** *(Continued)* **Troubleshooting a septic system.**

| Problem | Possible causes | Suggested remedies |
|---|---|---|
| Sewage is surfacing in yard. *(continued)* | 2. Water use is excessive. | a. Install water conservation devices, such as low-flush toilets, low-volume shower heads and faucet aerators.<br>b. Minimize use of the hot tub or whirlpool.<br>c. Monitor water use habits of household. Check for leaks in plumbing fixtures. |
| | 3. A seasonally high water table during rainy periods saturates the soil and limits its ability to accept wastewater from the house. | a. Install interceptor drains to lower or divert the high water table.<br>b. Use water conservation practices.<br>c. Modify absorption field system using "shallow placement of trenches" alternatives.<br>d. Monitor water use habits of household. Check for leaks in plumbing fixtures.<br>e. Divert dripline and downspout water from field.<br>f. Replumb sump pump out of septic system.<br>g. If the site permits, build an alternative system, such as a lagoon, low-pressure pipe system, or sand mound. |
| | 4. Solids are overflowing from the tank to the absorption field. | a. Pump the tank and check the baffles or tees.<br>b. Minimize garbage disposal use. |
| | 5. Pipe slopes or tank elevations are improper. | Rebuild absorption field on alternate site using correct pipe and trench or bed elevations. |
| | 6. Lift-station pump has failed or float switches are improperly set. | Clean septic tank, check pump (if used) and adjust float switches. |
| Sewage odors are present indoors or there are gurgling sounds in sink or tub drains. | 1. House sewer vent (soil stack) is plugged or frozen closed. | Clear plugged vents using a garden hose. Enlarge undersized vents. In winter, check for ice build-up on vents. |
| | 2. Water has been sucked from plumbing traps. | Replace water in traps by briefly running faucets. |
| | 3. Improper plumbing. | Have plumbing system checked by a professional. |
| | 4. Sewage backup in house. | Clean tank and check pump (if used). |
| Sewage odors are present outdoors. | 1. Sewage is surfacing in yard. (See also earlier problem section.) | Repair or replace absorption field. Clean tank and check pump (if used). |
| | 2. Inspection pipe caps damaged or removed. | Replace and secure damaged caps. |

*(Continued)*

## Chapter Twenty-six

Table 26.3 *(Continued)* **Troubleshooting a septic system.**

| Problem | Possible causes | Suggested remedies |
|---|---|---|
| Sewage odors are present outdoors. *(continued)* | 3. Source is from location other than homeowner's system. | Contact county health department sanitarian to investigate other potential sources. |
| Drinking water and/or surface waters are contaminated with bacteria. | 1. Septic system is too close to well, water table or fractured bedrock; cesspool or drywell is in use; or sewage discharges to surface or groundwater. | Replace with approved septic system on alternate site. |
| | 2. Well is improperly constructed. | Consult state-certified well driller or pump installer for options. |
| | 3. Water supply pipe from well is broken. | Repair piping and disinfect the entire plumbing system. |
| | 4. Source is from other than homeowner's system. | Contact county health department sanitarian to investigate other potential sources. |
| Lift-station alarm is activated. | 1. Circuit breaker is tripped. | Check electrical connections to pump and reset the breaker. |
| | 2. Pump has failed. | Make sure a professional replaces pump with proper size unit. |
| | 3. Controls are malfunctioning. | Check float switch controls and pump. |
| Distribution pipes and/or absorption field freezes in winter. | 1. System is improperly constructed. | Have system repaired or replaced by a qualified contractor. |
| | 2. There is foot or vehicle traffic over piping. | Keep people and vehicles off area. |
| | 3. The flow rate is too low from lack of use. | a. Have someone use water in house if you are away.<br>b. Operate septic tank as a holding tank.<br>c. Don't use antifreeze. |
| | 4. Lift-station check-valve is not working, or pump is not cycling properly. | Examine check-valve and/or replace it. Increase frequency of pump cycling. |

# Chapter 27

# Low-Cost Septic Systems

Making a habit of installing low-cost septic systems doesn't mean than you are cheating your customers. Actually, you may be doing them a very big favor. If you pass the savings from these systems onto the consumer, you have used your knowledge to save them money. What customer wouldn't appreciate this type of surprise?

Builders can face some very tough times during their careers. I know I have. Surviving economic downturns and recessions are just some of the obstacles to overcome. Increased competition is always a threat, and there are many other factors that can put a builder out of business. One way to improve your odds of survival is to hone your skills at saving money.

Making money is very nice, but it requires you to pay taxes. Saving money can be even better than making money, because the tax bite is not so vicious. I've never met anyone yet who enjoys paying taxes. Most of us do it, but few of us feel we get our money's worth.

The cost of living goes up almost constantly. This means that our incomes should rise with the increases in the cost of living. Did your income rise year before last? How much more money did you make last year? Are you on track to make more money this year? A lot of businesses, including building businesses, suffer from declining sales and income. Left alone, this type of pattern will eventually drive a business out of existence. If you are feeling a pinch in your pocketbook, you must look for new ways to make or save money.

One way for rural builders to reduce their cost of doing business comes in conjunction with septic systems. Some systems simply cost less than others to install. We've talked about this issue already. But for you to assure reduced septic costs time and time again, you must get involved. Spec builders have to look for good land. Custom builders sometimes have to look for alternatives to high-priced septic systems. If you can get into the swing of doing this right, you can make and save more money.

## Knowledge

Knowledge is one of the best tools that a builder can possess. If you know enough about what you're doing, you should be able to do it well. More knowledge in some work areas, such as septic systems, can give you a competitive edge. You've already taken a giant step in the right direction by reading this book. The fact that you are willing to invest your valuable time reading about one or two phases of your building business proves that you care about your business and hopefully about your customers.

Informed builders are better builders. This book is an excellent starting point for you, but it is not the only source of knowledge that you need to become more competitive. Seek other reading materials from local and government agencies. Agencies such as the United States Environmental Protection Agency can provide you with valuable reading material. Local and state agencies may also be able to offer some guidance on the issue of septic systems.

Your subcontractors are a natural source of information. If you have questions about your next septic job, who is better qualified to answer them than your septic engineer or installer? Ask questions. Become informed. Get on the right track, and you can see substantial savings over the coming years. These savings can help offset any economic drops in business production.

## Try Installing a System Yourself

If local regulations will allow you to install a septic system, try installing one yourself. Even if you don't make much money from the work, the hands-on experience will do a lot for you. Once you have put your feet in the trenches and your hands on the pipes, you will have a better understanding of how septic systems are installed. Reading about procedures is good, but carrying out the procedures is the best way to understand them. You might even discover that installing your own septic systems is an excellent way to boost your income from every rural house sold.

Custom builders are at the mercy of the market. So are spec builders, for that matter. In economies where building is kept at a minimum, making the most from every job is important. You might accomplish this by being your own septic installer. If this line of action doesn't suit you, there is still room to make and save more money by using your head.

## Choose Sites Selectively

Choose your building sites selectively whenever possible. Site selection is critical to a low-cost septic system. If you can avoid chamber and pump systems, you're money ahead. Earlier chapters in this book provided you with a host of suggestions for finding the best septic sites. Go back and read those chapters again. Highlight key points and refer to them when you are out in the field searching for land.

Building lots and land can be in short supply. Sometimes builders have to take lots that are not as desirable as one would like them to be. Any builder who has been in the business for a few years knows this. But some sites simply aren't worth the prices being asked for them. This doesn't necessarily mean that you shouldn't buy the lots. What it does mean is that you should negotiate for a lower price.

If you have a solid understanding of soil types, you can use perk tests to drive the cost of some building lots down. Sellers are reluctant to drop their prices for no apparent reason, but if you build a strong case, you may be able to influence sellers to lower their prices. I've done this many times and under varied circumstances. This approach is so effective that it is worth spending some time on the details.

## A Spec Builder

A spec builder has a wealth of options in choosing building sites. Since houses being built on speculation can be built almost anywhere, a builder doesn't have to accept the first building lot to come along. Unlike custom builders, who must build on the land that their customers own or want to buy, spec builders have a free hand to wheel and deal. They can look for the best land bargains available. As long as the building site shows good promise for resale, a spec builder doesn't have to put emotions in front of financial logic.

Let's assume that you are a spec builder. You have found three building sites. All of them appeal to you. The three building lots are similar in size and price. Location is not a problem with any of the lots. A spec house built on any one of them should sell well. You only want one of the lots, so you must decide which one to buy. All three lots will require a septic system and well. When all factors are considered, the lots seem to be equal in potential for a quick sale once a house is built. Which one will you buy?

In this scenario, there may not be much haggling to be done. Since all of the lots are about the same, you don't have a lot of room to work on the price. In this case, you might just pick the lot you like best and go with it. Or, you could try a price-lowering strategy. How can you do this? You can start by requesting septic designs for the various lots.

Once you have septic designs to review, you might find some differences in the land. One lot might require only a pipe-and-gravel septic system, while the other two call for a chamber system. If this were the case, you should choose the lot that is cleared for an inexpensive pipe-and-gravel system. With all other factors being equal, it would be foolish to buy a lot that required a more expensive septic system.

Now, let's change some of the criteria in our example. Assume that one of the three lots has a much better location than the other two. This one lot is the site you really want. But there is a problem. The good lot requires a chamber system for its septic disposal. Pipe-and-gravel systems are all you need on the other two lots. After doing some estimates, you discover that the better lot will require about $5,000 in additional site work due to the chamber system. This in effect puts the price of the better lot $5,000 higher than the other two, even though all three are priced the same. As much as you like the one lot, your budget numbers show that the area will not support a higher-priced house. This means that you will have to take the $5,000 out of your building profit. You don't like this idea, so you try a negotiation tactic.

After compiling cost estimates from three septic installers, you make the landowner an offer to buy the building site at a reduced cost. Your justification is in the expensive septic system required for the lot. The seller may not accept your offer, but there is a fair chance that you will get the land at a lower price. Since you are able to document and show the landowner a viable reason for making a low offer, you are in a better position to win the negotiations.

I have often been able to acquire land at reduced prices after showing sellers my reasons for making a low offer. If you can get the price of a building lot dropped to compensate for an expensive septic system, you have balanced the scales. It's true that you must still install a costly septic system, but the lower price of the land compensates you for it.

## Cost Overruns

Cost overruns are not uncommon among builders. When septic systems are involved, the chances of going over budget are increased. If you are building a house where the private sewer will be connected to a public sewer, there are fewer variables than there are with septic systems. Problems can come up with any type of work. Septic systems, however, seem to be especially prone to cost increases. There is really no good excuse for this. Bidding and installing a septic system is not a difficult task. It is far more difficult to anticipate the shift in lumber prices than it is to calculate the cost of installing a septic system.

Why do so many builders find themselves in financial straits when dealing with septic systems? The common denominator seems to be negligence. Too many builders make assumptions and then find out that their guesses were incorrect. I've seen this happen over and over again. And yes, it has happened to me, too.

How many times have you given customers a ballpark price? When was the last time that you took an educated guess at the cost of an electrical rough-in? Builders frequently plug numbers into their estimates based on experience. They also use square-foot pricing formulas. These approaches can be surprisingly accurate in many cases, but they should not be applied to septic installations.

Septic installations are like snowflakes; no two are identical. Many septic systems are similar enough to allow the use of an average cost when figuring a job. But for every nine that come out on the average, one will come out costing much more. This nine-out-of-ten statement is only an example. I don't know what the true percentages would work out to be. But I do know that if you bid low and have to pay high for a septic system, it could cost you thousands of dollars.

Get septic installers to give you firm price quotes before you make any commitments to customers. Don't accept estimates; demand quotes. A lot of contractors fall into trouble when they rely on price estimates. An estimate is not the same as a quote. If you have three solid quotes from reputable septic installers, there is no reason why you should suffer from cost overruns. The septic installer might lose money from a mistake in judgment, but you won't. Work only with firm, written quotes and you should not have to worry about coming in over budget on your next septic system.

## Shop Around

Before you accept the bid of a septic installer, shop around. You might be surprised at how much difference there can be in the prices offered by various contractors. I'm sure you already have a good idea of the potential price spreads. Any builder who has worked with subcontractors for a while has seen wide-ranging prices for identical work.

I've gotten bids in for jobs where the prices were so far apart that I assumed a mistake had been made by someone. In running down these bids, I've rarely found the great difference in price to be the result of a mistake. Some contractors simply ask for more money that others do. This has proved to be true in all the trades that I've subcontracted over the years.

The septic installer I now use is a great guy. He does fantastic work for fair prices. However, when I was searching for the perfect septic installer, I got some wild bids in. There have been times when one contractor has offered a bid price that was nearly double that of competitors. This is the exception rather than the rule, but vast differences do exist. It has not been unusual for me to see prices as much as 40 percent higher for identical work specifications.

As a bidding contractor, I know that it is unusual for contractors to arrive at identical prices for work. But, I consider any price that is more than 10 percent apart from a competitor's to be suspect. It might be too low or too high. Either way, I become concerned. I like to check my subcontractors out thoroughly before I give them any work. In doing this, I've often found a number of reasons for not using particular contractors. I've had the feeling that some were trying to take advantage of me. Sometimes I've sensed that a contractor didn't seem to want my work. And I've seen a lot of sloppy estimating. In my opinion, a contractor who cannot estimate a job properly is a risk as a subcontractor.

If you don't already have a septic installer that you know well and trust, find one. Solicit bids from several installers. I recommend getting at least five bids. The more prices you get, the easier it will be to predict who is on target. I'm not going to take up your time telling you how to deal with subcontractors, but you should be shopping your septic prices if you want to consistently produce profitable work. Even if you have an installer who is near-perfect, like mine, you should solicit outside bids from time to time to make sure that you are not paying too much for your septic work. I'm willing to pay a little extra to get the kind of service I get from my installer, but there has to be a limit as to how much extra cost is justified.

# Chapter 28

# Low-Cost Well Systems

How can you keep your costs down on well systems? There are many ways to reduce or limit what you pay for a well. Some of these methods have to do with field conditions. Others depend on your ability as a negotiator. Is it really possible to keep well expenses in check? Sure it is. You will have to exert a little effort, but the savings can be considerable.

In this chapter, we are going to look at a number of ways to reduce or hold your costs on well systems. This includes not only wells but the pump systems that are used with them. I plan to give you a number of examples so that you can see my suggestions in action.

At this point, you have learned a great deal about wells and pump systems. The foundation already laid gives us solid ground to work with in this chapter.

## Cutting Too Many Corners

Cutting too many corners to save money is dangerous. You can damage your reputation, offend customers, and suffer other consequences. I don't recommend providing substandard materials or workmanship on any job at any price. Even if a customer asks you to provide low-quality work, I suggest that you not do it. If you do intend to do it, get all of the facts of your agreement with the customer in writing. Let's look at some examples that illustrate the risks of working below normal standards.

## Using a Spring

Using a spring as a primary water source can put you at risk in many ways. Springs are a cheap source of water, but the water may not remain available at all times of the year. It's possible that the water source will dry up in the summer or freeze in the winter. As a builder of a new house, you have a warranty period to honor. If the spring you use for a water supply fails to produce water, you might have to provide the customer with a well at no charge. Paying for a well and a pump system out of your own pocket will not be a pleasant experience.

When you bid a house that requires a private water source, you should normally plan on a well system. You will have to determine what type of well system should be installed. We will talk more about this a little later. But, what should you do if a customer insists that you use a spring as a water source?

Let's say that you have a couple that wants you to build a custom home in the country. The house will need a private water supply. It so happens that the land where the house will be built has a good spring on it. In fact, the person who sold this couple the land told them that people had used the spring for drinking water for years. Based on representations made to them by others, the couple feels the spring will be adequate for their household water needs. They have instructed you to bid the job based on the spring as the only domestic water source. What are you going to do?

If you have enough experience, you might counsel the couple about the risks associated with using a spring as the only water source. Should you not have an adequate grasp of the facts pertaining to the use of various water sources, it would be appropriate to suggest that the couple talk to experts before they commit to using the spring.

You could at least explain the risk that the spring may run dry. Then you might mention, depending upon the geographic location, the risk of the spring freezing during winter months. Contamination of the water is another issue that you might bring up. The cost of enclosing the spring and making it acceptable as a source of potable water should definitely be discussed. Many people have misconceptions about the use of a spring as a domestic water source.

I would guess that a number of people would tell you that all that is needed to use a spring is to put a pipe in it and pump the water to the house. You now know this is not true. Some type of containment should be placed around a spring that will be used as a water source. It's very important to protect the spring from any entry of ground water. This process costs money. Granted, it doesn't cost nearly as much as digging or drilling a well, but it is an expense that some people will not be aware of.

In addition to the cost of a casing or lining for the spring, trenching and fencing may be needed. Springs located on the side of a hill should have trenched-in drainpipe installed on their uphill side. These slotted drainpipes collect surface water running down the hill and divert it away from the spring. Fencing can be needed to keep livestock from contaminating the spring. Since houses using springs for drinking water are usually in rural locations, livestock is a possibility that should be considered.

Once you, or some other expert, have explained all the facts involved in using a spring to your customers, they may need some time to think over their options. If the couple still wants to use the spring, you may have to succumb to their wishes if you want to build a house for the customers. But don't do this without some protection.

Assuming that your customers insist on using the spring, you should ask them to put their desires in writing. This written agreement is your protection. If the couple puts their orders to you in writing and assume all responsibility for the quantity and quality of water provided by the spring, you can limit your exposure to only the pump system. This type of action should be acceptable to any reasonable person. If the couple is unwilling to sign a statement of their specifications to you, there may be a good reason to avoid doing business with them. I hate to say it, but sometimes people set other people up to be taken advantage of. My personal experience has proved this to be true far too often.

What might happen if you don't get a liability waiver from the couple in our example? Maybe nothing, but a lot could happen. If you don't have anything in writing, the circumstances surrounding your work might be presented very differently in legal litigation. For example, the couple might tell a judge during a legal dispute that you recommended using the spring without ever advising them of the possibilities that might make it unsuitable.

If a judge were given a false impression of what actually went on between you and your customers, you might be found to be responsible for the lack of adequate or suitable drinking water. As an experienced builder, you may be judged to be much more knowledgeable than your customers on the subject of private water supplies. All in all, you might be ordered by the court to provide a conventional, suitable water source for the customer at your own expense. If this were the case, the cost you incur could easily run in excess of $4,000. That simple little liability waiver could save you this aggravation and expense.

**Cheap Pumps**

Cheap pumps are one way to reduce the cost of a well system. They may, however, be more trouble than they are worth. Some builders may feel that the pumps installed for new houses don't have a direct effect on their businesses. I disagree with this line of thinking.

Most builders don't install their own pump systems. They typically subcontract the work out to specialized contractors. In doing this, the builders put a buffer between themselves and the work being done by the subcontractors. Materials, like pumps, might be provided by builders or by the installers hired to put the pumps in. My experience has shown that most installers prefer to provide the materials that they will be working with. When this is the case, a pump failure ultimately rests on the shoulders of the installer who supplied the pump.

A builder who hires a pump contractor to supply and install a pump doesn't have to assume any cost in having the pump repaired or replaced while it is under warranty. The pump installer will have to make good on the work done. This doesn't mean, however, that the customer won't blame the builder for trouble with a pump. When a pump fails, people tend to get upset. If you want to avoid disgruntled customers, you should use the best materials available that fit into your budget.

There is not a lot of cost difference between a good pump and a questionable pump. The difference in cost between a nylon insert fitting and a brass insert fitting is minimal compared to the added benefits provided by the brass fitting. Using only one stainless-steel clamp per pipe joint will save a few dollars on a typical well installation, but the savings is not worth the risk in my opinion. If you cut every material corner that you can, the savings might reach a couple hundred dollars. This amount of money is nothing to be sneezed at, but why would you risk your reputation, your customer satisfaction, and the cost of callbacks for such a small savings? I just wouldn't do it. I'm all for saving money and reducing job costs, but I won't do it at the expense of my customers or my reputation.

**Smart Ways**

There are many smart ways to reduce the costs of wells and well systems. Many of the options don't have any effect on the quality of work or materials that you provide your customers with. These types of savings are well worth looking into. And this is what we are going to do right now.

## Well Selection

Well selection can have a lot to do with the cost of a job. You now know that driven wells are your least expensive option. Dug or bored wells are less expensive than drilled wells. Knowing this, you have some room to work on your budget, subject to local conditions.

When you do a site inspection, you have an opportunity to choose a location where a well will be best placed. There is more to choosing a well location than just a spot where the well will be out of the way. For example, a well that is placed 50 feet further from a house than it needs to be will increase the cost of a job. In the smallest way, the cost of extra trenching and pipe will be more. This is due directly to the additional length of piping needed to reach the home. A more costly aspect of this situation could be the upgrading of the needed pump.

If a well is at a distance and depth that is borderline between a half-horsepower pump and a three-quarter-horsepower pump, the extra 50 feet of distance would clearly call for the larger pump. This would increase the cost of the pump system. The combined cost of trenching, pipe, and increased pump size could amount to hundreds of dollars. Assuming that there is no real reason for extending the distance to the well, keeping it closer to the home will provide a cost savings without sacrifice to you or your customer.

## Shallow Wells

Shallow wells are much less expensive than drilled wells. If you have a choice between these two types of wells and feel that a shallow well will be adequate, there is no reason to spend extra money for a drilled well. While drilled wells do offer advantages over shallow wells, the additional cost might not be justified by the differences.

Depending on ground conditions, you may not have any choice in the type of well you choose for a job. Your customer might spec out a drilled well, regardless of geological conditions. If this happens, you will have little choice in the matter. It might be prudent to make your customer aware of the cost savings possible by using a shallow well, but I'd steer clear of recommending a shallow well.

If you have customers who specify a drilled well and you talk them into a shallow well, you could wind up in a world of trouble later in life. Since the people wanted a drilled well and you talked them into a shallow well to save them money, you would bear responsibility, in my opinion, if the shallow well did not perform to the customer's satisfaction. You are usually safe in recommending a superior product or service, but you can get into deep water fast by recommending something that is of a lesser quality.

## Driven Wells

Driven wells are the least expensive form of well that can be created. There are times when these wells are adequate for a full-time residence. But there are also times when driven wells can't keep up with the water demands placed on them. Should you install driven wells for new houses? Not normally.

If you are going to install a driven well for a customer, I feel that you should treat it in much the same way that we discussed for springs. Driven wells, in my opinion, should only be used for special circumstances. Trying to save money with the use of a driven well can be very risky.

## Pumps

Pumps are a necessary part of well systems. As a builder, you might select the type of pump to be used for a new home. Customers may often tell you what type of pumps they want. The decision as to what type of pump to use might be made by the pump installer. If you or your installer chooses a pump for a customer, you will be on the hook if something goes wrong with it. The only time when you will not have to bear the burden of responsibility for pump performance (outside of an improper installation) is when a customer gives you specific instructions on what type of pump to use. This means that you owe it to yourself to become aware of pump selection and applications.

You should know by now that the three pumps available to you for routine residential installations are single-pipe jet pumps, two-pipe pumps, and submersible pumps. I've stated that submersible pumps are my preference for deep wells. Can you save some money by installing a two-pipe pump for a deep well? It's very possible that you can. Should you take this approach? It really depends on your customers and your circumstances. Typically, I would recommend using a submersible pump, but there are some disadvantages.

## Disadvantages of Submersible Pumps

The disadvantages of submersible pumps are few. One possible disadvantage is the need to pull the pump out of a well if it fails. When a two-pipe pump fails, it is normally very accessible for repairs. This is comforting to some people, and it can result in lower repair costs.

I've heard that lightning is more prone to strike a submersible pump than a pump that is installed within a home. Whether or not this is true, I don't know. It has been my experience that more submersible pumps are affected by lightning than jet pumps. In fact, I don't know of a single instance when a jet pump has been hit by lightning. But I do have personal knowledge of several submersible pumps being struck. This may not seem like a big deal, but if a pump is not covered by lightning insurance, the cost of replacement for a customer can run several hundred dollars. Admittedly, lightning strikes might not be a good reason for choosing an above-ground pump, but it is a potential consideration.

Cost is another disadvantage of submersible pumps. In my opinion, submersible pumps are a good value, but their costs do tend to be higher than above-ground pumps. Money is always some part of the decision-making process for builders. Even if a submersible is worth every penny of its cost, such a pump may not be needed.

## Two-Pipe Pumps

Two-pipe pumps can be used for deep wells. These pumps require two pipes, and that is a disadvantage in two ways. The cost of a second pipe causes the overall price of a pump system to increase. Perhaps more importantly, the need for a second pipe creates twice the risk of pipe failure. There are tradeoffs both pro and con for two-pipe pumps.

Submersible pumps can move more water more efficiently than two-pipe pumps. This may not matter to a homeowner. Having a pump sitting in a closet produces some risk to a home. If something goes wrong with the pump system, the home could be flooded. Noise is another factor to consider when deciding between a submersible pump and an aboveground pump. Pumps installed in basements or the common living space of a home can be noisy. Submersible pumps cannot be heard when they run. Jet pumps can be heard. This can be a concern for some homeowners.

I've had homeowners complain about being able to hear water running down a drain located in some wall of their homes. If people will complain about the relatively quiet draining of water, they are certainly likely to complain about the running noise of an inside pump. If you consider installing a pump system close to living space to reduce the cost of a water distribution system, thereby saving some money, consider the noise factor carefully.

## Single-Pipe Pumps

Single-pipe pumps are very limited in their uses. If water has to be lifted more than about 25 feet, single-pipe jet pumps become questionable in their ability to perform. This is not to say that a jet pump can't pull water up to 30 feet. It is also possible that a jet pump won't be able to raise water more than 20 feet. The elevation above sea level is a factor in how well a jet pump, which works on a vacuum basis, will perform.

Installing a single-pipe pump and expecting it to lift water more than 25 feet is a risk. While you can save some money with a single-pipe pump, you must not install this type of pump under questionable circumstances. Common sense goes a long way in day-to-day decisions.

Circumstances will often dictate whether or not a single-pipe pump can be used. Just as you can't drive a well point through bedrock, you can't expect a simple jet pump to pull water up from a 100-foot-deep well. The main question you will normally have to consider will be a decision between jet pumps and submersible pumps.

## Pressure Tanks

Pressure tanks should be used on all residential well systems. The size of a tank is relevant to the pump it is working with. You already know that larger pressure tanks reduce the running time of a pump over a period of years. However, a super-large tank is not usually necessary. As long as the tank being used is sized in conjunction with the pump it's working with, it is not necessary to oversize the tank.

Never try to save money by eliminating the use of a pressure tank. I'm quite sure you will regret any action along this line. Pressure tanks that hold between 20 to 40 gallons generally work out very well. Larger tanks don't hurt anything, but their additional cost may not be worthwhile. Before you can justify an extra-large tank, you must have some special considerations to take into account.

## Shopping Prices

Shopping prices can really pay off. Some suppliers charge more than others. This should come as no surprise. Competitors in most fields charge varying prices. If you go to buy gas, you can find a multitude of dealers who are offering essentially the same gasoline for prices that are nowhere near the same. Do all banks charge the same interest fees on loans? Of course not. It stands to reason that some prices will be higher than others for identical equipment.

Being a builder, you may not buy your well pumps and associated equipment directly. You may buy it indirectly through your pump installer. Either way, shopping pays off. You can shop individual material prices or you can shop subcontractor prices.

It doesn't hurt you or your customers to look for lower prices. Keeping a good perspective is, however, important. Going with the low bidder on a labor-and-material bid can be very risky. This is not the case when you are looking only at material prices. If you get three prices on an identical pump, you should have nothing to lose by going with the lowest price. It is not until you factor in the quality of workmanship, dependability, and similar human traits that prices become difficult to rely on.

Experience goes a long way in knowing what subcontractors to work with. Until you have a lengthy track record with subcontractors, you don't know what to expect from them. Regardless of what is promised, you can't be sure of what you will get. Really, you can never be sure. However, after doing several jobs with the same subs, you begin to get comfortable with them. This in itself can be dangerous. If you get lulled into a sense of security, you can come up short at some point.

Most installers prefer to provide their own materials. There are two good reasons for this. One reason is that when an installer is supplying material for a job, you will not be depended upon to put the proper materials on the job site. This saves you time, but it can cost you money. Most subcontractors who supply their own materials mark up the costs. This isn't all bad if you look at all sides of the deal.

If you assume that the mark-up on a well system is $100, it may very well be worthwhile to you to let the pump installer supply all materials. A mark-up of $300 is a little harder to swallow. Since there are no set rules on how much mark-up may be added to the cost of products, you have to educate yourself. If you don't, there will be no way for you to know how much you are being charged for materials that you would be able to buy yourself.

Some material suppliers won't sell pumps and related materials to builders for the same prices that they charge plumbers and pump installers. This may not seem fair, but it does happen. If you check material prices, you might find that your pump installer can have a strong mark-up on the products being offered without having much effect on your costs. Let me give you an example of what I'm talking about.

Assume that you need a complete pump setup for a deep well. You will be installing a submersible pump and a stand-type pressure tank. For the sake of this example, assume that the wholesale cost for your plumber is $750. Further assume that the plumber adds a hefty markup to the wholesale price and gives you a figure of $900. Is this too much for you to pay? It depends.

Let's say that you can't buy direct from plumbing wholesalers. This is basically the case in Maine. Plumbing wholesalers in my area will sell only to licensed plumbers. A builder can't buy from them at all, at least in theory. I know of a few nonlicensed people who deal with these suppliers, but most can't.

If you have to go to a building-supply center or hardware supplier to buy a pump, you might get a 10-percent discount from the retail price. Even with your discount, the price might be more than what your plumber is charging you. I know this can be true, because I've seen it from both sides of the table. As both a licensed builder and a master plumber, I have better buying power than most builders. There are times when I can buy something through my plumbing company at wholesale prices, mark it up 30 percent for my building company, and still be paying less for it than what it would cost me with my builder's discount at many stores. In my personal circumstances, this doesn't mean much on a paper trade. But, it could mean a lot to you.

Think about what I've just told you. It might be just as cost-effective for you to pay your plumber an inflated price as it would for you to pay your supplier a discounted price. This makes everyone happy. Your plumber makes some extra money, and you don't lose any money; you might even save a few bucks. This is the type of thing you have to weigh carefully.

**Read the Specs**

When you price out materials, read the specs carefully. You should also pay close attention to the proposals installers supply you with. A few little gimmicks in a proposal can make a big difference in the bid price.

Two well pumps that appear identical could be very different. Motor size is one factor to consider, but so is the production rate of the pump. There are many good brands of pumps available. The brand you choose is likely to depend on the dealers in your area. I don't have a particular favorite, but there are two name brands that I have a lot of confidence in. If I'm calling the shots, one of the two of these brands is used on my jobs.

As a plumber, I've worked on many different brands of pumps. Some are easier to get parts for than others. It seems that some brands are more prone to failure than others, but this might not be true. Just because my plumbers repair a lot of pumps from one particular manufacturer doesn't necessarily mean that the manufacturer produces bad pumps. It could simply be that the brand being serviced is much more popular than other brands. If you have 25 of Brand-A pumps in an area and only 2 of Brand-X and 5 of Brand-B, it stands to reason that there should be more failures reported with Brand-A pumps. It's purely a numbers game.

I do believe that some pumps are better than others. There is no question that some pumps are easier to work on than others. Each manufacturer usually has something special to offer as a feature or benefit. You can save some money by purchasing no-name pumps, and they might do a fine job. However, you will normally be better off to stick with major brands that are used regularly in your area. If nothing else, parts should be easier to come by.

Most architects specify materials very clearly. Homeowners rarely do. Depending upon who is providing you with job specifications, you may know exactly which materials to bid, or you might have no idea of what is wanted. If you have questions, ask them. The type and brand of pump you figure into your bid could be enough to sway the decision on who wins a job. Don't take chances on this. Get a clear understanding of what is wanted, and bid the job accordingly.

Watch the proposals given to you by subcontractors for little phrases which may come back to haunt you. Let's say, for example, that you spec out a Brand-D submersible pump for the installer to bid. If two contractors follow your instructions and bid the work with a Brand-D pump and one contractor bids the job with a Brand-X pump, the bid with the different pump won't be comparable to the others. This kind of tactic is used from time to time by contractors striving to be the low bidder. Some of these contractors are very good at hiding their true intentions. You may never know that the type of pump used wasn't the type you specified.

A tricky contractor can use a common phrase to fool you. Being a contractor, I assume you are familiar with the old "or equal" expression. Who's to say what is equal? When I get a quote or proposal with the "or equal" phrase, I strike through it. If contractors won't commit to using the materials I specify, they don't work for me.

How many builders hang around and watch an entire pump installation? Not many. Do you know how easy it would be for me to switch pump types on you? If you're like most builders, it would be very easy. I could even substitute a used pump without your knowing about it. These types of switches are very easy when submersible pumps are being installed.

Once a pump goes down into a well, it doesn't normally surface again until something is wrong with its performance. So if I stick a cheap pump in your well and charge you for an expensive one, I've made a few extra dollars. If I'm really tricky, I'll have one of the good pumps set out in the open when I arrive on the job and as I'm working. At the last moment, I'll switch to a used or cheap pump. If you happen to stop by to look over the materials or to chat, you see the good pump or maybe just a box for a good pump. What you don't see is the inexpensive pump I've installed in place of the pump you think you are getting.

I'm not saying that pumps are switched on the job very often, but it can happen. And really, who's going to know the difference until the pump fails? By then, the warranty has probably expired and the ripoff contractor may be long gone. I don't want to scare you, but it seems to me that you should be aware of what could happen on an unsupervised job.

## Do Your Own Installations

If local authorities will allow it, it might be worth it to you to do your own installations. The installation of a pump system is not very complicated. If you have basic mechanical skills and can follow instructions, you can install a pump and pipe up the accessories. Doing this work yourself can save you hundreds of dollars.

Some areas will allow individuals who are not working under the supervision of a master plumber to install pump systems. The regulations generally require the installer to stop after the installation of the pressure tank. In other words, you might be allowed to install the pump and related accessories, but not the water distribution pipes of a house.

If you can get a limited license or if no license is required, there is an opportunity for you to pick up extra profit from each job you do. In my area, the labor to install a submersible pump system, not counting the cost of trenching, is usually around $750. Since the work can be done in about half a day by an experienced installer, this isn't a bad rate of pay. I don't know if you would be allowed to do your own installations, but it's worth considering.

## Negotiations

Negotiations with your well and pump installers are one option for getting lower prices. When a bid price is submitted to a general contractor from a subcontractor, there is often some room in it for haggling. Simply asking for a lower price may be all that's required to get one. If your subcontractors have figured the job right and won't budge on their numbers, there is another card you can play.

Wells don't generally have to be installed at any particular time or phase of construction. Since most contractors hit dead days, you may be able to use these broken days to your advantage. Subcontractors may be willing to give you a lower price if you don't pin them down to an exact installation date. If a sub knows that your job is ready and waiting when a dead day comes along, the sub can work on your job. This is much better than sitting in the shop waiting for the phone to ring. Since your job is being treated as fill-in work, you might see a lower price.

I've often used the fill-in procedure to lower my costs with subcontractors. Rarely have I encountered trouble from doing this. However, make sure that some time limit is set on when the work will be done. This limit should be explained in your written agreement. For example, you might note that the well will be drilled within 45 days from the date of your agreement. This protects you and still gives the well driller a wide latitude within which to work. Creative ideas like this can definitely work to your advantage.

### Volume Deals

Volume deals can pay off for you by the end of a year. If you are in a position to offer your well and pump installers several jobs in close proximity of each other, discounts should be available to you.

When I was building at peak volume, I was producing about 60 houses a year. It was common for me to start 10 houses in 10 days. Knowing that I had these houses coming up, I was able to negotiate some very good volume discounts with all my subs. All the plumbing for the houses was done by my own piece workers, so discounts weren't a factor in the pump installations. The well work, however, was done by subcontractors, and I was able to get price breaks by giving them several contracts to take care of at the same time. Between offering multiple contracts and a flexible production schedule, my well installers gave me great prices. Not all builders are in a position to offer multiple contracts, but if you are, it should be worth a discount.

# Chapter 29

# Suggestions for Bidding Jobs

To be a successful contractor, you must protect yourself during the bidding phase. This can be done in many ways. One of the most important steps is to put everything in writing. Verbal statements rarely hold water in court. In order to have solid footing in a legal battle, you need written documentation.

You've seen some of the many risks associated with wells and septic systems. When you assume responsibility for installing these systems, you are putting yourself at risk. This, however, is part of your job. But the risk factor can be kept in check with a little thought and preparation on your part.

What type of proposal do you use? Is it one of those generic, fill-in-the-blank forms that you can order from a catalog? I know that generic proposals and contracts are used by a lot of contractors. Just because the forms are used in large numbers doesn't mean that they are good forms. You should have your attorney draft documents for you to use.

Even if you have the best forms in the world, they are not much good unless you use them. How many times have you given customers a quote over the phone? Do you have any record of your conversation? I doubt it, and even if you did, it probably wouldn't be enough to save you in a court battle. Your bids should be given in writing. If you have to give a price over the phone, follow it up with a written proposal.

Confusion is one of the biggest reasons why contractors get into arguments with their customers. If your intentions are spelled out in black and white, there is much less room for confusion to set in. Until you and your customer understand each other's plans completely, there is the risk of a conflict.

As a builder, you have to take responsibility for a lot of various trades. You are probably more familiar with some of the trades than you are with others. This is only natural. Regardless of how much you know, or think you know, about a trade other than your own, you can't afford to make representations without documentation. And you may not be able to create solid documentation without some help from experts in the fields of work with which you are dealing.

Look at the last contract you signed with your well driller. Pay particular attention to the disclaimers. Now do the same thing with the last contract you signed for a septic installer. Compare those contracts with the proposals you presented to your customers for those jobs. Does your bid package detail the same disclaimers that are found in the proposals from your subcontractors? If it doesn't, you are assuming too much risk.

Many contractors take a casual approach to bidding jobs. I've seen bids come into my office for major work in which only a couple of paragraphs described the work and terms being offered. This not only makes for a poor proposal in terms of sales appeal, but it leaves a lot of room for confusion and confrontation.

Your formal quote to a customer is a serious communication. If you leave details out of your quote that will later be put into a contract, you may run into resistance with your customers when you ask them to sign your contract. I believe it is best to provide complete details of every offer in the quote. If there will be exclusions, they should be spelled out in the quote. Any substitution of materials should also be put in clear language. Your quote is your sales presentation, so it should be accurate and enticing.

As a general contractor or builder, I'm sure that you've received numerous quotes and estimates from subcontractors. You must have seen some pretty skimpy ones from time to time. My experience has shown that a majority of bids are poorly prepared. Customers take accuracy, neatness, and thoroughness into consideration when trying to decide on whom to award a contract to. But getting the job is not your only concern.

Once you have won a bid, you want to make money and keep your customer happy. Detailed quotes and contracts are a good step toward reaching your goal. If pertinent details are omitted from either a quote or a contract, you might very well have a disgruntled customer on your hands. It's possible that you will even be dragged into court. I can give you an example of this eventuality to consider.

One of my carpentry subcontractors was telling me a story about a house he built recently for a customer. As part of the contractor's contract, he was supposed to have his site subcontractor provide a certain amount of loam for a job. During the construction process, the site contractor and the owner of the property had a discussion. They agreed jointly to use sand in place of loam for much of the fill work, as sand would be cheaper. A written change order was not issued to document this new agreement. The site contractor performed the services that he and the homeowner had agreed to.

After the house was built, the homeowner refused to pay the subcontractor for some of the work that had been done. One of the phases being contested was the site work. The owner wouldn't pay up, so the site contractor took him to court. As the general contractor, my friend was also required to make an appearance in court.

I'm told that the court hearing didn't last very long. After some verbal exchanges and accusations, the judge referred to the only written agreement between the parties. The contract called for a certain amount of loam. When the judge asked if the described amount of loam had been installed, the site contractor had to admit that it had not. He tried to qualify his answer by telling the court about the verbal agreement to substitute sand for a portion of the loam, but the judge wouldn't hear it. Based on the written agreement, the court ruled in favor of the homeowner.

If everything I was told was true, and I believe it was, the site contractor got a raw deal. I can't fault the judge for going by the only real evidence available, but the homeowner took advantage of the site contractor. If you want to avoid this type of problem, make sure that all of your agreements are in writing. This pertains to all aspects of your deal, including any exclusions.

### Disclosure

Every well proposal I have ever seen has had some form of disclosure. A common statement is that the driller will not guarantee the quantity or quality of the water produced by a well. This is one simple sentence in the quote or contract, but it can have a huge impact. Let's talk about this for a few minutes.

Assume that you are building a new house for a young couple. The house is in a rural location and will require both a well and a septic system. As the builder, you are responsible for all aspects of the job, including the well. Further assume that your agreement with the well driller does contain some version of the standard disclosure about quantity and quality. Now let's say that the well driller hits water and everything is fine until about six months later, when the dry season rolls around.

One day, as you are going about your business, the homeowner calls and complains that the house you built has no water. After troubleshooting the situation, you discover that the well has run dry. Being a builder, you have to warranty the house for one year. After only six months, the well has dried up. What's going to happen?

The homeowner is probably going to expect you to dig or drill a deeper well. If you go back to your well driller with this request as warranty work, you will probably be told to forget it. The driller can simply produce the contract you signed, the one with the innocent little sentence in it, and prove that the quantity of water was not guaranteed. Where does this leave you? Sort of between a rock and a hard spot. You've got to give the customer a better well, and your well driller doesn't have to do any of the work as warranty work. Under the circumstances, you are going to have to pay for the new well out of your own pocket.

You can avoid this type of situation by putting the same disclosure in your contract with the homeowner. If the homeowner is getting the same deal as you are, and the facts are set down in writing and signed, you're in a much better position to avoid paying for a new well.

The use of disclosures and liability waivers can go a long way in protecting you and your business. If you have specific terms written clearly into any agreement you make with a customer, you are much less likely to get into arguments. This is not only good protection but good public relations as well.

### Danger Spots

What are some of the danger spots that you should be extra cautious about as you are bidding jobs with wells and septic systems? There are many circumstances that can put you at risk. Some cannot be avoided, but others can be. With the use of disclosures, exclusions, and liability waivers, you can put up a good shield around you and your business. To expand on this, let's look at some of the specific areas of concern as a bidding contractor. We will start with septic systems.

### Septic Systems

Septic systems are fairly simple to install. Yet there can be a great deal of risk when bidding jobs involving private sewage systems. To help protect yourself and your company, you need to address some of these key issues in your proposals and contracts.

## Septic Permits

Septic permits are typically required before the septic system can be installed. This is normally not a problem, but it can be. If you are bidding a job where the customer wants or needs a private waste-disposal system, you should make your price and work subject to the issuance of a septic permit. Then, if for any reason a permit cannot be obtained, you have an out.

Having a request for a septic permit denied is not common, but it can occur. It could be due to poor soil conditions, but there is another major reason why the request might be turned down. There are some areas where a public sewer has been installed in recent years. Assuming the building lot is large enough to accommodate a septic system, a property owner might prefer having a private system. This would eliminate the tap fee charged by the sewer authorities for connecting to the public sewer. It would also do away with monthly usage fees for the public sewer. However, even if other houses in the area are using private systems, the jurisdiction issuing permits might not allow the installation of another private system when a public sewer is available. This is a common practice. You could walk into a nightmare if your quotes and contracts don't cover this possibility.

## A Specific Design

When you write up a quote or contract to install a septic system, you should name a specific type of design. If your quote is based on a pipe-and-gravel system, say so in your quote. Reference the septic design drawing that you used to calculate your pricing. If for some reason the specified design isn't approved, you're at much less risk. An open clause that states that you will install a septic system for a certain price is much more dangerous. The more specific you are in your language, the less likely you are to have a problem.

## Clearing

The clearing of an area is often needed for the installation of a septic field. Who's responsible for this work? If your proposal says you will install a septic system and does not exclude clearing, it could (and probably should) be interpreted that you will take care of all aspects of the job, including clearing.

If you are going to clear land for a septic system, detail in your proposal what work will be done. How large an area will be cleared? The answer to this question should be a part of your proposal. Will you remove stumps, rocks, and other leftover debris? If not, exclude this work. Otherwise, stipulate what you will do with the debris. Make your work description very clear.

### Access

Access to a septic site can be a problem. You might have to cut down trees to get trucks and equipment to the site. If your customer is not aware that trees outside the septic area will be removed, you might wind up with an angry client on your hands. If the customer comes out to inspect your progress and is shocked to see favorite trees gone without notice, you could be in trouble. If an access path must be made, be specific about where it will be and what will be done.

I've had occasions when the only practical way to get equipment to a septic site was by crossing over the land of adjoining owners. This, obviously, is not something that should be done without written permission. If your customer tells you that the neighbors won't mind a bit if you use their driveway or land for access, don't believe it until you have it in writing.

### Rock

Rock can make the normal installation of a septic system impossible. You should address this issue, as well as other underground obstacles, somewhere in your proposal and contract. I leave it up to you and your lawyer to work out the exact wording, but make sure that you have some type of protection in the event that unseen obstacles prevent you from doing the work you are proposing at the price you are quoting.

### Materials

The availability and price of materials are another issue that you might want to cover in your protective paperwork. I don't believe this is a big issue with septic systems, but it wouldn't hurt to have some type of clause in your agreements to deal with rising prices and materials that are not readily available.

## The Sewer

Who is going to run the sewer from the house to the septic tank? Septic installers often run the sewer to within 5 feet of a home's foundations. From there a plumber takes over. This, however, is not always the case. Sometimes plumbers do it all. If you are bidding all phases of a job, this is not such a big deal. However, if you are bidding the septic work and not the plumbing work, the question of who is responsible for the sewer could become an issue. Digging the trench for a sewer can take some time. There is also the labor and material needed to install the pipe to be considered. Keep the sewer installation in mind when you are bidding your next job.

## Landscaping

Are you going to take care of landscaping the septic area? Will you be doing the finish grade work? Who will seed and straw the area? All of these questions should be answered in your proposal and contract. Even the type of dirt used to cover the excavated site should be detailed.

## Perimeter Trees

Perimeter trees can interfere with septic systems. If trees are left standing too close to a drain field, their roots may invade it. If you feel that perimeter trees should be taken down, stress your point in your proposal. Owners who prohibit you from following your professional instincts on this issue should be asked to sign a liability waiver that releases you from any damage done by the perimeter trees.

## Wells

Wells, like septic systems, can present bidding contractors with some problems. Certain clauses often found in the contracts offered by well drillers can set a builder up for trouble. If you don't provide some protection for yourself in the contracts signed between you and your customers, you could lose thousands of dollars. A lot of houses depend on wells for their potable water, so you had better be prepared for bidding on them. Let me give you a few pointers on what to look out for.

## 298    Chapter Twenty-nine

### By the Job

Will you be paying your well driller by the job or by the foot? Both payment methods are usually available. Depending upon circumstances, there can be dramatic differences in the overall cost of producing a well, depending upon which pricing method is used. Since this is a serious subject, let me give you a little background on how I have worked with wells in the past.

All of the well drillers I have worked with have always given me two payment options. I can pay so much for each foot of depth in a well, or I can pay a flat rate that will not change no matter how deep the well is. The flat-rate price also guarantees me water, so if a driller hits a dry hole, I'm not paying to have the second well sunk. Under these conditions, which way would you go?

The flat-rate price for a well is usually pretty steep. Well drillers usually look at historical data for existing wells in the area. They calculate what they believe to be the worst-case scenario, and base their prices on that. The per-foot price often works out to be cheaper, sometimes as much as $1,000 less, but there is additional risk involved for the person paying the bill.

In the past, I've always gambled on my wells. Even though the guaranteed deals are safe, I wanted to save the extra money if I could. Just like the well drillers, I would do my own research on area well depths. Sometimes I would just talk to owners of adjoining properties to see how deep their wells were. This gave me a good idea, but no guarantee, of how deep my wells would have to be. Using this strategy, I've saved lots of money over the years. I could have been caught on some of them with extra-deep wells or even a dry hole, but I never have.

When I built my most recent personal home, I opted for a guaranteed price. This was the first time in all of my building activity that I ever went with a flat-rate fee. I'm not even sure myself why I did it, but I did. And, am I ever glad that I did.

Most wells in my area are between 250 and 300 feet deep. If I were estimating my costs on a per-foot basis, I would have used the 300-foot figure. I have river frontage on my land, and I thought the water vein might be even closer. My neighbors live some distance away (about half a mile), so their wells are not a great barometer to use, but they are better than nothing. I figured my well would not be deeper than 300 feet and maybe as shallow as 200 feet. Yet, something gnawed at me about this. For whatever reason, I decided to go with a fixed price. My well wound up being 404 feet deep. Even the driller was shocked at the depth. In this case, the driller lost and I won. If I'd chosen a per-foot price, the cost would have been well above my budgeted amount.

Which pricing method should you use? If you want to be safe, go with the guaranteed price. It's up to you to decide which way you will go, but you should put something in your proposal to identify which conditions you are giving your customer. You might say that the cost of the well will be a specific amount and that the amount is guaranteed. The price might scare your customers, but they will rest easy knowing that there will be no surprises.

You could say that your price is based on so much per foot for the well. If the depth is less, you will credit the difference to the customer. Should the depth be more than what your estimated cost is, the customer will pay the extra. This approach might frighten your customers more than the higher fixed price.

As a matter of practice, I give my customers the same option that the well drillers give me. I let them choose between the per-foot and the fixed-price method. Once they have made a decision, I memorialize it in our agreement. This method has always worked very well for me.

I've been lucky with my wells. Few have run deeper than expected, and I've never run up against a dry hole. Other contractors I know have not been so fortunate. Having to pay for two or three wells to get one is no bargain, so be careful on this issue.

### Drilled or Dug

Will the well you provide be drilled or dug? You could simply put in your agreement that you will provide a well. Most customers wouldn't question this approach, but some might. However, after the well is installed, you could get some grief if you installed a dug well and the customer thought a drilled well was being installed. You should identify clearly the type of well your price includes.

Dug wells are cheaper than drilled wells, but they sometimes run dry. Drilled wells rarely run out of water, due to their extreme depth. If your customer is looking to cut corners and the local ground conditions will allow it, dug and jet-pump wells are the least expensive way to go while still maintaining a viable hope for a steady water supply.

Driven wells, in my opinion, should not be considered a worthwhile option for full-time homes. These little wells can produce good water, but the quantity is limited. While a driven well makes sense for a weekend cottage or fishing camp, I don't feel it is suitable for the demands of full-time use.

Drilled wells are the most expensive to install, but they are also the most dependable. A submersible pump and a drilled well are the way to go in my opinion. I've owned houses that were served by dug wells and by drilled wells. In my opinion, the drilled well can't be beat.

Well preference should be left to the customer's discretion. However, you need to make it clear in your paperwork what is wanted, what your prices are based on, and how problems that may arise will be handled.

### Water Quantity

Will your customer be guaranteed a certain quantity of water from a well? Your well driller will probably not make this commitment, so you shouldn't either. I've never seen a driller's proposal that didn't exclude the quantity of water. This is a touchy subject, but it is one you should deal with before a problem pops up. If your customer comes back to you because of a lack of water quantity, you are unlikely to have anywhere to turn. Think about this.

How can you get your customer to accept the fact that you will not guarantee a quantity of water? There is a method that I've used for years that has never let me down. Get several quotes from well drillers. If you're not afraid to show your customers the prices that you are paying, show them the actual quotes. In my experience, every quote has stipulated that the well company would not be responsible for quantity. Once you prove to the customer that it is an industry standard to exclude quantity, you should have made your point. If not, get more estimates. Once you have collected a handful of quotes, all of them excluding quantity, your customer should give in.

### Flow Rate

I've had customers demand that I guarantee them a minimum flow rate of recovery for their wells. No driller I have ever dealt with was willing to do this. Therefore, I've never done it. This problem is similar to the issue of quantity. If you can't find a driller who will give you a guarantee, you shouldn't offer such a commitment to a customer. Even if I found a driller who would make such a claim, I don't think I would be comfortable doing so.

### Water Quality

Water quality is another issue that most well drillers will not guarantee. The perspective of most drillers is that, once water is hit, their job is done. Most drillers will go deep enough to provide a reasonable rate of recovery, but they will not refund your money if the water smells like rotten eggs. Water with a high sulfur content does stink, and some wells are full of this disgusting water.

Sulfur is not the only disagreeable element found in water. Iron can be a big problem. It stains plumbing fixtures and leaves a black buildup in toilet tanks, water heaters, storage tanks, and so forth. Hard water is very common in some areas, and it makes washing dishes, clothes, and hair difficult. Soap does not perform well when mixed with hard water. Acid can be so dominant in drinking water that it will eat holes in copper pipes and upset the stomachs of some people. There are other types of mineral inclusions that can disappoint homeowners. While sulfur is not too common, other types of mineral conditions are.

The types of water conditions we are discussing don't usually make water unsafe to drink. But, smelling rotten eggs every time you put a glass of water to your mouth certainly isn't something most people want to live with. For this reason, you need to cover the bases on water quality in your quotes and contracts. Again, I would use the quotes of well drillers as evidence that a guarantee of quality just isn't standard procedure within the industry.

### Location

Location might be an issue that will come to haunt you with the installation of a well. If the only place to put a well is smack dab in someone's front lawn, you might have a problem. Dug wells have a large diameter and usually consist of a concrete casing and top. Having one of these for a lawn ornament will not get a person's home on the cover of a fashionable magazine. Drilled wells are not as conspicuous as dug wells, but they are still not things of beauty. The 6-inch steel casing of a drilled well is easier to camouflage than the 3-foot diameter of a dug well, but you'd better clear the location with your customer before you make a firm commitment.

I suggest that you locate the proposed well site during your site visit. Have the customer agree to the location and identify some landmarks and measurements to detail the location on paper. Include some option for yourself in case the proposed location proves unsuitable.

### Access

Well-drilling rigs are big. It takes a lot of room to move them around. Tree limbs or entire trees might have to be removed to allow access for these big rigs. Make sure you talk this situation over with your customers in advance. Have the customers agree to whatever your needs are for getting a well truck in and out.

## Permission

Getting permission from a governing body to install a well is normally not a big deal. However, you could run into a problem similar to the one we talked about with septic systems. If a public water supply is available, it may be mandatory to hook up to it. This is not an uncommon situation. You will seldom run into this type of a problem, but don't assume that it couldn't come up.

## The Trench

Who is going to be responsible for the trench needed from the well to the house? Well drillers will sometimes take care of the trench if they are installing the pump system, but they don't normally provide a trench in their drilling price. If you have full responsibility for all aspects and costs of building a home, you will have to get a price on the trench from somewhere, and there will be no need to bother the customer with details of who will install it. However, if you are acting as the general contractor on only parts of the construction process, the trench could be an expensive issue to have to settle at some point.

## The Pump System

The pump system falls into a category similar to the trench. If you're taking care of the whole job, the customer doesn't have to know if the pump system will be installed by a plumber or by a well driller. But if your job is segmented, you might have to determine who is assuming responsibility for the pump system. This work involves a good deal of material and labor, so it is not a cheap item to make a mistake on.

## A Spec Sheet

A detailed spec sheet should accompany all of your proposals and contracts. This is just good business for any contractor. Many of the risk elements that we have discussed can be covered in the spec sheet. For example, you should list in the specifications for a well system all of the types of materials that will be used. This would start with the type of well being priced. The type of casing and grouting should also be listed. You would then note the brand and size of the pump being supplied.

Reference should be made to the brand, type, and capacity of the pressure tank. The more details you put on your specifications sheet, the better.

As a part of my company policy, I require customer to sign more than just my contracts. I ask them to sign the specifications sheets and any blueprints or drawings being used. This eliminates any possibility that the customer will claim that the documents were never shown to them. There have been occasions when this practice saved me a lot of trouble. If a customer starts complaining about something that is covered in the documents, a quick reminder of their signatures settles them down fast.

**What You Don't Know**

What you don't know can definitely hurt you when bidding jobs. If you are not familiar with septic systems and wells, spend some time studying them. You have to be able to talk intelligently about these subjects even though you are not expected to be an expert on the topic. If a customer asks you the difference between a one-pipe jet pump and a two-pipe jet pump, you had better have a good answer. What would you say if a customer wanted your advice on whether to use a jet pump or a submersible pump? You should be prepared to answer such questions with solid answers. Would you recommend using a garbage disposer with a septic tank?

A lot of builders don't know much about wells or septic systems. This is understandable but not acceptable. You owe it to your customers and yourself to become educated on the services you offer. Failure to do this can hurt your reputation and your bank account. Since you are reading this book, you are obviously interested in learning more about rural water and waste systems. I applaud you for this.

Chapter

# 30

# Cost Overruns

You can make more money by avoiding cost overruns. I'm sure you know this, but you may not know how to avoid going over budget. A lot of contractors have trouble keeping their spending within the confines of a budget. When was the last time that you had a job run into higher costs than you were expecting? Do you perform job costs for all your work? If you're not doing job costing, you can't know how well your estimates worked out compared to the real cost of a job. It may be surprising to you, but a lot of contractors don't maintain regular job-costing procedures.

Before you can stay on budget, you have to create a budget to follow. This is something else that far too many contractors are lax in doing. Many contractors work up a lump-sum figure and work only from it. They may know if they make or lose money, but they don't have enough information to know where the wins and losses come from. This is bad business.

## Setting Up a Job Budget

Setting up a job budget is not difficult to do. You don't need a degree in accounting to manage the task. It does take a little time to establish a framework of prices, but it is time well spent. Most successful contractors never start a job without a budget. You shouldn't either.

When I'm getting ready to bid on a job, I break my bid work down into phases. My breakdowns are done in tight categories. Some contractors lump all of their excavation and grading work into one category. I don't. My procedure calls for a category to meet every major aspect of a job. For example, I would have a category for site clearing. Another category would be reserved for roadwork. A different category would cover digging footings and foundation holes. Rough grading would have its own section in my budget. So would final grading. Basically, every phase of a job would be broken down. Since we are talking about wells and septic systems, let's use these two aspects of a job as our working example.

## Septic Systems

The manner in which you break down the costs for a septic system can vary. If you are hiring a subcontractor to provide all labor and materials to install a septic system, you might plug in only one price. This, of course, would be the price quoted to you by your septic installer.

If you are supplying materials for your septic contractor, the cost breakdown would be more extensive. You could lump all material costs into one category, but I'd go further than that. I would list all of the major components of the system separately. For example, I'd have space for crushed stone, concrete products, chambers (if needed), pipe, dirt, and so forth.

My reasons for going into extensive breakdowns are numerous. For one thing, I'm not as likely to overlook an expense if my worksheet for a budget details all major expenses. Using a septic system as an example, I might forget to figure in the cost of a permit if a slot didn't exist for the price of a permit. It's conceivable that I would fail to figure a price for stone if a blank line or box doesn't exist for the cost. By having a detailed worksheet, I can go through it systematically and not worry about what I might be leaving out. This is good protection against cost overruns.

Once I start a job, I can use my budget to track the profit picture of my work. When an invoice is delivered for crushed stone, I can compare it with the figure I have in my budget. If the stone cost less than what I had figured, I'm in great shape. When the cost is on target, I'm okay. A price that is higher than my bid figure raises a red flag. There may not be much I can do about it on the job being dealt with, but it tells me to find out what I did wrong so it won't happen again.

When the dust settles, I can review my estimated figures and compare them with the true costs of materials. It's easy for me to job-cost any phase of a job, because I have each phase broken down independently. After a few jobs, I can compare notes and see if I'm experiencing any recurring problems. For example, if I'm habitually missing my estimated figures for fill dirt, I know this is an area I need to work on. When I'm coming in under budget on concrete materials constantly, I might decide to lower my prices a little if competition is tough. The budgets that I create serve many purposes.

### Wells

When I have a house that will need a well, I break my budget down into several categories. One of the categories is for the direct well work, either the drilling or the boring. A second category covers the pump. Pressure tanks have a category of their own. Gauges, fittings, pipe, and miscellaneous items are lumped together. Trenching gets its own category. And, of course, labor has a section in the budget. I have used my extensive breakdowns for about 15 years, and I have always found them helpful.

### Take-Offs

If you are figuring the cost of a job accurately, you must include take-offs. To complete an accurate take-off, you must do extensive breakdowns of what materials and labor will be needed for a particular phase of work. You are probably accustomed to doing this with your building materials. Since wells and septic systems might be beyond your scope of detailed knowledge, preparing a full take-off for such a system might be more than you can handle. If you're using subcontractors for the work, you're not the one who has to worry about take-offs for wells and septic systems. All you have to do is write or type in the prices given to you by your subs. This makes it that much easier to come up with budget numbers.

If you are figuring your own materials with a take-off, you can use the take-off as a part of your budget. For example, if you decide that 200 feet of PE pipe will be needed when doing your take-off, you can use that information in breaking down your budget. Putting together a budget from take-offs is simple and effective. You need to create some type of working budget. Take-offs don't have to be a part of this administrative task, but you do need some way of coming up with firm numbers. If you cheat and cut corners in arriving at your budget figures, you will only be hurting yourself. Trying to save time can wind up costing you money and maybe a lot of it.

## Accuracy

The accuracy of any budget you build must be dependable. If the numbers aren't solid, they will do you very little good. Everyone makes mistakes from time to time. The odds of you making an error in a complex take-off are high, especially if you are not doing the take-off in an undisturbed atmosphere. Even transferring numbers from a subcontractor's bid package to your budget sheet can result in mistakes. Sometimes a seven looks like a one. Forgetting to add a zero in a price can make a huge difference. If you depend on a computerized spreadsheet to add up your figures, you might never notice a simple mistake until it is far too late. After you have created your budget, go back over it and double-check its accuracy.

## Sorting Through Bids

Sorting through bids can be a tedious job. In order to get a clear picture of what each bid includes, it is necessary to spend adequate time pouring over proposals and bid packages. During my many years as a builder, I've seen a number of strange proposals and bid packages. Some of the prices I've reviewed turned out to be way out of line. If I had been too busy to study prices given to me, my negligence could have cost me a bundle. This can be true of prices provided by both subcontractors and suppliers.

If you have much experience as a builder, you have undoubtedly had occasions when suppliers have botched up your price quotes. I've had suppliers omit all interior doors from my requests for prices. Hey, if you leave out all the interior doors for a house, your bid price is going to look pretty good. But, a cheap bid price that's wrong doesn't do anyone much good. It's essential that you take an active role in checking and double-checking all quotes given to you. Let's put this into perspective for wells and septic systems.

## Well Prices

Well prices are generally fairly consistent. Some trades seem to vary in their prices much more than others. Well contractors with whom I've dealt have normally been close competitors. This makes it easier in some ways and more difficult in others.

Contractors who install wells typically offer two types of pricing. The options are normally either a per-foot price or a flat-rate fee. There are pros and cons to both of these types of pricing structures. When well installers quote a flat-rate fee, it is usually the highest price an installer expects a well to cost. It is often possible to beat flat-rate fees by choosing a per-foot price. But picking a per-foot price structure is risky. Flat-rate fees are guaranteed; per-foot prices can run away with your cash.

When you are reviewing bid prices from well installers, you have to look closely. Most of the prices will be fairly clear and concise. But, there can be hidden charges or wording that might leave you out on a limb. To expand on this, let's start with a per-foot price structure.

When you get a quote for a per-foot well, there will be no total price provided. The well installer will charge so much per foot for every foot of depth that is required for a suitable well. Casing for the well will also be priced on a per-foot basis, with no reference to how many feet will be needed. This leaves you in an open-ended situation. You know how much the well will cost per foot, but you have no idea of how many feet you will be billed for. Under this circumstance, you are at risk in giving a customer a flat-rate quote. Are you going to make your price to the customer based on a per-foot basis, or are you going to make your best guess and give a firm price? Banks and other lenders will require a firm price if they are supplying construction financing.

I've only used flat-rate well pricing once in my long building career. With all the wells I've had installed, I always came out better by gambling on the per-foot price. Then when I built my most recent personal home, I opted for a guaranteed price. It's a good thing I did. My well turned out to be much deeper than I or my well driller thought it would be. To be safe, you should work with guaranteed prices. If you are willing to gamble, you can quote flat rates to your customers and play the odds of per-foot pricing with your well installer. This can net you more money out of a job, but it can also cause you to pay for cost overruns out of your own pocket.

Our main interest right now is in comparing price bids. In the case of well installers, this is pretty easy to do. If all your bidders are bidding the same type of well and quoting per-foot prices, you can look at their prices and compare apples for apples. But, there are some opportunities for unethical installers to take advantage of you. They may dig or drill deeper than necessary. The amount of casing installed may not be reflected accurately on your bill. A per-foot pricing basis is similar to a labor-and-material pricing structure. Both are risky. If you like to work with known figures, stick with guaranteed prices.

## Guaranteed Prices

Guaranteed prices are often available from well installers. These prices tend to be on the high side, but they are dependable. If you have three well contractors all bidding identical specifications with guaranteed pricing, there is not much effort required in evaluating their prices. All you have to do is look for the low number. In theory, you can't lose money on a guaranteed deal. There will not be any cost overrun for you to suffer from. If anyone loses, it's the well installer. This approach is a conservative, safe one.

## Reading Between the Lines

Reading between the lines of a quote from a well company is sometimes necessary. Not all quotes are quite what they appear to be. A busy builder can make assumptions and find big trouble arising from a lack of details. Let me expound upon this.

Assume that you have received quotes for installing a well. If you're new to wells, you might think that the price you are given includes all the necessary work and materials for a pump system. This will not normally be the case. A lot of well installers do install pump systems, but I've never seen one include the pump work in their well price. A well price normally includes nothing more than drilling a well, installing a casing, and capping the casing. I suppose a really slick con artist could try to sell you a well without the casing and cap included. This has never happened to my knowledge, but it won't hurt for you to confirm exactly what is included in all prices you receive.

Pump systems are generally treated as a different job from well installations. Even when the same contractor is doing both jobs, the work is normally priced separately. It should be. This way you can compare the well installers' pump prices with those of your plumbers. If the pump work and well work are both mixed in together, you have no way of comparing the pump prices given to you by plumbers to those given to you by well installers.

In my experience, grouting has always been included in the prices given to me for well installations. This is an aspect of the job that I suppose could be left out of a proposal offered by a well installer. The installer might insist that you were planning to grout the well with your own crews. Again, I haven't seen this happen, but it is a potential risk if you want to look at all the pessimistic possibilities.

Your first job as a general contractor is to understand all the bid packages that you are working with. If you have questions, call the subcontractors and get answers to your concerns. Don't make assumptions;

they generally only serve to hurt you. Get all the facts before you award a contract.

All the well installers I have dealt with in the past have been good ones. I've never had a bad experience with a well installer. I can't say this about many of the trades I've worked with, but well installers have been good to me. Still, you can't allow your defenses to become too lax. Read each quote carefully. Has anything been disclaimed? It's common for water quantity and quality to be disclaimed, but there shouldn't be any other caveats.

If you are having a well installer bid on your pump package, make sure the materials planned for use are specified. Don't accept just the name of the pump to be used. Get a model number and a description sheet of the pump. Compare flow rates, horsepower, voltage, and other specifics with the pumps competitive bidders are listing. Find out what type of pressure tank is being supplied and what its capacity is. What type of pipe will be installed? Are nylon fittings going to be used, or will they be brass? Will the installer double-clamp all connections? Create a checklist of questions to ask installers. Don't attempt to choose the best bidder until you have all the facts to work with.

Trenching and backfilling are both parts of a pump installation. This is work that most installers don't include in their prices. Some companies do factor in this expense, but many don't. If you receive a bunch of bids and assume that the trenching is included, you can be in for a rude awakening. The cost of a backhoe or excavator and operator can become quite expensive, especially when it's not in your budget. Confirm who is responsible for digging and filling in the well trench before you make a decision on whom to award a job to.

**Septic Prices**

Septic prices can involve a lot of material and money. Unlike wells, where per-foot prices are common, most septic systems are priced with flat-rate fees. A septic design is handed out to a number of septic installers, who give you quotes. They are generally comprehensive, but there can be some hooks hidden in them.

If you have a septic system that requires a pump system, who is paying for the pump, the controls, and related pumping materials? Who is going to install this equipment? Are you going to have to pay a plumber to install the pump and an electrician to wire the system? Not all septic contractors include the price for labor and material required to make a pump system operational. Making the mistake of thinking these prices are included in a septic price can cost you thousands of dollars.

Most septic installers will provide prices that include all work and material associated with a designed system. But you can't count on this. You have to clarify the situation before you commit to accepting a bid.

It has been my experience that most septic installers provide their own permits and materials. There have been times when I've bought septic permits and septic materials for installers. Looking at a septic price and thinking it includes everything when it actually only includes the cost of labor can be disastrous to your profit from a job. Make sure you understand what each bid includes.

Excavation can be a major expense in putting a septic system together. This work is normally included in the bids of septic contractors. So is the cost of stone, pipe, and fill dirt. Septic tanks and distribution boxes are commonly provided by septic installers. The prices you get should be turnkey numbers, but check them out.

One aspect of septic systems that does seem to vary is the installation of a sewer between a septic tank and the sanitary plumbing from a house. Sometimes plumbers do this work; at other times septic installers do it. This work involves trenching, pipe, fittings, and labor. The amount of money is rarely a fortune, but it's enough that you won't want to pay it out of your own pocket. Find out which one of your subcontractors is bidding the sewer work.

Final grading over a septic system is normally done by the site contractor. This person might also be your septic installer. Many site contractors do install septic systems. Again, however, you can't afford to make assumptions. Determine exactly who is going to handle the final grading.

**Suppliers**

When you will be supplying your own materials for either pump systems or septic systems, you will have to get prices from material suppliers. This work can be frustrating. Some suppliers just don't seem concerned enough about getting a person's business to meet the requirements set forth in a bid request.

When I built my most recent home, I sent bid packages out to seven major suppliers. Of those seven, only five responded with prices. Two of the suppliers must have decided that my job wasn't worth bidding. Now I'm not talking about a bid of a few hundred dollars. Tens of thousands of dollars in materials were on my list. Even so, two out of seven suppliers chose not to bid the job. Why? I don't know. It must be nice to be so independently wealthy that you don't need to bid work.

Most suppliers will bid jobs, but they may not do it in the way you request them to. I like my bids broken down into itemized lists. A lot of suppliers prefer to throw out a lump-sum figure at contractors. If you don't have itemized pricing to work with, you are at a disadvantage. It's possible to look at the bottom-line figures of lump-sum bids and choose a low bidder, but you could be wasting money. Let me tell you what I mean.

Take a pump installation as an example. If you get three lump-sum bids, you can easily see which supplier is offering the overall lowest price. But is one supplier more expensive on the well pump than another? Could you buy a pressure tank for less from one place than you could from another? You won't know without an itemized list of prices.

It's rare when one supplier offers the best prices on all types of materials. This is true of building materials, plumbing materials, and probably most other materials. A supplier who can give you a great price on dimensional lumber might have lousy prices on siding or roof shingles. You probably already know this, but in case you don't, it's true.

Shopping individual prices can save you a lot of money. I'm not one to nickel and dime a supplier, but I do believe in shopping various phases of a job. I might buy a pump from one place, pipe from another, and a pressure tank from yet another. If my only extra effort is two additional phone calls and I can save a significant amount of money, I'm all for it. Now I wouldn't buy some of my fittings here and some of them there. That gets too confusing. But, within reason, selective shopping of itemized materials is good business.

Some contractors rely on suppliers to do take-offs for them. I don't have a problem with letting suppliers provide a take-off for a job, but I certainly don't trust a take-off that is prepared by a supplier to be accurate. This type of take-off can be quite good, but it can be way off base. I've seen it in both my building and my plumbing business.

Suppliers don't always have the time or the proper personnel to produce quality take-offs. Miscounting a few copper fittings is no big deal, but forgetting to include the cost of a pressure tank is a whole other issue. If you use take-offs prepared by suppliers, I suggest that you check them over very closely. Many times these lists will have omissions.

I generally provide my suppliers with a bid list. After making a take-off, I spec out the brand names I want and ask all suppliers to bid the work as specified. This is the only way I know of to get a true comparison of prices. If suppliers are allowed to make their own decisions in the types of materials used for a bid, you might get five quotes with none of them being comparable.

When I recommend specifying material, I'm not talking only about big-ticket items. Many contractors will spec out a particular pump and pressure tank but fail to give specifications for valves, fittings, pipe, and other elements of a job. The difference in cost between a cheap gate valve and a name-brand valve can be $15 or more. With enough fittings and accessories, a supplier can lower the cost of a job considerably by pricing materials of a lesser quality. If you don't spec out a job in detail, you won't know what your prices are based on.

Once you get bids in from suppliers, read them over thoroughly. I can't count the number of times that suppliers have left whole categories of materials off bids being made. Go down the bid lists item by item and make sure that everything that you wanted a price on is listed. Simply looking at the total on the bottom of a bid sheet can be very deceiving.

### Once a Job Starts

Once a job starts, you have to stay on your toes to remain within your budget. How often do you compare your monthly invoices from suppliers with the prices that you have been quoted for materials being billed? Ideally, you should check each invoice against quoted prices. There are many times when the prices you are charged will not agree with the terms of a quote. How often is this likely to happen? It happens just about every month in my businesses.

When you get a quote from a supplier, it is natural to assume that you will be billed at the quoted price when the invoice arrives. I have found, however, that the prices billed are often higher than the quoted prices. Is this a computer error? Did someone forget to lock in the quoted prices? Do suppliers try to take advantage of contractors? I don't know why it happens, but it does. Sometimes the difference is only a few pennies, but sometimes the discrepancy is hundreds of dollars.

I try to check each invoice that is tied to a quoted job with the quoted prices. Sometimes an invoice slips through without me seeing it, but this is rare. During my checks, I often find problems with my billing. The trouble doesn't come from just one or two suppliers; it seems to happen with a lot of them. There are some suppliers whom I've never caught in a mistake, but they are the exception rather than the rule.

Keeping tabs on your billing invoices from suppliers is a simple way to avoid cost overruns. If a supplier bills you at a price higher than the written, quoted price, you can ask for, and usually receive, the lower price. Even small price increases can add up over the course of a year. Paying $5 too much for this and $10 too much for that on job after job can cause you to lose thousands of dollars.

It takes time to go over invoices. Sometimes the price discrepancies you find will not be worth the time it takes to find and correct them. But, there can be some big differences in prices. On the last house I built, there was a $700 problem with one of my invoices. For this amount of money, I'm willing to scan invoices for errors.

## Extras

How do you handle customers who want extras on a job? If someone asks you to add an in-line filter to the job while putting in the pump system, what do you do? A change order should be written and signed by all parties to reflect the change. This is the best way to avoid confusion and possible payment problems. Yet far too few contractors take the time to use change orders.

It is not uncommon for customers to request additional work and get it without ever being billed for it. This is a sure way to bust your budget. When proper records are not kept of changes on a job, the billing for additional work can easily slip through the cracks. Many times mechanics do little extras on jobs and forget to turn in work orders. Anytime additional work is done, it needs to be recorded and billed.

## Written Records

Written records kept during a job can pay off later. There are different ways of keeping track of what goes into a job. You might just keep delivery tickets and time cards as reference points. Some contractors, like me, use day sheets to list the materials used on any given day. This procedure is more accurate than piling up delivery tickets.

When my crews go into the field for small jobs, they make written records of all materials used on their job. This is done each day that they work on a job. If they work two jobs in the same day, they turn in two different logs. Employee time is also kept in a journal. The time is broken down by phases. If 2 hours are spent putting a pump in and 1 hour in backfilling a trench, the day log will reflect it. This type of paperwork makes it easy for me to track production, inventory, estimates, and job costs.

## As a Job is Winding Down

As a job is winding down, you should start to gather your job-costing data. This might mean a trip into the field to count materials used, or you might be able to rely on paperwork created during the course of a job. One way or another, you need to know how much labor and material went into a job. This might be as simple as jotting down numbers from bills you have received from subcontractors. If you are not supplying any of the labor from payroll people and your subcontractors are providing all materials, your paperwork will be kept at a minimum.

By the time a job is finished, you should be geared up to perform a full job-cost report. It may be necessary to wait for final billing to come in from your suppliers before putting a final job-cost together. This doesn't stop you from starting your work. The longer you wait to produce a job-cost report, the more likely you are to neglect doing the report at all.

## Pulling It All Together

Once you have all the data needed for a job cost, you can start pulling it all together for a clear view of how well you budgeted the job. Only an accurate job-cost report can prove whether you made or lost money on a job. Finding out that you've lost money on a job is never pleasant. But discovering financial losses on one job can help you to avoid them on future jobs.

Job costs give you power, profit power. A good job cost will show you what you did right and where you went wrong. You can determine if your estimating skills are good when you review a job-cost report. There is a wealth of knowledge available to contractors who use job-costing techniques properly.

To stay on budget, you have to track all your financial activities. This should be done during the course of a job and after a job is finished. Until you can account for all your costs of doing a job, you can't be sure how profitable your business is. Since profit is a prime reason for being in business, it's very important to keep track of the money you are making or losing.

## A Money Diet

You may have to put your company on a money diet. Sometimes businesses become fat with overhead expenses. This puts strain on a company. Reducing overhead can increase profits. If your business is too fat with overhead, put it on a diet. How does this apply to cost overruns? Well, I'll tell you.

Overhead reduction might not seem to have anything to do with cost overruns on jobs. In fact, it doesn't, assuming that a contractor calculates all overhead into job quotes. This is often done incorrectly, if at all. But overhead expenses that are not accounted for can ruin the projected profit of a job.

Do you have field supervisors on your payroll? If you do, how do you bill out their time? How many employees do you have? Are some of them administrative employees? Do you account for their cost in your job quotes? How you factor in the cost of employees, insurance, rent, utilities, and other expenses can clearly affect the profit percentages on your jobs.

Employees cost employers a lot more than their hourly rate of pay when all associated expenses are factored into a total hourly cost. Someone who is being paid $10 per hour might be costing an employer $14 an hour. Many factors influence the total cost of employees. Paid vacation, company-provided insurance coverage, and a host of other expenses contribute to the total cost of an employee.

Not all builders have employees. Many contractors work exclusively with subcontractors. When this is the case, figuring overhead expenses is a little easier. I don't plan to get into a long discussion on company overhead. But, I do want you to be aware that any overhead that is not factored into a job quote can cause you to make less money than you were hoping to.

**Here We Are**

Well, here we are, at the end of our time together. You've been given a lot of information on wells and septic systems. I trust you've found it easy to understand and beneficial. Builders like yourself who take the time to increase their knowledge tend to be more successful. I wish you the very best of luck in all your endeavors.

# Chapter 31

# Worksite Safety

Worksite safety doesn't get as much attention as it should. Far too many people are injured on jobs every year. Most of the injuries could be prevented, but they are not. Why is this? People are in a hurry to make a few extra bucks, so they cut corners. This happens with plumbing contractors and piece workers. It even affects hourly plumbers who what to shave 15 minutes off their workday, so that they can head back to the shop early.

Based on my field experience, most accidents occur as a result of negligence. Plumbers try to cut a corner, and they wind up getting hurt. This has proved true with my personal injuries. I've only suffered two serious on-the-job injuries, and both of them were a direct result of my carelessness. I knew better than to do what I was doing when I was hurt, but I did it anyway. Well, sometimes you don't get a second chance, and the life you affect may not be your own. So, let's look at some sensible safety procedures that you can implement in your daily activity.

## Very Dangerous

Plumbing can be a very dangerous trade. The tools of the trade have potential to be killers. Requirements of the job can place you in positions where a lack of concentration could result in serious injury or death. The fact that plumbing can be dangerous is no reason to rule out the trade as your profession. Driving can be extremely dangerous, but few people never set foot in an automobile out of fear.

Fear is generally a result of ignorance. When you have a depth of knowledge and skill, fear begins to subside. As you become more accomplished at what you do, fear is forgotten. While it is advisable to learn to work without fear, you should never work without respect. There is a huge difference between fear and respect.

If, as a plumber, you are afraid to climb up on a roof to flash a pipe, you are not going to last long in the plumbing trade. However, if you scurry up the roof recklessly, you could be injured severely, perhaps even killed. You must respect the position you are putting yourself in. If you are using a ladder to get on the roof, you must respect the outcome of what a mistake could have. Once you are on the roof, you must be conscious of footing conditions, and the way that you negotiate the pitch of the roof. If you pay attention, are properly trained, and don't get careless, you are not likely to get hurt.

Being afraid of a roof will limit or eliminate your plumbing career. Treating your trip to and from the roof like a walk into your living room could be deadly. Respect is the key. If you respect the consequences of your actions, you are aware of what you are doing and the odds for a safe trip improve.

Many young plumbers are fearless in the beginning. They think nothing of darting around on a roof or jumping down in a sewer trench. As their careers progress, they usually hear about or see on-the-job accidents. Someone gets buried in the cave-in of a trench. Somebody falls off a roof. A metal ladder being set up hits a power line. A careless plumber steps into a flooded basement and is electrocuted because of submerged equipment. The list of possible job-related injuries is a long one.

Millions of people are hurt every year in job-related accidents. Most of these people were not following solid safety procedures. Sure, some of them were victims of unavoidable accidents, but most were hurt by their own hand, in one way or another. You don't have to be one of these statistics.

In over twenty-five years of plumbing, I have only been hurt seriously on the job twice. Both times were my fault. I got careless. In one of the instances, I let loose clothing and a powerful drill work together to chew up my arm. In the other incident, I tried to save myself the trouble of repositioning my stepladder while drilling holes in floor joists. My desire to save a minute cost me torn stomach muscles and months of pain from a twisting drill.

My accidents were not mistakes; they were stupidity. Mistakes are made through ignorance. I wasn't ignorant of what could happen to me.

I knew the risk I was taking, and I knew the proper way to perform my job. Even with my knowledge, I slipped up and got hurt. Luckily, both of my injuries healed, and I didn't pay a lifelong price for my stupidity.

During my long plumbing career I have seen a lot of people get hurt. Most of these people have been helpers and apprentices. Of all the on-the-job accidents I have witnessed, every one of them could have been avoided. Many of the incidents were not extremely serious, but a few were.

As a plumber, you will be doing some dangerous work. You will be drilling holes, running threading machines, snaking drains, installing roof flashings, and a lot of other potentially dangerous jobs. Hopefully, your employer will provide you with quality tools and equipment. If you have the right tool for the job, you are off to a good start in staying safe.

Safety training is another factor you should seek from your employer. Some plumbing contractors fail to tell their employees how to do their jobs safely. It is easy for someone, like an experienced plumber, who knows a job inside and out to forget to inform an inexperienced person of potential danger.

For example, a plumber might tell you to break up the concrete around a pipe to allow the installation of a closet flange and never consider telling you to wear safety glasses. The plumber will assume you know the concrete is going to fly up in your face as it is chiseled up. However, as a rookie, you might not know about the reaction concrete has when hit with a cold chisel. One swing of the hammer could cause extreme damage to your eyesight.

Simple jobs, like the one in the example, are all it takes to ruin a career. You might be really on your toes when asked to scoot across an I-beam, but how much thought are you going to give to tightening the bolts on a toilet? The risk of falling off the I-beam is obvious. Having chips of china, from a broken toilet where the nuts were turned one time too many, flying into your eyes is not so obvious. Either way, you can have a work-stopping injury.

Safety is a serious issue. Some job sites are very strict in the safety requirements maintained. But a lot of jobs have no written rules of safety. If you are working on a commercial job, supervisors are likely to make sure you abide by the rules of the Occupational Safety and Health Administration (OSHA). Failure to comply with OSHA regulations can result in stiff financial penalties. However, if you are working in residential plumbing, you may never set foot on a job where OSHA regulations are observed.

In all cases, you are responsible for your own safety. Your employer and OSHA can help you to remain safe, but in the end, it is up to you. You are the one who has to know what to do and how to do it. And not only do you have to take responsibility for your own actions, you also have to watch out for the actions of others. It is not unlikely that you could be injured by someone else's carelessness. Now that you have had the primer course, let's get down to the specifics of job-related safety.

As we move into specifics, you will find the suggestions in this chapter broken down into various categories. Each category will deal with specific safety issues related to the category. For example, in the section on tool safety, you will learn procedures for working safely with tools. As you move from section to section, you may notice some overlapping of safety tips. For example, in the section on general safety, you will see that it is wise to work without wearing jewelry. Then you will find jewelry mentioned again in the tool section. The duplication is done to pinpoint definite safety risks and procedures. We will start into the various sections with general safety.

## General Safety

General safety covers a lot of territory. It starts from the time you get into the company vehicle and carries you right through to the end of the day. Much of the general safety recommendations involve the use of common sense. Now, let's get started.

### Vehicles

Many plumbers are given company trucks for their use in getting to and from jobs. You will probably spend a lot of time loading and unloading company trucks. And, of course, you will spend time either riding in or driving them. All of these areas can threaten your safety.

If you will be driving the truck, take the time to get used to how it handles. Loaded plumbing trucks don't drive like the family car. Remember to check the vehicle's fluids, tires, lights, and related equipment. Many plumbing trucks are old and have seen better days. Failure to check the vehicle's equipment could result in unwanted headaches. Also, remember to use the seat belts; they do save lives.

Apprentices are normally charged with the duty of unloading the truck at the job site. There are a lot of ways to get hurt in doing this job. Many plumbing trucks use roof racks to haul pipe and ladders. If you are unloading these items, make sure they will not come into contact with low-hanging electrical wires. Copper pipe and aluminum ladders make very good electrical conductors, and they will carry the power surge through you on the way to the ground. If you are unloading heavy items, don't put your body in awkward positions. Learn the proper ways for lifting, and never lift objects inappropriately. If the weather is wet, be careful climbing on the truck. Step bumpers get slippery, and a fall can impale you on an object or bang up your knee.

When it is time to load the truck, observe the same safety precautions you did in unloading. In addition to these considerations, always make sure your load is packed evenly and well secured. Be especially careful of any load you attach to the roof rack, and always double check the cargo doors on trucks with utility bodies.

It will not only be embarrassing to lose your load going down the road, it could be deadly. I have seen a one-piece fiberglass tub\shower unit fly out of the back of a pick-up truck as the truck was rolling up an interstate highway. As a young helper, I lost a load of pipe in the middle of a busy intersection. In that same year, the cargo doors on the utility body of my truck flew open as I came off a ramp, onto a major highway. Tools scattered across two lanes of traffic. These types of accidents don't have to happen. It's your job to make sure they don't.

## Clothing

Clothing is responsible for a lot of on-the-job injuries. Sometimes it is the lack of clothing that causes the accidents, and there are many times when too much clothing creates the problem. Generally, it is wise not to wear loose fitting clothes. Shirttails should be tucked in, and short-sleeve shirts are safer than long-sleeved shirts when operating some types of equipment.

Caps can save you from minor inconveniences, like getting glue in your hair, and hard hats provide some protection from potentially damaging accidents, like having a steel fitting dropped on your head. If you have long hair, keep it up and under a hat.

Good footwear is essential in the trade. Normally a strong pair of hunting-style boots will be best. The thick soles provide some protection from nails and other sharp objects you may step on. Boots with steel toes can make a big difference in your physical well being. If you are going to be climbing, wear foot gear with a flexible sole that grips well. Gloves can keep your hands warm and clean, but they can also contribute to serious accidents. Wear gloves sparingly, depending upon the job you are doing.

## Jewelry

On the whole, jewelry should not be worn in the workplace. Rings can inflict deep cuts in your fingers. They can also work with machinery to amputate fingers. Chains and bracelets are equally dangerous, probably more so.

## Eye And Ear Protection

Eye and ear protection is often overlooked. An inexpensive pair of safety glasses can prevent you from spending the rest of your life blind. Ear protection reduces the effect of loud noises, such as jackhammers and drills. You may not notice much benefit now, but in later years you will be glad you wore it. If you don't want to lose your hearing, wear ear protection when subjected to loud noises.

## Pads

Kneepads not only make a plumber's job more comfortable, they help to protect the knees. Some plumbers spend a lot of time on their knees, and pads should be worn to ensure that they can continue to work for many years.

The embarrassment factor plays a significant role in job-related injuries. People, especially young people, feel the need to fit in and to make a name for themselves. Plumbing is sort of a macho trade. There is no secret that plumbers often fancy themselves as strong human specimens. Many plumbers are strong. The work can be hard and in doing it, becoming strong is a side benefit. But you can't allow safety to be pushed aside for the purpose of making a macho statement.

All too many people believe that working without safety glasses, ear protection, and so forth makes them tough. That's just not true; it may make them appear dumb, and it may get them hurt, but it does not make them look tough. If anything, it makes them look stupid or inexperienced.

Don't fall into the trap so many young plumbers do. Never let people goad you into bad safety practices. Some plumbers are going to laugh at your kneepads. Let them laugh, you will still have good knees when they are hobbling around on canes. I'm dead serious about this issue. There is nothing sissy about safety. Wear your gear in confidence, and don't let the few jokesters get to you.

## Tool Safety

Tool safety is a big issue in plumbing. Anyone in the plumbing trade will work with numerous tools. All of these tools are potentially dangerous, but some of them are especially hazardous. This section is broken down by the various tools used on the job. You cannot afford to start working without the basics in tool safety. The more you can absorb on tool safety, the better off you will be.

The best starting point is reading all literature available from the manufacturers of your tools. The people that make the tools provide some good safety suggestions with them. Read and follow the manufacturers' recommendations.

The next step in working safely with your tools is to ask questions. If you don't understand how a tool operates, ask someone to explain it to you. Don't experiment on your own, the price you pay could be much too high.

Common sense is irreplaceable in the safe operation of tools. If you see an electrical cord with cut insulation, you should have enough common sense to avoid using it. In addition to this type of simple observation, you will learn some interesting facts about tool safety. Now, let me tell you what I've learned about tool safety over the years.

There are some basic principals to apply to all of your work with tools. We will start with the basics, and then we will move on to specific tools. Here are the basics:

- Keep body parts away from moving parts.

- Don't work with poor lighting conditions.

- Be careful of wet areas when working with electrical tools.

- If special clothing is recommended for working with your tools, wear it.

- Use tools only for their intended purposes.

- Get to know your tools well.

- Keep your tools in good condition.

Now, let's take a close look at the tools you are likely to use. Plumbers use a wide variety of hand tools and electrical tools. They also use specialty tools. So let's see how you can use all these tools without injury.

**Torches**

Torches are often used by plumbers on a daily basis. Some plumbers use propane torches, and others use acetylene torches. In either case, the containers of fuel for these torches can be very dangerous. The flames produced from torches can also do a lot of damage.

When working with torches and tanks of fuel, you must be very careful. Don't allow your torch equipment to fall on a hard surface, the valves may break. Check all your connections closely, a leak allowing fuel to fill an area could result in an explosion when you light the torch.

Always pay attention to where your flame is pointed. Carelessness with the flame could start unwanted fires. This is especially true when working near insulation and other flammable substances. If the flame is directed at concrete, the concrete may explode. Since moisture is retained in concrete, intense heat can cause the moisture to force the concrete to explode.

It's not done often enough, but you should always have a fire extinguisher close by when working with a torch. If you have to work close to flammable substances, use a heat shield on the torch. When your flame is close to wood or insulation, try to remove the insulation or wet the flammable substance, before applying the flame. When you are done using the torch, make sure the fuel tank is turned off and the hose is drained of all fuel. Use a striker to light your torch. The use of a match or cigarette lighter puts your hand too close to the source of the flame.

Fast Facts

- Don't allow torch equipment fall on a hard surface.

- Check torch connections carefully.

- Pay attention to where the flame of your torch is pointed.

- Concrete can explode when heated.

- Use a heat shield when using a torch near potentially flammable materials.

- Turn off your torch's fuel tank when not in use.

- Use a striker to light your touch.

**Lead Pots And Ladles**

Lead pots and ladles offer their own style of potential danger. Plumbers today don't use molten lead as much as they used to, but hot lead is still used. It is used to make joints with cast-iron pipe and fittings.

When working with a small quantity of lead, many plumber heat the lead in a ladle. They melt the lead with their torch and pour it straight from the ladle. When larger quantities of lead are needed, lead pots are used. The pots are filled with lead and set over a flame. All of this type of work is dangerous.

Never put wet materials in hot lead. If the ladle is wet or cold when it is dipped into the pot of molten lead, the hot lead can explode. Don't add wet lead to the melting pot if it contains molten lead. Before you pour the hot lead in a waiting joint, make sure the joint is not wet. As another word of caution, don't leave a working lead pot where rain dripping off a roof can fall into it.

Obviously, molten lead is hot, so you shouldn't touch it. Be very careful to avoid tipping the pot of hot lead over. I remember one accident when a pot of hot lead was knocked over on a plumber's foot. Let me just say the scene was terrifying.

**Drills And Bits**

Drills have been my worst enemy in plumbing. The two serious injuries I have received were both related to my work with a right-angle drill. The drills used most by plumbers are not the little pistol-grip, hand-held types of drills most people think of. The day-to-day drilling done by plumbers involves the use of large, powerful right-angle drills. These drills have enormous power when they get in a bind. Hitting a nail or a knot in the wood being drilled can do a lot of damage. You can break fingers, lose teeth, suffer head injuries, and a lot more. As with all electrical tools, you should always check the electrical cord before using your drill. If the cord is not in good shape, don't use the drill.

Always know what you are drilling into. If you are doing new-construction work it is fairly easy to look before you drill. However, drilling in a remodeling job can be much more difficult. You cannot always see what you are getting into. If you are unfortunate enough to drill into a hot wire, you can get a considerable electrical shock.

The bits you use in a drill are part of the safe operation of the tool. If your drill bits are dull, sharpen them. Dull bits are much more dangerous than sharp ones. When you are using a standard plumber's bit to drill through thin wood, like plywood, be careful. Once the worm driver of the bit penetrates the plywood fully, the big teeth on the bit can bite and jump, causing you to lose control of the drill. If you will be drilling metal, be aware that the metal shavings will be sharp and hot.

**Power Saws**

Plumbers don't use power saws as much as carpenters, but they do use them. The most common type of power saw used by plumbers is the reciprocating saw. These saws are used to cut pipe, plywood, floor joists, and a whole lot more. In addition to reciprocating saws, plumbers use circular saws and chop saws. All of the saws have the potential for serious injury.

Reciprocating saws are reasonably safe. Most models are insulated to help avoid electrical shocks if a hot wire is cut. The blade is typically a safe distance from the user, and the saws are pretty easy to hold and control. However, the brittle blades do break. This could result in an eye injury.

Circular saws are used by plumbers occasionally. The blades on these saws can bind and cause the saws to kick back. Chop saws are sometimes used to cut pipe. If you keep your body parts out of the way and wear eye protection, chop saws are not unusually dangerous.

**Hand-Held Drain Cleaners**

Hand-held drain cleaners don't get a lot of use from plumbers that do new-construction work, but they are a frequently-used tool of service plumbers. Most of these drain cleaners resemble, to some extent, a straight, hand-held drill. There are models that sit on stands, but most small snakes are hand-held. These small-diameter snakes are not nearly as dangerous as their big brothers, but they do deserve respect. These units carry all the normal hazards of an electric tool, but there is more.

The cables used for small drain cleaning jobs are usually very flexible. They are basically springs. The heads attached to the cables take on different shapes and looks. When you look at these thin cables, they don't look dangerous, but they can be. When the cables are being fed down a drain and turning, they can hit hard stoppages and go out of control. The cable can twist and kink. If your finger, hand, or arm is caught in the cable, injury can be the result. To avoid this, don't allow excessive cable to exist between the drain pipe and the machine.

**Large Sewer Machines**

Large sewer machines are much more dangerous than small ones. These machines have tremendous power and their cables are capable of removing fingers. Broken bones, severe cuts, and assorted other injuries are also possible with big snakes.

One of the most common conflicts with large sewer machines is with the cables. When the cutting heads hit roots or similar hard items in the pipe, the cable goes wild. The twisting, thrashing cable can do a lot of damage. Again, limiting excess cable is one of the best protections possible. Special sleeves are also available to contain unruly cables.

Most big machines can be operated by one person, but it is wise to have someone standing by in case help is needed. Loose clothing results in many drain machine accidents. The use of special mitts can help reduce the risk of hand injuries. Electrical shocks are also possible when doing drain cleaning.

**Power Pipe Threaders**

Power pipe threaders are very nice to have if you are doing much work with threaded pipe. However, these threading machines can grind body parts as well as they thread pipe. Electric threaders are very dangerous in the hands of untrained people. It is critical to keep fingers and clothing away from the power mechanisms. The metal shavings produced by pipe threaders can be very sharp, and burrs left on the threaded pipe can slash your skin. The cutting oil used to keep the dies from getting too hot can make the floor around the machine slippery.

**Air-Powered Tools**

Air-powered tools are not used often by plumbers. Jackhammers are probably the most used air-powered tools for plumbers. When using tools with air hoses, check all connections carefully. If you experience a blow-out, the hose can spiral wildly out of control.

## Powder-Actuated Tools

Powder-actuated tools are used by plumbers to secure objects to hard surfaces, like concrete. If the user is properly trained, these tools are not too dangerous. However, good training, eye protection, and ear protection are all necessary. Misfires and chipping hard surfaces are the most common problem with these tools.

## Ladders

Ladders are used frequently by plumbers, both stepladders and extension ladders. Many ladder accidents are possible. You must always be aware of what is around you when handling a ladder. If you brush against a live electrical wire with a ladder you are carrying, your life could be over. Ladders often fall over when the people using them are not careful. Reaching too far from a ladder can be all it takes to take a fall.

When you set up a ladder, or rolling scaffolds make sure it is set up properly. The ladder should be on firm footing, and all safety braces and clamps should be in place. When using an extension ladder, many plumbers use a rope to tie rungs together where the sections overlap. The rope provides an extra guard against the ladder's safety clamps failing and the ladder collapsing. When using an extension ladder, be sure to secure both the base and the top. I had an unusual accident on a ladder that I would like to share with you.

I was on a tall extension ladder, working near the top of a commercial building. The top of my ladder was resting on the edge of the flat roof. There was metal flashing surrounding the edge of the roof, and the top of the ladder was leaning against the flashing. There was a picket fence behind me and electrical wires entering the building to my right. The entrance wires were a good ways away, so I was in no immediate danger. As I worked on the ladder, a huge gust of wind blew around the building. I don't know where it came from; it hadn't been very windy when I went up the ladder.

The wind hit me and pushed me and the ladder sideways. The top of the ladder slid easily along the metal flashing, and I couldn't grab anything to stop me. I knew the ladder was going to go down, and I didn't have much time to make a decision. If I pushed off of the ladder, I would probably be impaled on the fence. If I rode the ladder down, it might hit the electrical wires and fry me. I waited until the last minute and jumped off of the ladder.

I landed on the wet ground with a thud, but I missed the fence. The ladder hit the wires and sparks flew. Fortunately, I wasn't hurt and electricians were available to take care of the electrical problem. This was a case where I wasn't really negligent, but I could have been killed. If I had secured the top of the ladder, my accident wouldn't have happened.

**Screwdrivers And Chisels**

Eye injuries and puncture wounds are common when working with screwdrivers and chisels. When the tools are use properly and safety glasses are worn, few accidents occur.
 The key to avoiding injury with most hand tools is simply to use the right tool for the job. If you use a wrench as a hammer or a screwdriver as a chisel, you are asking for trouble.
 There are, of course, other types of tools and safety hazards found in the plumbing trade. However, this list covers the ones that result in the most injuries. In all cases, observe proper safety procedures and utilize safety gear, such as eye and ear protection.

**Co-Worker Safety**

Co-worker safety is the last segment of this chapter. I am including it because workers are frequently injured by the actions of co-workers. This section is meant to protect you from others and to make you aware of how your actions might affect your co-workers.
 Most plumbers find themselves working around other people. This is especially true on construction jobs. When working around other people, you must be aware of their actions, as well as your own. If you are walking out of a house to get something off the truck and a roll of roofing paper gets away from a roofer, you could get an instant headache.
 If you don't pay attention to what is going on around you, it is possible to wind up in all sorts of trouble. Cranes lose their loads some times, and such a load landing on you is likely to be fatal. Equipment operators don't always see the plumber kneeling down for a piece of pipe. It's not hard to have a close encounter with heavy equipment. While we are on the subject of equipment, let me bore you with another war story.
 One day I was in a sewer ditch, connecting the sewer from a new house to the sewer main. The section of ditch that I was working in was only about four feet deep. There was a large pile of dirt near the edge

of the trench; it had been created when the ditch was dug. The dirt wasn't laid back like it should have been; it was piled up.

As I worked in the ditch, a backhoe came by. The operator had no idea I was in the ditch. When he swung the backhoe around to make a turn, the small scorpion-type bucket on the back of the equipment hit the dirt pile.

I had stood up when I heard the hoe approaching, and it was a good thing I had. When the equipment hit the pile of dirt, part of the mound caved in on me. I tried to run, but it caught both of my legs and the weight drove me to the ground. I was buried from just below my waist. My head was okay, and my arms were free. I was still holding my shovel.

I yelled, but nobody heard me. I must admit, I was a little panicked. I tried to get up and couldn't. After a while, I was able to move enough dirt with the shovel to crawl out from under the dirt. I was lucky. If I had been on my knees making the connection or grading the pipe, I might have been smothered. As it was, I came out of the ditch no worse for the wear. But, boy, was I mad at the careless backhoe operator. I won't go into the details of the little confrontation I had with him.

That accident is a prime example of how other workers can hurt you and never know they did it. You have to watch out for yourself at all times. As you gain field experience, you will develop a second nature for impending co-worker problems. You will learn to sense when something is wrong or is about to go wrong. But you have to stay alive and healthy long enough to get that experience.

Always be aware of what is going on over your head. Avoid working under other people and hazardous overhead conditions. Let people know where you are, so you won't get stranded on a roof or in an attic when your ladder is moved or falls over.

You must also remember that your actions could harm co-workers. If you are on a roof to flash a pipe and your hammer gets away from you, somebody could get hurt. Open communication between workers is one of the best ways to avoid injuries. If everyone knows where everyone else is working, injuries are less likely. Primarily, think and think some more. There is no substitute for common sense. Try to avoid working alone, and remain alert at all times.

**Chapter**

# 32

# First Aid

Everyone should invest some time in learning the basics of first-aid. You never know when having skills in first-aid treatments may save your life. Plumbers live what can be a dangerous life. On-the-job injuries are not uncommon. Most injuries are fairly minor, but they often require treatment. Do you know the right way to get a sliver of copper out of your hand? If your helper suffers from an electrical shock when a drill cord goes bad, do you know what to do? Well, many plumbers don't possess good first-aid skills.

Before we get too far into this chapter, there are a few points are want to make. First of all, I'm not a medical doctor or any type of trained medical-care person. I've taken first-aid classes, but I'm certainly not an authority on medical issues. The suggestions that I will give you in this chapter are for informational purposes only. This book is not a substitute for first-aid training offered by qualified professionals.

My intent here is to make you aware of some basic first-aid procedures that can make life on the job much easier. But, I want you to understand that I'm not advising you to use my advice to administer first-aid. Hopefully, this chapter will show you the types of advantages you can gain from taking first-aid classes. Before you attempt first-aid on anyone, including yourself, you should attend a structured, approved first-aid class. I'm going to give you information that is as accurate as I can make it, but don't assume that my words are enough. Take a little time to seek professional training in the art of first-aid.

You may never use what you learn, but the one time it is needed, you will be glad you made the effort to learn what to do. With this said, let's jump right into some tips on first-aid.

### Open Wounds

Open wounds are a common problem for plumbers. Many tools and materials used by plumbers can create open wounds. What should you do if you or one of your workers is cut?

- Stop the bleeding as soon as possible.

- Disinfect and protect the wound from contamination.

- You may have to take steps to avoid shock symptoms.

- Once the patient is stable, seek medical attention for severe cuts.

When a bad cut is encountered, the victim may slip into shock. A loss of consciousness could result from a loss of blood. Death from extreme bleeding is also a risk. As a first-aid provider, you must act quickly to reduce the risk of serious complications.

### Bleeding

To stop bleeding, direct pressure is normally a good tactic. This may be as crude as clamping your hand over the wound, but a cleaner compression is desirable. Ideally, a sterile material should be placed over the wound and secured, normally with tape (even if it's duct tape). Whenever possible, wear rubber gloves to protect yourself from possible disease transfer if you are working on someone else. Thick gauze used as a pressure material can absorb blood and begin to allow the clotting process to start.

Bad wounds may bleed right through the compress material. If this happens, don't remove the blood-soaked material. Add a new layer of material over it. Keep pressure on the wound. If you are not prepared with a first-aid kit, you could substitute gauze and tape with strips cut from clothing that can be tied in place over the wound.

When you are dealing with a bleeding wound, it is usually best to elevate it. If you suspect a fractured or broken bone in the area of the wound, elevation may not be practical. When we talk about elevating a wound, it simply means to raise the wound above the level of the victim's heart. This helps the blood flow to slow, due to gravity.

**Super Serious**

Super serious bleeding might not stop even after a compression bandage is applied and the wound is elevated. When this is the case, you must resort to putting pressure on the main artery which is producing the blood. Constricting an artery is not an alternative for the steps that we have discussed previously.

Putting pressure on an artery is serious business. First, you must be able to locate the artery, and you should not keep the artery constricted any longer than necessary. You may have to apply pressure for awhile, release it, and then apply it again. It's important that you do not restrict the flow of blood in arteries for long periods of time. I hesitate to go into too much detail on this process, as I feel it is a method that you should be taught in a controlled, classroom situation. However, I will hit the high spots. But remember, these words are not a substitute for professional training from qualified instructors.

Open arm wounds are controlled with the brachial artery. The location of this artery is in the area between the biceps and triceps, on the inside of the arm. It's about halfway between the armpit and the elbow. Pressure is created with the flat parts of your fingertips. Basically, you are holding the victim's wrist with one hand and closing off the artery with your other hand. Pressure exerted by your fingers pushes the artery against the arm bone and restricts blood flow. Again, don't attempt this type of first-aid until you have been trained properly in the execution of the procedure.

Severe leg wounds may require the constriction of the femoral artery. This artery is located in the pelvic region. Normally, bleeding victims are placed on their backs for this procedure. The heel of a hand is placed on the artery to restrict blood flow. In some cases, fingertips are used to apply pressure. I'm uncomfortable with going into great detail on these procedures, because I don't want you to rely solely on what I'm telling you. It's enough that you understand that knowing when and where to apply pressure to arteries can save lives and that you should seek professional training in these techniques.

## Tourniquets

Tourniquets get a lot of attention in movies, but they can do as much harm as they do good if not used properly. A tourniquet should only be used in a life-threatening situation. When a tourniquet is applied, there is a risk of losing the limb to which the restriction is applied. This is obviously a serious decision and one that must be made only when all other means of stopping blood loss have been exhausted.

Unfortunately, plumbers might run into a situation where a tourniquet is the only answer. For example, if a worker allowed a power saw to get out of control, a hand might be severed or some other type of life-threatening injury could occur. This would be cause for the use of a tourniquet. Let me give you a basic overview of what's involved when a tourniquet is used.

Tourniquets should be at least two inches wide. A tourniquet should be placed at a point that is above a wound, between the bleeding and the victim's heart. However, the binding should not encroach directly on the wound area. Tourniquets can be fashioned out of many materials. If you are using strips of cloth, wrap the cloth around the limb that is wounded and tie a knot in the material. Use a stick, screwdriver, or whatever else you can lay your hands on to tighten the binding.

Once you have made a commitment to apply a tourniquet, the wrapping should be removed only by a physician. It's a good idea to note the time that a tourniquet is applied, as this will help doctors later in assessing their options. As an extension of the tourniquet treatment, you will most likely have to treat the patient for shock.

## Infection

Infection is always a concern with open wounds. When a wound is serious enough to require a compression bandage, don't attempt to clean the cut. Keep pressure on the wound to stop bleeding. In cases of sever wounds, be on the look out for shock symptoms, and be prepared to treat them. Your primary concern with a serious open wound is to stop the bleeding and gain professional medical help as soon as possible.

Lesser cuts, which are more common than deep ones, should be cleaned. Regular soap and water can be used to clean a wound before applying a bandage. Remember, we are talking about minor cuts and scrapes at this point. Flush the wound generously with clean water. A piece of sterile gauze can be used to pat the wound dry. Then a clean, dry bandage can be applied to protect the wound while in transport to a medical facility.

Fast Facts

- Use direct pressure to stop bleeding.

- Wear rubber gloves to prevent direct contact with a victim's blood.

- When feasible, elevate the part of the body that is bleeding.

- Extremely serious bleeding can require you to put pressure on the artery supplying the blood to the wound area.

- Tourniquets can do more harm than good.

- Tourniquets should be at least two inches wide.

- Tourniquets should be placed above the bleeding wound, between the bleeding and the victim's heart.

- Tourniquets should not be applied directly on the wound area.

- Tourniquets should only be removed by trained medical professionals.

- If you apply a tourniquet, note the time that you apply the tourniquet.

- When a bleeding wound requires a compression bandage, don't attempt to clean the wound. Simply apply compression quickly.

- Watch victims with serious bleeding for symptoms of shock.

- Lesser bleeding wounds should be cleaned before being bandaged.

### Splinters And Such

Splinters and such foreign objects often invade the skin of plumbers. Getting these items out cleanly is best done by a doctor, but there are some on-the-job methods that you might want to try. A magnifying glass and a pair of tweezers work well together when removing embedded objects, such as splinters and slivers of copper tubing. Ideally, tweezers being used should be sterilized either over an open flame, such as the flame of your torch, or in boiling water.

Splinters and slivers that are submerged beneath the skin can often be lifted out with the tip of a sterilized needle. The use of a needle in conjunction with a pair of tweezers is very effective in the removal of most simple splinters. If you are dealing with something that has gone extremely deep into tissue, it is best to leave the object alone until a doctor can remove it.

### Eye Injuries

Eye injuries are very common on construction and remodeling jobs. Most of these injuries could be avoided if proper eye protection was worn, but far too many workers don't wear safety glasses and goggles. This sets the stage for eye irritations and injuries.

Before you attempt to help someone who is suffering from an eye injury, you should wash your hands thoroughly. I know this is not always possible on construction sites, but cleaning your hands is advantageous. In the meantime, keep the victim from rubbing the injured eye. Rubbing can make matters much worse.

Never attempt to remove a foreign object from someone's eye with the use of a rigid device, such as a toothpick. Cotton swabs that have been wetted can serve well as a magnet to remove some types of invasion objects. If the person you are helping has something embedded in an eye, get the person to a doctor as soon as possible. Don't attempt to remove the object yourself.

When you are investigating the cause of an eye injury, you should pull down the lower lid of the eye to determine if you can see the object causing trouble. A floating object, such as a piece of sawdust trapped between an eye and an eyelid can be removed with a tissue, a damp cotton swab, or even a clean handkerchief. Don't allow dry cotton material to come into contact with an eye.

If looking under the lower lid doesn't a source of discomfort, check under the lower lid. Clean water can be used to flush out many eye contaminants without much risk of damage to the eye. Objects that cannot be removed easily should be left alone until a physician can take over.

- Wash your hands, if possible, before treating eye injuries.

- Don't rub an eye wound.

- Don't attempt to remove embedded items from an eye.

- Clean water can be used to flush out some eye irritants.

**Scalp Injuries**

Scalp injuries can be misleading. What looks like a serious wound can be a fairly minor cut. On the other hand, what appears to be only a cut can involve a fractured skull. If you or someone around you sustains a scalp injury, such as having a hammer fall on your head from an overhead worker, take it seriously. Don't attempt to clean the wound. Expect profuse bleeding.

If you don't suspect a skull fracture, raise the victim's head and shoulders to reduce bleeding. Try not to bend the neck. Put a sterile bandage over the wound, but don't apply excessive pressure. If there is a bone fracture, pressure could worsen the situation. Secure the bandage with gauze or some other material that you can wrap around it. Seek medical attention immediately.

**Facial Injuries**

Facial injuries can occur on plumbing jobs. I've seen helpers let their right-angle drills get away from them with the result being hard knocks to the face. On one occasion, I remember a tooth being lost, and split lips and tongues that have been bitten are common when a drill goes on a rampage.

Extremely bad facial injuries can cause a blockage of the victim's air passages. This, of course, is a very serious condition. It's critical that air passages be open at all times. If the person's mouth contains broken teeth or dentures, remove them. Be careful not to jar the individual's spine if you have reason to believe there may be injury to the back or neck.

Conscious victims should be positioned, when possible, so that secretions from the mouth and nose will drain out. Shock is a potential concern in severe facial injuries. For most on-the-job injuries, plumbers should be treated for comfort and sent for medical attention.

## Nose Bleeds

Nose bleeds are not usually difficult to treat. Typically, pressure applied to the side of the nose where bleeding is occurring will stop the flow of blood. Applying cold compresses can also help. If external pressure is not stopping the bleeding, use a small, clean pad of gauze to create a dam on the inside of the nose. Then, apply pressure on the outside of the nose. This will almost always work. If it doesn't, get to a doctor.

## Back Injuries

There is really only one thing that you need to know about back injuries. Don't move the injured party. Call for professional help and see that the victim remains still until help arrives. Moving someone who has suffered a back injury can be very risky. Don't do it unless there is a life-threatening cause for your action, such as a person trapped in a fire or some other type of deadly situation.

## Legs And Feet

Legs and feet sometimes become injured on job sites. The worst case of this type that I can remember was when a plumber knocked a pot of molten lead over on his foot. It sends shivers up my spine just to recall that incident. Anyway, when someone suffers a minor foot or leg injury, you should clean and cover the wound. Bandages should be supportive without being constrictive. The appendage should be elevated above the victim's heart level when possible. Prohibit the person from walking. Remove boots and socks so that you can keep an eye on the person's toes. If the toes begin to swell or turn blue, loosen the supportive bandages.

## Blisters

Blisters may not seem like much of an emergency, but they can sure take the steam out of a helper or plumber. In most cases, blisters can be covered with a heavy gauze pad to reduce pain. It is generally recommended to leave blisters unbroken. When a blister breaks, the area should be cleaned and treated as an open wound. Some blisters tend to be more serious than others. For example, blisters in the palm of a hand or on the sole of a foot should be look at by a doctor.

## Hand Injuries

Hand injuries are common in the plumbing trade. Little cuts are the most frequent complaint. Getting flux in a cut is an eye-opening experience, so even the smallest break in the skin should be covered. Serious hand injuries should be elevated. This tends to reduce swelling. You should not try to clean really bad hand injuries. Use a pressure bandage to control bleeding. If the cut is on the palm of a hand, a roll of gauze can be squeezed by the victim to slow the flow of blood. Pressure should stop the bleeding, but if it doesn't, seek medical assistance. As with all injuries, use common sense on whether or not professional attention is needed after first-aid is applied.

## Shock

Shock is a condition that can be life threatening even when the injury responsible for a person going into shock is not otherwise fatal. We are talking about traumatic shock, not electrical shock. Many factors can lead to a person going into shock. A serious injury is a common cause, but many other causes exist. There are certain signs of shock which you can look for.

If a person's skin turns pale or blue and is cold to the touch, it's a likely sign of shock. Skin that becomes moist and clammy can indicate shock is present. A general weakness is also a sign of shock. When a person is going into shock, the individual's pulse is likely to exceed 100 beats per minute. Breathing is usually increased, but it may be shallow, deep, or irregular. Chest injuries usually result in shallow breathing. Victims who have lost blood may be thrashing about as they enter into shock. Vomiting and nausea can also signal shock.

As a person slips into deeper shock, the individual may become unresponsive. Look at the eyes, they may be widely dilated. Blood pressure can drop, and in time, the victim will lose consciousness. Body temperature will fall, and death will be likely if treatment is not rendered.

There are three main goals when treating someone for shock. Get the person's blood circulating well. Make sure an adequate supply of oxygen is available to the individual, and maintain the person's body temperature.

When you have to treat a person for shock, you should keep the victim lying down. Cover the individual so that the loss of body heat will be minimal. Get medical help as soon as possible. The reason it's best to keep a person lying down is that the individual's blood should circulate better. Remember, if you suspect back or neck injuries, don't move the person.

People who are unconscious should be placed on one side so that fluids will run out of the mouth and nose. It's also important to make sure that air passages are open. A person with a head injury may be laid out flat or propped up, but the head should not be lower than the rest of the body. It is sometimes advantageous to elevate a person's feet when they are in shock. However is there is any difficulty in breathing or if pain increases when the feet are raised, lower them.

Body temperature is a big concern with shock patients. You want to overcome or avoid chilling. However, don't attempt to add additional heat to the surface of the person's body with artificial means. This can be damaging. Use only blankets, clothes, and other similar items to regain and maintain body temperature.

Avoid the temptation to offer the victim fluids, unless medical care is not going to be available for a long time. Avoid fluids completely if the person is unconscious or is subject to vomiting. Under most job-site conditions, fluids should not be administered.

Checklist of shock symptoms

- Skin that is pale, blue, or cold to the touch

- Skin that is moist and clammy

- General weakness

- Pulse rate in excess of 100 beats per minute

- Increased breathing

- Shallow breathing

- Thrashing

- Vomiting and nausea

- Unresponsive action

❏ Widely dilated eyes

❏ A drop in blood pressure

## Burns

Burns are not real common among plumbers, but they can occur in the workplace. There are three types of burns that you may have to deal with. First-degree burns are the least serious. These burns typically come from overexposure to the sun, which construction workers often suffer from, quick contact with a hot object, like the tip of a torch, and scalding water, which could be the case when working with a boiler or water heater.

Second-degree burns are more serious. They can come from a deep sunburn or from contact with hot liquids and flames. A person who is affected by a second-degree burn may have a red or mottled appearance, blisters, and a wet appearance of the skin within the burn area. This wet look is due to a loss of plasma through the damaged layers of skin.

Third-degree burns are the most serious. They can be caused by contact with open flames, hot objects, or immersion in very hot water. Electrical injuries can also result in third-degree burns. This type of burn can look similar to a second-degree burn, but the difference will be the loss of all layers of skin.

### Treatment

Treatment for most job-related burns can be administered on the job site and will not require hospitalization. First-degree burns should be washed with or submerged in cold water. A dry dressing can be applied if necessary. These burns are not too serious. Eliminating pain is the primary goal with first-degree burns.

Second-degree burns should be immersed in cold (but not ice) water. The soaking should continue for at least one hour and up to two hours. After soaking, the wound should be layered with clean cloths that have been dipped in ice water and wrung out. Then the wound should be dried by blotting, not rubbing. A dry, sterile gauze should then be applied. Don't break open any blisters. It is also not advisable to use ointments and sprays on severe burns. Burned arms and legs should be elevated, and medical attention should be acquired.

Bad burns, the third-degree type, need quick medical attention. First, don't remove a burn victim's clothing, skin might come off with it. A thick, sterile dressing can be applied to the burn area. Personally, I would avoid this if possible. A dressing might stick to the mutilated skin and cause additional skin loss when the dressing is removed. When hands are burned, keep them elevated above the victim's heart. The same goes for feet and legs. You should not soak a third-degree burn in cold water, it could induce more shock symptoms. Don't use ointments, sprays, or other types of treatments. Get the burn victim to competent medical care as soon as possible.

## Heat-Related Problems

Heat-related problems can include heat stroke and heat exhaustion. Cramps are also possible when working in hot weather. There are people who don't consider heat stroke to be serious. They are wrong. Heat stroke can be life threatening. People affected by heat stroke can develop body temperatures in excess of 106 degrees F. Their skin is likely to be hot, red, and dry. You might think sweating would take place, but it doesn't. Pulse is rapid and strong, and victims can sink into an unconscious state.

If you are dealing with heat stroke, you need to lower the person's body temperature quickly. There is a risk, however, of cooling the body too quickly once the victim's temperature is below 102 degrees F. You can lower body temperature with rubbing alcohol, cold packs, cold water on clothes or in a bathtub of cold water. Avoid the use of ice in the cooling process. Fans and air conditioned space can be used to achieve your cooling goals. Get the body temperature down to at least 102 degrees and then go for medical help.

### Cramps

Cramps are not uncommon among workers during hot spells. A simple massage can be all it takes to cure this problem. Salt-water solutions are another way to control cramps. Mix one teaspoonful of salt per glass of water and have the victim drink half a glass about every 15 minutes.

### Exhaustion

Heat exhaustion is more common than heat stroke. A person affected by heat exhaustion is likely to maintain a fairly normal body temperature. But, the person's skin may be pale and clammy. Sweating may be very noticeable, and the individual will probably complain of being tired and weak. Headaches, cramps, and nausea may accompany the symptoms. In some cases, fainting might occur.

The salt-water treatment described for cramps will normally work with heat exhaustion. Victims should lie down and elevate their feet about a foot off the floor or bed. Clothing should be loosened, and cool, wet cloths can be used to add comfort. If vomiting occurs, get the person to a hospital for intravenous fluids.

We could continue talking about first-aid for a much longer time. However, the help I can give you here for medical procedures is limited. You owe it to yourself, your family, and the people you work with to learn first-aid techniques. This can be done best by attending formal classes in your area. Most towns and cities offer first-aid classes on a regular basis. I strongly suggest that you enroll in one. Until you have some hands-on experience in a classroom and gain the depth of knowledge needed, you are not prepared for emergencies. Don't get caught short. Prepare now for the emergency that might never happen.

# Chapter 33

# Mathematics for the Trade

| | |
|---|---|
| A or a | Area, acre |
| AWG | American Wire Gauge |
| B or b | Breadth |
| bbl | Barrels |
| bhp | Brake horsepower |
| BM | Board measure |
| Btu | British thermal units |
| BWG | Birmingham Wire Gauge |
| B & S | Brown and Sharpe Wire Gauge (American Wire Gauge) |
| C of g | Center of gravity |
| cond | Condensing |
| cu | Cubic |
| cyl | Cylinder |
| D or d | Depth, diameter |
| dr | Dram |
| evap | Evaporation |
| F | Coefficient of friction; Fahrenheit |
| F or f | Force, factor of safety |
| ft (or ') | Foot |
| ft lb | Foot pound |
| fur | Furlong |
| gal | Gallon |
| gi | Gill |
| ha | Hectare |
| H or h | Height, head of water |
| HP | horsepower |
| IHP | Indicated horsepower |
| in (or ") | Inch |
| L or l | Length |
| lb | Pound |
| lb/sq in. | Pounds per square inch |
| mi | Mile |
| o.d. | Outside diameter (pipes) |
| oz | Ounces |
| pt | Pint |
| P or p | Pressure, load |
| psi | Pounds per square inch |
| R or r | Radius |
| rpm | Revolutions per minute |
| sq ft | Square foot |
| sq in. | Square inch |
| sq yd | Square yard |
| T or t | Thickness, temperature |
| temp | Temperature |
| V or v | Velocity |
| vol | Volume |
| W or w | Weight |
| W. I. | Wrought iron |

Figure 33.1  Abbreviations.

Circumference of a circle = π × diameter or 3.1416 × diameter
Diameter of a circle = circumference × 0.31831
Area of a square = length × width
Area of a rectangle = length × width
Area of a parallelogram = base × perpendicular height
Area of a triangle = ½ base × perpendicular height
Area of a circle = π radius squared or diameter squared × 0.7854
Area of an ellipse = length × width × 0.7854
Volume of a cube or rectangular prism = length × width × height
Volume of a triangular prism = area of triangle × length
Volume of a sphere = diameter cubed × 0.5236 (diameter × diameter × diameter × 0.5236)
Volume of a cone = π × radius squared × ⅓ height
Volume of a cylinder = π × radius squared × height
Length of one side of a square × 1.128 = the diameter of an equal circle
Doubling the diameter of a pipe or cylinder increases its capacity 4 times
The pressure (in lb/sq in.) of a column of water = the height of the column (in feet) × 0.434
The capacity of a pipe or tank (in U.S. gallons) = the diameter squared (in inches) × the length (in inches) × 0.0034
1 gal water = 8½ lb = 231 cu in.
1 cu ft water = 62½ lb = 7½ gal

**Figure 33.2** Useful formulas.

| | | |
|---|---|---|
| Sine | $\sin = \dfrac{\text{side opposite}}{\text{hypotenuse}}$ | |
| Cosine | $\cos = \dfrac{\text{side adjacent}}{\text{hypotenuse}}$ | |
| Tangent | $\tan = \dfrac{\text{side opposite}}{\text{side adjacent}}$ | |
| Cosecant | $\csc = \dfrac{\text{hypotenuse}}{\text{side opposite}}$ | |
| Secant | $\sec = \dfrac{\text{hypotenuse}}{\text{side adjacent}}$ | |
| Cotangent | $\cot = \dfrac{\text{side adjacent}}{\text{side opposite}}$ | |

**Figure 33.3** Trigonometry.

| Pentagon | 5 sides |
| Hexagon | 6 sides |
| Heptagon | 7 sides |
| Octagon | 8 sides |
| Nonagon | 9 sides |
| Decagon | 10 sides |

**Figure 33.4** Polygons.

---

The capacity of pipes is as the square of their diameters. Thus, doubling the diameter of a pipe increases its capacity four times. The area of a pipe wall may be determined by the following formula:

$$\text{Area of pipe wall} = 0.7854 \times [(\text{o.d.} \times \text{o.d.}) - (\text{i.d.} \times \text{i.d.})]$$

**Figure 33.5** Piping.

---

The approximate weight of a piece of pipe may be determined by the following formulas:

Cast-iron pipe: weight = $(A^2 - B^2) \times \text{length} \times 0.2042$
Steel pipe: weight = $(A^2 - B^2) \times \text{length} \times 0.2199$
Copper pipe: weight = $(A^2 - B^2) \times \text{length} \times 0.2537$
A = outside diameter of the pipe in inches
B = inside diameter of the pipe in inches

**Figure 33.6** Determining pipe weight.

---

The formula for calculating expansion or contraction in plastic piping is:

$$L = Y \times \frac{T - F}{10} \times \frac{L}{100}$$

L = expansion in inches
Y = constant factor expressing inches of expansion per 100°F temperature change per 100 ft of pipe
T = maximum temperature (°F)
F = minimum temperature (°F)
L = length of pipe run in feet

**Figure 33.7** Expansion in plastic piping.

| | |
|---|---|
| *Parallelogram* | Area = base × distance between the two parallel sides |
| *Pyramid* | Area = ½ perimeter of base × slant height + area of base |
| | Volume = area of base × ⅓ of the altitude |
| *Rectangle* | Area = length × width |
| *Rectangular prism* | Volume = width × height × length |
| *Sphere* | Area of surface = diameter × diameter × 3.1416 |
| | Side of inscribed cube = radius × 1.547 |
| | Volume = diameter × diameter × diameter × 0.5236 |
| *Square* | Area = length × width |
| *Triangle* | Area = one-half of height times base |
| *Trapezoid* | Area = one-half of the sum of the parallel sides × the height |
| *Cone* | Area of surface = one-half of circumference of base × slant height + area of base |
| | Volume = diameter × diameter × 0.7854 × one-third of the altitude |
| *Cube* | Volume = width × height × length |
| *Ellipse* | Area = short diameter × long diameter × 0.7854 |
| *Cylinder* | Area of surface = diameter × 3.1416 × length + area of the two bases |
| | Area of base = diameter × diameter × 0.7854 |
| | Area of base = volume ÷ length |
| | Length = volume ÷ area of base |
| | Volume = length × area of base |
| | Capacity in gallons = volume in inches ÷ 231 |
| | Capacity of gallons = diameter × diameter × length × 0.0034 |
| | Capacity in gallons = volume in feet × 7.48 |
| *Circle* | Circumference = diameter × 3.1416 |
| | Circumference = radius × 6.2832 |
| | Diameter = radius × 2 |
| | Diameter = square root of = (area ÷ 0.7854) |
| | Diameter = square root of area × 1.1233 |

**Figure 33.8** Area and other formulas.

The formulas for pipe radiation of heat are as follows:

$$L = \frac{144}{OD \times 3.1416} \times R \div 12$$

D = outside diameter (OD) of pipe
L = length of pipe needed in feet
R = square feet of radiation needed

Figure 33.9  Formulas for pipe radiation of heat.


Temperature may be expressed according to the Fahrenheit (F) scale or the Celsius (C) scale. To convert °C to °F or °F to °C, use the following formulas:

°F = 1.8 × °C + 32
°C = 0.55555555 × °F − 32
°C = °F − 32 ÷ 1.8
°F = °C. × 1.8 + 32

Figure 33.10  Temperature conversion.


To figure the final temperature when two different temperatures of water are mixed together, use the following formula:

$$\frac{(A \times C) + (B \times D)}{A + B}$$

A = weight of lower temperature water
B = weight of higher temperature water
C = lower temperature
D = higher temperature

Figure 33.11  Computing water temperature.


### Radiation

3 ft of 1-in. pipe equal 1 ft² R.
2⅓ lineal ft of 1¼-in. pipe equal 1 ft² R.
Hot water radiation gives off 150 Btu/ft² R/hr.
Steam radiation gives off 240 Btu/ft² R/hr.
On greenhouse heating, figure ⅔ ft² R/ft² glass.
1 ft² of direct radiation condenses 0.25 lb water/hr.

Figure 33.12  Radiant heat facts.

| −100°–30° | | |
|---|---|---|
| °C | Base temperature | °F |
| −73 | −100 | −148 |
| −68 | −90 | −130 |
| −62 | −80 | −112 |
| −57 | −70 | −94 |
| −51 | −60 | −76 |
| −46 | −50 | −58 |
| −40 | −40 | −40 |
| −34.4 | −30 | −22 |
| −28.9 | −20 | −4 |
| −23.3 | −10 | 14 |
| −17.8 | 0 | 32 |
| −17.2 | 1 | 33.8 |
| −16.7 | 2 | 35.6 |
| −16.1 | 3 | 37.4 |
| −15.6 | 4 | 39.2 |
| −15.0 | 5 | 41.0 |
| −14.4 | 6 | 42.8 |
| −13.9 | 7 | 44.6 |
| −13.3 | 8 | 46.4 |
| −12.8 | 9 | 48.2 |
| −12.2 | 10 | 50.0 |
| −11.7 | 11 | 51.8 |
| −11.1 | 12 | 53.6 |
| −10.6 | 13 | 55.4 |
| −10.0 | 14 | 57.2 |

| 31°–71° | | |
|---|---|---|
| °C | Base temperature | °F |
| −0.6 | 31 | 87.8 |
| 0 | 32 | 89.6 |
| 0.6 | 33 | 91.4 |
| 1.1 | 34 | 93.2 |
| 1.7 | 35 | 95.0 |
| 2.2 | 36 | 96.8 |
| 2.8 | 37 | 98.6 |
| 3.3 | 38 | 100.4 |
| 3.9 | 39 | 102.2 |
| 4.4 | 40 | 104.0 |
| 5.0 | 41 | 105.8 |
| 5.6 | 42 | 107.6 |

**Figure 33.13** Temperature conversion.

| Vacuum in inches of mercury | Boiling point |
|---|---|
| 29 | 76.62 |
| 28 | 99.93 |
| 27 | 114.22 |
| 26 | 124.77 |
| 25 | 133.22 |
| 24 | 140.31 |
| 23 | 146.45 |
| 22 | 151.87 |
| 21 | 156.75 |
| 20 | 161.19 |
| 19 | 165.24 |
| 18 | 169.00 |
| 17 | 172.51 |
| 16 | 175.80 |
| 15 | 178.91 |
| 14 | 181.82 |
| 13 | 184.61 |
| 12 | 187.21 |
| 11 | 189.75 |
| 10 | 192.19 |
| 9 | 194.50 |
| 8 | 196.73 |
| 7 | 198.87 |
| 6 | 200.96 |
| 5 | 202.25 |
| 4 | 204.85 |
| 3 | 206.70 |
| 2 | 208.50 |
| 1 | 210.25 |

**Figure 33.14** Boiling points of water based on pressure.

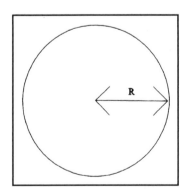

**Figure 33.15** Radius of a circle.

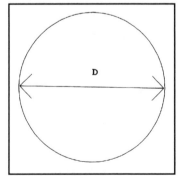

**Figure 33.16** Diameter of a circle.

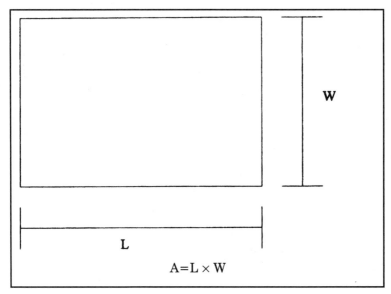

**Figure 33.17** Area of a rectangle.

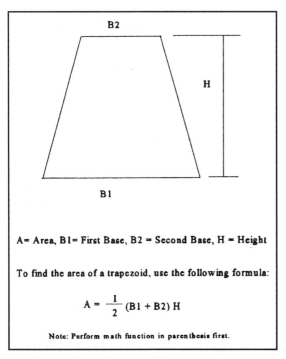

A = Area, B1 = First Base, B2 = Second Base, H = Height

To find the area of a trapezoid, use the following formula:

$$A = \frac{1}{2}(B1 + B2)H$$

Note: Perform math function in parenthesis first.

**Figure 33.18** Area of a trapezoid.

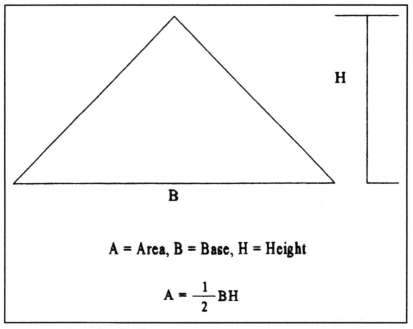

Figure 33.19  Area of a triangle.

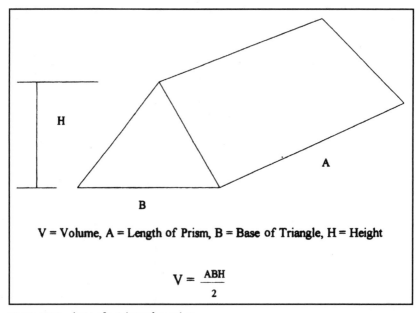

Figure 33.20  Area of a triangular prism.

Mathematics for the Trade 357

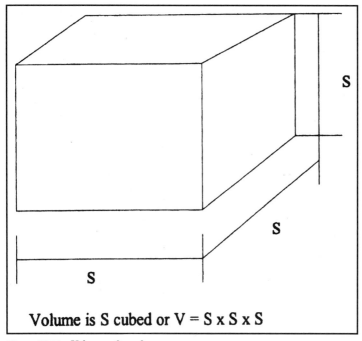

Figure 33.21  Volume of a cube.

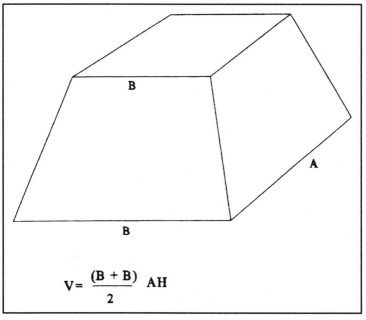

Figure 33.22  Volume of a trapezoidal prism.

| Set | Travel | Set | Travel | Set | Travel |
|---|---|---|---|---|---|
| 2 | 2.828 | ¼ | 15.907 | ½ | 28.987 |
| ¼ | 3.181 | ½ | 16.261 | ¾ | 29.340 |
| ½ | 3.531 | ¾ | 16.614 | 21 | 29.694 |
| ¾ | 3.888 | 12 | 16.968 | ¼ | 30.047 |
| 3 | 4.242 | ¼ | 17.321 | ½ | 30.401 |
| ¼ | 4.575 | ½ | 17.675 | ¾ | 30.754 |
| ½ | 4.949 | ¾ | 18.028 | 22 | 31.108 |
| ¾ | 5.302 | 13 | 18.382 | ¼ | 31.461 |
| 4 | 5.656 | ¼ | 18.735 | ½ | 31.815 |
| ¼ | 6.009 | ½ | 19.089 | ¾ | 32.168 |
| ½ | 6.363 | ¾ | 19.442 | 23 | 32.522 |
| ¾ | 6.716 | 14 | 19.796 | ¼ | 32.875 |
| 5 | 7.070 | ¼ | 20.149 | ½ | 33.229 |
| ¼ | 7.423 | ½ | 20.503 | ¾ | 33.582 |
| ½ | 7.777 | ¾ | 20.856 | 24 | 33.936 |
| ¾ | 8.130 | 15 | 21.210 | ¼ | 34.289 |
| 6 | 8.484 | ¼ | 21.563 | ½ | 34.643 |
| ¼ | 8.837 | ½ | 21.917 | ¾ | 34.996 |
| ½ | 9.191 | ¾ | 22.270 | 25 | 35.350 |
| ¾ | 9.544 | 16 | 22.624 | ¼ | 35.703 |
| 7 | 9.898 | ¼ | 22.977 | ½ | 36.057 |
| ¼ | 10.251 | ½ | 23.331 | ¾ | 36.410 |
| ½ | 10.605 | ¾ | 23.684 | 26 | 36.764 |
| ¾ | 10.958 | 17 | 24.038 | ¼ | 37.117 |
| 8 | 11.312 | ¼ | 24.391 | ½ | 37.471 |
| ¼ | 11.665 | ½ | 24.745 | ¾ | 37.824 |
| ½ | 12.019 | ¾ | 25.098 | 27 | 38.178 |
| ¾ | 12.372 | 18 | 25.452 | ¼ | 38.531 |
| 9 | 12.726 | ¼ | 25.805 | ½ | 38.885 |
| ¼ | 13.079 | ½ | 26.159 | ¾ | 39.238 |
| ½ | 13.433 | ¾ | 26.512 | 28 | 39.592 |
| ¾ | 13.786 | 19 | 26.866 | ¼ | 39.945 |
| 10 | 14.140 | ¼ | 27.219 | ½ | 40.299 |
| ¼ | 14.493 | ½ | 27.573 | ¾ | 40.652 |
| ½ | 14.847 | ¾ | 27.926 | 29 | 41.006 |
| ¾ | 15.200 | 20 | 28.280 | ¼ | 41.359 |
| 11 | 15.554 | ¼ | 28.635 | ½ | 41.713 |

**Figure 33.23** Set and travel relationships in inches for 45° offsets.

| Inches | Decimal of an inch | Inches | Decimal of an inch |
|---|---|---|---|
| 1/64 | .015625 | 33/64 | .515625 |
| 1/32 | .03125 | 17/32 | .53125 |
| 3/64 | .046875 | 35/64 | .546875 |
| 1/16 | .0625 | 9/16 | .5625 |
| 5/64 | .078125 | 37/64 | .578125 |
| 3/32 | .09375 | 19/32 | .59375 |
| 7/64 | .109375 | 39/64 | .609375 |
| 1/8 | .125 | 5/8 | .625 |
| 9/64 | .140625 | 41/64 | .640625 |
| 5/32 | .15625 | 21/32 | .65625 |
| 11/64 | .171875 | 43/64 | .671875 |
| 3/16 | .1875 | 11/16 | .6875 |
| 13/64 | .203125 | 45/64 | .703125 |
| 7/32 | .21875 | 23/32 | .71875 |
| 15/64 | .234375 | 47/64 | .734375 |
| 1/4 | .25 | 3/4 | .75 |
| 17/64 | .265625 | 49/64 | .765625 |
| 9/32 | .28125 | 25/32 | .78125 |
| 19/64 | .296875 | 51/64 | .796875 |
| 5/16 | .3125 | 13/16 | .8125 |
| 21/64 | .328125 | 53/64 | .828125 |
| 11/32 | .34375 | 27/32 | .84375 |
| 23/64 | .359375 | 55/64 | .859375 |
| 3/8 | .375 | 7/8 | .875 |
| 25/64 | .390625 | 57/64 | .890625 |
| 13/32 | .40625 | 22/32 | .90625 |
| 27/64 | .421875 | 59/64 | .921875 |
| 7/16 | .4375 | 15/16 | .9375 |
| 29/64 | .453125 | 61/64 | .953125 |
| 15/32 | .46875 | 31/32 | .96875 |
| 31/64 | .484375 | 63/64 | .984375 |
| 1/2 | .5 | 1 | 1 |

**Figure 33.24** Simple offsets.

Figure 33.25  Calculated 45° offsets.

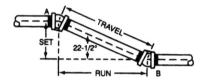

Figure 33.26  Decimal equivalents of fractions of an inch.

| To find side* | When known side is | Multiply Side | For 60° ells by | For 45° ells by | For 30° ells by | For 22½° ells by | For 11¼° ells by | For 5⅝° ells by |
|---|---|---|---|---|---|---|---|---|
| T | S | S | 1.155 | 1.414 | 2.000 | 2.613 | 5.125 | 10.187 |
| S | T | T | .866 | .707 | .500 | .383 | .195 | .098 |
| R | S | S | .577 | 1.000 | 1.732 | 2.414 | 5.027 | 10.158 |
| S | R | R | 1.732 | 1.000 | .577 | .414 | .198 | .098 |
| T | R | R | 2.000 | 1.414 | 1.155 | 1.082 | 1.019 | 1.004 |
| R | T | T | .500 | .707 | .866 | .924 | .980 | .995 |

*S = set, R = run, T = travel.

Figure 33.27  Multipliers for calculating simple offsets.

| Metric | U.S. |
|---|---|
| 144 in.² | 1 ft² |
| 9 ft² | 1 yd² |
| 1 yd² | 1296 in.² |
| 4840 yd² | 1 a |
| 640 a | 1 mi² |

Figure 33.28  Square measure.

| | |
|---|---|
| 1 cm² | 0.1550 in.² |
| 1 dm² | 0.1076 ft² |
| 1 ms2 | 1.196 yd² |
| 1 A (are) | 3.954 rd² |
| 1 ha | 2.47 a (acres) |
| 1 km² | 0.386 mi² |
| 1 in.² | 6.452 cm² |
| 1 ft² | 9.2903 dm² |
| 1 yd² | 0.8361 m² |
| 1 rd² | 0.2529 A (are) |
| 1 a (acre) | 0.4047 ha |
| 1 mi² | 2.59 km² |

Figure 33.29  Square measures of length and area.

| | |
|---|---|
| 100 mm² | 1 cm² |
| 100 cm² | 1 dm² |
| 100 dm² | 1 m² |

**Figure 33.30** Metric square measure.

| | | |
|---|---|---|
| 144 in.² | 1 ft² | 272.25 ft² |
| 9 ft² | 1 yd² | |
| 30¼ yd² | 1 rd² | 272.25 ft² |
| 160 rd² | 1 a | 4840 yd² |
| | | 43,560 ft²) |
| 640 a | 1 m² | 3,097,600 yd² |
| 36 mi² | 1 township | |

**Figure 33.31a** Square measures.

| Square feet | Square meters |
|---|---|
| 1 | 0.925 |
| 2 | 0.1850 |
| 3 | 0.2775 |
| 4 | 0.3700 |
| 5 | 0.4650 |
| 6 | 0.5550 |
| 7 | 0.6475 |
| 8 | 0.7400 |
| 9 | 0.8325 |
| 10 | 0.9250 |
| 25 | 2.315 |
| 50 | 4.65 |
| 100 | 9.25 |

**Figure 33.31b** Square feet to square meters.

| Fraction | Square root |
|---|---|
| 1/8 | .3535 |
| 1/4 | .5000 |
| 3/8 | .6124 |
| 1/2 | .7071 |
| 5/8 | .7906 |
| 3/4 | .8660 |
| 7/8 | .9354 |

**Figure 33.32** Square roots of fractions.

| Fraction | Cube root |
|---|---|
| 1/8 | .5000 |
| 1/4 | .6300 |
| 3/8 | .7211 |
| 1/2 | .7937 |
| 5/8 | .8550 |
| 3/4 | .9086 |
| 7/8 | .9565 |

**Figure 33.33** Cube roots of fractions.

| Number | Cube | Number | Cube | Number | Cube |
|---|---|---|---|---|---|
| 1 | 1 | 36 | 46,656 | 71 | 357,911 |
| 2 | 8 | 37 | 50.653 | 72 | 373,248 |
| 3 | 27 | 38 | 54,872 | 73 | 389,017 |
| 4 | 64 | 39 | 59,319 | 74 | 405,224 |
| 5 | 125 | 40 | 64,000 | 75 | 421,875 |
| 6 | 216 | 41 | 68,921 | 76 | 438,976 |
| 7 | 343 | 42 | 74,088 | 77 | 456,533 |
| 8 | 512 | 43 | 79,507 | 78 | 474,552 |
| 9 | 729 | 44 | 85,184 | 79 | 493,039 |
| 10 | 1,000 | 45 | 91,125 | 80 | 512,000 |
| 11 | 1,331 | 46 | 97,336 | 81 | 531,441 |
| 12 | 1,728 | 47 | 103,823 | 82 | 551,368 |
| 13 | 2,197 | 48 | 110,592 | 83 | 571,787 |
| 14 | 2,477 | 49 | 117,649 | 84 | 592,704 |
| 15 | 3,375 | 50 | 125,000 | 85 | 614,125 |
| 16 | 4,096 | 51 | 132,651 | 86 | 636,056 |
| 17 | 4,913 | 52 | 140,608 | 87 | 658,503 |
| 18 | 5,832 | 53 | 148,877 | 88 | 681,472 |
| 19 | 6,859 | 54 | 157,464 | 89 | 704,969 |
| 20 | 8,000 | 55 | 166,375 | 90 | 729,000 |
| 21 | 9,621 | 56 | 175,616 | 91 | 753,571 |
| 22 | 10,648 | 57 | 185,193 | 92 | 778,688 |
| 23 | 12,167 | 58 | 195,112 | 93 | 804,357 |
| 24 | 13,824 | 59 | 205,379 | 94 | 830,584 |
| 25 | 15,625 | 60 | 216,000 | 95 | 857,375 |
| 26 | 17,576 | 61 | 226,981 | 96 | 884,736 |
| 27 | 19,683 | 62 | 238,328 | 97 | 912,673 |
| 28 | 21,952 | 63 | 250,047 | 98 | 941,192 |
| 29 | 24,389 | 64 | 262,144 | 99 | 970,299 |
| 30 | 27,000 | 65 | 274,625 | 100 | 1,000,000 |
| 31 | 29,791 | 66 | 287,496 | | |
| 32 | 32,768 | 67 | 300,763 | | |
| 33 | 35,937 | 68 | 314,432 | | |
| 34 | 39,304 | 69 | 328,500 | | |
| 35 | 42,875 | 70 | 343,000 | | |

**Figure 33.34** Cubes of numbers.

Mathematics for the Trade 363

| Number | Square root | Number | Square root | Number |
|---|---|---|---|---|
| 1 | 1.00000 | 36 | 6.00000 | 71 |
| 2 | 1.41421 | 37 | 6.08276 | 72 |
| 3 | 1.73205 | 38 | 6.16441 | 73 |
| 4 | 2.00000 | 39 | 6.24499 | 74 |
| 5 | 2.23606 | 40 | 6.32455 | 75 |
| 6 | 2.44948 | 41 | 6.40312 | 76 |
| 7 | 2.64575 | 42 | 6.48074 | 77 |
| 8 | 2.82842 | 43 | 6.55743 | 78 |
| 9 | 3.00000 | 44 | 6.63324 | 79 |
| 10 | 3.16227 | 45 | 6.70820 | 80 |
| 11 | 3.31662 | 46 | 6.78233 | 81 |
| 12 | 3.46410 | 47 | 6.85565 | 82 |
| 13 | 3.60555 | 48 | 6.92820 | 83 |
| 14 | 3.74165 | 49 | 7.00000 | 84 |
| 15 | 3.87298 | 50 | 7.07106 | 85 |
| 16 | 4.00000 | 51 | 7.14142 | 86 |
| 17 | 4.12310 | 52 | 7.21110 | 87 |
| 18 | 4.24264 | 53 | 7.28010 | 88 |
| 19 | 4.35889 | 54 | 7.34846 | 89 |
| 20 | 4.47213 | 55 | 7.41619 | 90 |
| 21 | 4.58257 | 56 | 7.48331 | 91 |
| 22 | 4.69041 | 57 | 7.54983 | 92 |
| 23 | 4.79583 | 58 | 7.61577 | 93 |
| 24 | 4.89897 | 59 | 7.68114 | 94 |
| 25 | 5.00000 | 60 | 7.74596 | 95 |
| 26 | 5.09901 | 61 | 7.81024 | 96 |
| 27 | 5.19615 | 62 | 7.87400 | 97 |
| 28 | 5.29150 | 63 | 7.93725 | 98 |
| 29 | 5.38516 | 64 | 8.00000 | 99 |
| 30 | 5.47722 | 65 | 8.06225 | 100 |
| 31 | 5.56776 | 66 | 8.12403 | |
| 32 | 5.65685 | 67 | 8.18535 | |
| 33 | 5.74456 | 68 | 8.24621 | |
| 34 | 5.83095 | 69 | 8.30662 | |
| 35 | 5.91607 | 70 | 8.36660 | |

**Figure 33.35** Square roots of numbers.

| Number | Square | Number | Square | Number | Square |
|---|---|---|---|---|---|
| 1 | 1 | 36 | 1296 | 71 | 5041 |
| 2 | 4 | 37 | 1369 | 72 | 5184 |
| 3 | 9 | 38 | 1444 | 73 | 5329 |
| 4 | 16 | 39 | 1521 | 74 | 5476 |
| 5 | 25 | 40 | 1600 | 75 | 5625 |
| 6 | 36 | 41 | 1681 | 76 | 5776 |
| 7 | 49 | 42 | 1764 | 77 | 5929 |
| 8 | 64 | 43 | 1849 | 78 | 6084 |
| 9 | 81 | 44 | 1936 | 79 | 6241 |
| 10 | 100 | 45 | 2025 | 80 | 6400 |
| 11 | 121 | 46 | 2116 | 81 | 6561 |
| 12 | 144 | 47 | 2209 | 82 | 6724 |
| 13 | 169 | 48 | 2304 | 83 | 6889 |
| 14 | 196 | 49 | 2401 | 84 | 7056 |
| 15 | 225 | 50 | 2500 | 85 | 7225 |
| 16 | 256 | 51 | 2601 | 86 | 7396 |
| 17 | 289 | 52 | 2704 | 87 | 7569 |
| 18 | 324 | 53 | 2809 | 88 | 7744 |
| 19 | 361 | 54 | 2916 | 89 | 7921 |
| 20 | 400 | 55 | 3025 | 90 | 8100 |
| 21 | 441 | 56 | 3136 | 91 | 8281 |
| 22 | 484 | 57 | 3249 | 92 | 8464 |
| 23 | 529 | 58 | 3364 | 93 | 8649 |
| 24 | 576 | 59 | 3481 | 94 | 8836 |
| 25 | 625 | 60 | 3600 | 95 | 9025 |
| 26 | 676 | 61 | 3721 | 96 | 8216 |
| 27 | 729 | 62 | 3844 | 97 | 9409 |
| 28 | 784 | 63 | 3969 | 98 | 9604 |
| 29 | 841 | 64 | 4096 | 99 | 9801 |
| 30 | 900 | 65 | 4225 | 100 | 10000 |
| 31 | 961 | 66 | 4356 | | |
| 32 | 1024 | 67 | 4489 | | |
| 33 | 1089 | 68 | 4624 | | |
| 34 | 1156 | 69 | 4761 | | |
| 35 | 1225 | 70 | 4900 | | |

Figure 33.36  Squares of numbers.

| Diameter | Circumference | Diameter | Circumference |
| --- | --- | --- | --- |
| 1/8 | 0.3927 | 10 | 31.41 |
| 1/4 | 0.7854 | 10½ | 32.98 |
| 3/8 | 1.178 | 11 | 34.55 |
| 1/2 | 1.570 | 11½ | 36.12 |
| 5/8 | 1.963 | 12 | 37.69 |
| 3/4 | 2.356 | 12½ | 39.27 |
| 7/8 | 2.748 | 13 | 40.84 |
| 1 | 3.141 | 13½ | 42.41 |
| 1 1/8 | 3.534 | 14 | 43.98 |
| 1 1/4 | 3.927 | 14½ | 45.55 |
| 1 3/8 | 4.319 | 15 | 47.12 |
| 1 1/2 | 4.712 | 15½ | 48.69 |
| 1 5/8 | 5.105 | 16 | 50.26 |
| 1 3/4 | 5.497 | 16½ | 51.83 |
| 1 7/8 | 5.890 | 17 | 53.40 |
| 2 | 6.283 | 17½ | 54.97 |
| 2 1/4 | 7.068 | 18 | 56.54 |
| 2 1/2 | 7.854 | 18½ | 58.11 |
| 2 3/4 | 8.639 | 19 | 56.69 |
| 3 | 9.424 | 19½ | 61.26 |
| 3 1/4 | 10.21 | 20 | 62.83 |
| 3 1/2 | 10.99 | 20½ | 64.40 |
| 3 3/4 | 11.78 | 21 | 65.97 |
| 4 | 12.56 | 21½ | 67.54 |
| 4 1/2 | 14.13 | 22 | 69.11 |
| 5 | 15.70 | 22½ | 70.68 |
| 5½ | 17.27 | 23 | 72.25 |
| 6 | 18.84 | 23½ | 73.82 |
| 6½ | 20.42 | 24 | 75.39 |
| 7 | 21.99 | 24½ | 76.96 |
| 7½ | 23.56 | 25 | 78.54 |
| 8 | 25.13 | 26 | 81.68 |
| 8½ | 26.70 | 27 | 84.82 |
| 9 | 28.27 | 28 | 87.96 |
| 9½ | 29.84 | 29 | 91.10 |
|   |   | 30 | 94.24 |

**Figure 33.37** Circumference of a circle.

| Diameter | Area | Diameter | Area |
|---|---|---|---|
| 1/8 | 0.0123 | 10 | 78.54 |
| 1/4 | 0.0491 | 10½ | 86.59 |
| 3/8 | 0.1104 | 11 | 95.03 |
| 1/2 | 0.1963 | 11½ | 103.86 |
| 5/8 | 0.3068 | 12 | 113.09 |
| 3/4 | 0.4418 | 12½ | 122.71 |
| 7/8 | 0.6013 | 13 | 132.73 |
| 1 | 0.7854 | 13½ | 143.13 |
| 1⅛ | 0.9940 | 14 | 153.93 |
| 1¼ | 1.227 | 14½ | 165.13 |
| 1⅜ | 1.484 | 15 | 176.71 |
| 1½ | 1.767 | 15½ | 188.69 |
| 1⅝ | 2.073 | 16 | 201.06 |
| 1¾ | 2.405 | 16½ | 213.82 |
| 1⅞ | 2.761 | 17 | 226.98 |
| 2 | 3.141 | 17½ | 240.52 |
| 2¼ | 3.976 | 18 | 254.46 |
| 2½ | 4.908 | 18½ | 268.80 |
| 2¾ | 5.939 | 19 | 283.52 |
| 3 | 7.068 | 19½ | 298.60 |
| 3¼ | 8.295 | 20 | 314.16 |
| 3½ | 9.621 | 20½ | 330.06 |
| 3¾ | 11.044 | 21 | 346.36 |
| 4 | 12.566 | 21½ | 363.05 |
| 4½ | 15.904 | 22 | 380.13 |
| 5 | 19.635 | 22½ | 397.60 |
| 5½ | 23.758 | 23 | 415.47 |
| 6 | 28.274 | 23½ | 433.73 |
| 6½ | 33.183 | 24 | 452.39 |
| 7 | 38.484 | 24½ | 471.43 |
| 7½ | 44.178 | 25 | 490.87 |
| 8 | 50.265 | 26 | 530.93 |
| 8½ | 56.745 | 27 | 572.55 |
| 9 | 63.617 | 28 | 615.75 |
| 9½ | 70.882 | 29 | 660.52 |
|   |   | 30 | 706.86 |

Figure 33.38  Area of a circle.

| COLUMN | 1 | | 2 | 3 | 4 | 5 | 6 | 7 | 8 | 9 | 10 |
|---|---|---|---|---|---|---|---|---|---|---|---|
| Line | | Description | Lb per sq in thru each section (psi) | Gal per min through section | Length of section (feet) | Trial pipe size (inches) | Equivalent length of fittings and valves (feet) | Total equivalent length Col. 4 and Col. 6 (100 feet) | Friction loss per 100 feet of total (psi) | Friction loss in equivalent length Col. 7 × Col. 8 (psi) | Excess pressure over friction losses (psi) |
| a | Service and cold water distribution piping[a] | Minimum pressure available at main | 55.00 | | | | | | | | |
| b | | Highest pressure required at a fixture (Section 604.3) | 15.00 | | | | | | | | |
| c | | Meter loss 2" meter | 11.00 | | | | | | | | |
| d | | Tap in main loss 2" tap (Table E103A) | 1.61 | | | | | | | | |
| e | | Static head loss 21 × 0.43 psi | 9.03 | | | | | | | | |
| f | | Special fixture loss backflow preventer | 9.00 | | | | | | | | |
| g | | Special fixture loss—Filter | 0.00 | | | | | | | | |
| h | | Special fixture loss—Other | 0.00 | | | | | | | | |
| i | | Total overall losses and requirements (sum of Lines b through h) | 45.64 | | | | | | | | |
| j | | Pressure available to overcome pipe friction (Line a minus Lines b to h) | 9.36 | | | | | | | | |
| | Designation | | FU | | | | | | | | |
| | Pipe section (from diagram) | AB | 294 | 108.0 | 54 | 2½ | 12 | 0.66 | 3.3 | 2.18 | |
| | Cold water distribution piping | BC | 264 | 108.0 | 8 | 2½ | 2.5 | 0.105 | 3.2 | 0.34 | |
| | | CD | 132 | 77.0 | 13 | 2½ | 8 | 0.21 | 1.9 | 0.40 | |
| | | CF | 132 | 77.0 | 150 | 2½ | 12 | 1.62 | 1.9 | 3.08 | |
| | | DE | 132 | 77.0 | 150 | 2½ | 14.5 | 1.645 | 1.9 | 3.12 | |
| k | Total pipe friction losses (cold) | | | | | | 9.36 | 6.24 | 6.24 | | 3.12 |
| 1 | Difference (Line j minus Line k) | | | | | | | | | | |
| | Pipe section (from diagram) | A'B' | 294 | 108.0 | 54 | 2½ | 9.6 | 0.64 | 3.3 | 2.1 | |
| | Hot water distribution piping | B'C' | 24 | 38.0 | 8 | 2 | 9.0 | 0.17 | 1.4 | 0.24 | |
| | | C'D' | 12 | 28.6 | 13 | 1½ | 5 | 0.18 | 3.2 | 0.58 | |
| | | C'F'b | 12 | 28.6 | 150 | 1½ | 14 | 1.64 | 3.2 | 5.25 | |
| | | D'E'b | 12 | 28.6 | 150 | 1½ | 7 | 1.57 | 3.2 | 5.02 | |
| k | Total pipe friction losses (hot) | | | | | | 9.36 | 7.94 | 7.94 | | 1.42 |
| 1 | Difference (Line j minus Line k) | | | | | | | | | | |

For SI: 1 inch = 25.4 mm, 1 foot = 304.8 mm, 1 psi = 6.895 kPa, 1 gpm = 3.785 L/m.

a. To be considered as pressure gain for fixtures below main (to consider separately, omit from "i" and add to "j").
b. To consider separately, in k use C–F only if greater loss than above.

**Figure 33.39** This table can help resolve any problems that might arise in relation to pipe sizing. *(Courtesy of International Code Council, Inc. and International Plumbing Code 2000).*

## LOAD VALUES ASSIGNED TO FIXTURES[a]

| FIXTURE | OCCUPANCY | TYPE OF SUPPLY CONTROL | LOAD VALUES, IN WATER SUPPLY FIXTURE UNITS (wsfu) | | |
|---|---|---|---|---|---|
| | | | Cold | Hot | Total |
| Bathroom group | Private | Flush tank | 2.7 | 1.5 | 3.6 |
| Bathroom group | Private | Flush valve | 6.0 | 3.0 | 8.0 |
| Bathtub | Private | Faucet | 1.0 | 1.0 | 1.4 |
| Bathtub | Public | Faucet | 3.0 | 3.0 | 4.0 |
| Bidet | Private | Faucet | 1.5 | 1.5 | 2.0 |
| Combination fixture | Private | Faucet | 2.25 | 2.25 | 3.0 |
| Dishwashing machine | Private | Automatic | | 1.4 | 1.4 |
| Drinking fountain | Offices, etc. | 3/8" valve | 0.25 | | 0.25 |
| Kitchen sink | Private | Faucet | 1.0 | 1.0 | 1.4 |
| Kitchen sink | Hotel, restaurant | Faucet | 3.0 | 3.0 | 4.0 |
| Laundry trays (1 to 3) | Private | Faucet | 1.0 | 1.0 | 1.4 |
| Lavatory | Private | Faucet | 0.5 | 0.5 | 0.7 |
| Lavatory | Public | Faucet | 1.5 | 1.5 | 2.0 |
| Service sink | Offices, etc. | Faucet | 2.25 | 2.25 | 3.0 |
| Shower head | Public | Mixing valve | 3.0 | 3.0 | 4.0 |
| Shower head | Private | Mixing valve | 1.0 | 1.0 | 1.4 |
| Urinal | Public | 1" flush valve | 10.0 | | 10.0 |
| Urinal | Public | 3/4" flush valve | 5.0 | | 5.0 |
| Urinal | Public | Flush tank | 3.0 | | 3.0 |
| Washing machine (8 lbs.) | Private | Automatic | 1.0 | 1.0 | 1.4 |
| Washing machine (8 lbs.) | Public | Automatic | 2.25 | 2.25 | 3.0 |
| Washing machine (15 lbs.) | Public | Automatic | 3.0 | 3.0 | 4.0 |
| Water closet | Private | Flush valve | 6.0 | | 6.0 |
| Water closet | Private | Flush tank | 2.2 | | 2.2 |
| Water closet | Public | Flush valve | 10.0 | | 10.0 |
| Water closet | Public | Flush tank | 5.0 | | 5.0 |
| Water closet | Public or private | Flushometer tank | 2.0 | | 2.0 |

For SI: 1 inch = 25.4 mm, 1 pound = 0.454 kg.

[a] For fixtures not listed, loads should be assumed by comparing the fixture to one listed using water in similar quantities and at similar rates. The assigned loads for fixtures with both hot and cold water supplies are given for separate hot and cold water loads and for total load, the separate hot and cold water loads being three-fourths of the total load for the fixture in each case.

Figure 33.40 Every fixture involved in plumbing has a load value. They are determined here. *(Courtesy of International Code Council, Inc. and International Plumbing Code 2000).*

| SUPPLY SYSTEMS PREDOMINANTLY FOR FLUSH TANKS | | | SUPPLY SYSTEMS PREDOMINANTLY FOR FLUSH VALVES | | |
|---|---|---|---|---|---|
| Load | Demand | | Load | Demand | |
| (Water supply fixture units) | (Gallons per minute) | (Cubic feet per minute) | (Water supply fixture units) | (Gallons per minute) | (Cubic feet per minute) |
| 1 | 3.0 | 0.04104 | | | |
| 2 | 5.0 | 0.0684 | | | |
| 3 | 6.5 | 0.86892 | | | |
| 4 | 8.0 | 1.06944 | | | |
| 5 | 9.4 | 1.256592 | 5 | 15.0 | 2.0052 |
| 6 | 10.7 | 1.430376 | 6 | 17.4 | 2.326032 |
| 7 | 11.8 | 1.577424 | 7 | 19.8 | 2.646364 |
| 8 | 12.8 | 1.711104 | 8 | 22.2 | 2.967696 |
| 9 | 13.7 | 1.831416 | 9 | 24.6 | 3.288528 |
| 10 | 14.6 | 1.951728 | 10 | 27.0 | 3.60936 |
| 11 | 15.4 | 2.058672 | 11 | 27.8 | 3.716304 |
| 12 | 16.0 | 2.13888 | 12 | 28.6 | 3.823248 |
| 13 | 16.5 | 2.20572 | 13 | 29.4 | 3.930192 |
| 14 | 17.0 | 2.27256 | 14 | 30.2 | 4.037136 |
| 15 | 17.5 | 2.3394 | 15 | 31.0 | 4.14408 |
| 16 | 18.0 | 2.90624 | 16 | 31.8 | 4.241024 |
| 17 | 18.4 | 2.459712 | 17 | 32.6 | 4.357968 |
| 18 | 18.8 | 2.513184 | 18 | 33.4 | 4.464912 |
| 19 | 19.2 | 2.566656 | 19 | 34.2 | 4.571856 |
| 20 | 19.6 | 2.620128 | 20 | 35.0 | 4.6788 |
| 25 | 21.5 | 2.87412 | 25 | 38.0 | 5.07984 |
| 30 | 23.3 | 3.114744 | 30 | 42.0 | 5.61356 |
| 35 | 24.9 | 3.328632 | 35 | 44.0 | 5.88192 |
| 40 | 26.3 | 3.515784 | 40 | 46.0 | 6.14928 |
| 45 | 27.7 | 3.702936 | 45 | 48.0 | 6.41664 |
| 50 | 29.1 | 3.890088 | 50 | 50.0 | 6.684 |

**Figure 33.41** This table will let a user estimate demand. *(Courtesy of International Code Council, Inc. and International Plumbing Code 2000).*

| SUPPLY SYSTEMS PREDOMINANTLY FOR FLUSH TANKS | | | SUPPLY SYSTEMS PREDOMINANTLY FOR FLUSH VALVES | | |
|---|---|---|---|---|---|
| Load | Demand | | Load | Demand | |
| (Water supply fixture units) | (Gallons per minute) | (Cubic feet per minute) | (Water supply fixture units) | (Gallons per minute) | (Cubic feet per minute) |
| 60 | 32.0 | 4.27776 | 60 | 54.0 | 7.21872 |
| 70 | 35.0 | 4.6788 | 70 | 58.0 | 7.75344 |
| 80 | 38.0 | 5.07984 | 80 | 61.2 | 8.181216 |
| 90 | 41.0 | 5.48088 | 90 | 64.3 | 8.595624 |
| 100 | 43.5 | 5.81508 | 100 | 67.5 | 9.0234 |
| 120 | 48.0 | 6.41664 | 120 | 73.0 | 9.75864 |
| 140 | 52.5 | 7.0182 | 140 | 77.0 | 10.29336 |
| 160 | 57.0 | 7.61976 | 160 | 81.0 | 10.82808 |
| 180 | 61.0 | 8.15448 | 180 | 85.5 | 11.42964 |
| 200 | 65.0 | 8.6892 | 200 | 90.0 | 12.0312 |
| 225 | 70.0 | 9.3576 | 225 | 95.5 | 12.76644 |
| 250 | 75.0 | 10.0260 | 250 | 101.0 | 13.50168 |
| 275 | 80.0 | 10.6944 | 275 | 104.5 | 13.96956 |
| 300 | 85.0 | 11.3628 | 300 | 108.0 | 14.43744 |
| 400 | 105.0 | 14.0364 | 400 | 127.0 | 16.97736 |
| 500 | 124.0 | 16.57632 | 500 | 143.0 | 19.11624 |
| 750 | 170.0 | 22.7256 | 750 | 177.0 | 23.66136 |
| 1,000 | 208.0 | 27.80544 | 1,000 | 208.0 | 27.80544 |
| 1,250 | 239.0 | 31.94952 | 1,250 | 239.0 | 31.94952 |
| 1,500 | 269.0 | 35.95992 | 1,500 | 269.0 | 35.95992 |
| 1,750 | 297.0 | 39.70296 | 1,750 | 297.0 | 39.70296 |
| 2,000 | 325.0 | 43.446 | 2,000 | 325.0 | 43.446 |
| 2,500 | 380.0 | 50.7984 | 2,500 | 380.0 | 50.7984 |
| 3,000 | 433.0 | 57.88344 | 3,000 | 433.0 | 57.88344 |
| 4,000 | 535.0 | 70.182 | 4,000 | 525.0 | 70.182 |
| 5,000 | 593.0 | 79.27224 | 5,000 | 593.0 | 79.27224 |

For SI: 1 gpm = 3.785 L/m, 1 cfm = 0.4719 L/s.

**Figure 33.42** The table for estimating demand for flush tanks and valves continues. (*Courtesy of International Code Council, Inc. and International Plumbing Code 2000*).

## LOSS OF PRESSURE THROUGH TAPS AND TEES IN POUNDS PER SQUARE INCH (psi)

| GALLONS PER MINUTE | 5/8 | 3/4 | 1 | SIZE OF TAP OR TEE (inches) 1 1/4 | 1 1/2 | 2 | 3 |
|---|---|---|---|---|---|---|---|
| 10  | 1.35 | 0.64 | 0.18  | 0.08 |      |      |      |
| 20  | 5.38 | 2.54 | 0.77  | 0.31 | 0.14 |      |      |
| 30  | 12.1 | 5.72 | 1.62  | 0.69 | 0.33 | 0.10 |      |
| 40  |      | 10.2 | 3.07  | 1.23 | 0.58 | 0.18 |      |
| 50  |      | 15.9 | 4.49  | 1.92 | 0.91 | 0.28 |      |
| 60  |      |      | 6.46  | 2.76 | 1.31 | 0.40 |      |
| 70  |      |      | 8.79  | 3.76 | 1.78 | 0.55 | 0.10 |
| 80  |      |      | 11.5  | 4.90 | 2.32 | 0.72 | 0.13 |
| 90  |      |      | 14.5  | 6.21 | 2.94 | 0.91 | 0.16 |
| 100 |      |      | 17.94 | 7.67 | 3.63 | 1.12 | 0.21 |
| 120 |      |      | 25.8  | 11.0 | 5.23 | 1.61 | 0.30 |
| 140 |      |      | 35.2  | 15.0 | 7.12 | 2.20 | 0.41 |
| 150 |      |      |       | 17.2 | 8.16 | 2.52 | 0.47 |
| 160 |      |      |       | 19.6 | 9.30 | 2.92 | 0.54 |
| 180 |      |      |       | 24.8 | 11.8 | 3.62 | 0.68 |
| 200 |      |      |       | 30.7 | 14.5 | 4.48 | 0.84 |
| 225 |      |      |       | 38.8 | 18.4 | 5.6  | 1.06 |
| 250 |      |      |       | 47.9 | 22.7 | 7.00 | 1.31 |
| 275 |      |      |       |      | 27.4 | 7.70 | 1.59 |
| 300 |      |      |       |      | 32.6 | 10.1 | 1.88 |

For SI: 1 inch = 25.4 mm, 1 psi = 6.895 kPa, 1 gpm = 3.785 L/m.

**Figure 33.43** Pressure can be lost in taps and tees. This examines the numbers. *(Courtesy of International Code Council, Inc. and International Plumbing Code 2000).*

## ALLOWANCE IN EQUIVALENT LENGTH OF PIPE FOR FRICTION LOSS IN VALVES AND THREADED FITTINGS (feet)

| FITTING OR VALVE | PIPE SIZES (inches) | | | | | | | |
|---|---|---|---|---|---|---|---|---|
| | ½ | ¾ | 1 | 1¼ | 1½ | 2 | 2½ | 3 |
| 45-degree elbow | 1.2 | 1.5 | 1.8 | 2.4 | 3.0 | 4.0 | 5.0 | 6.0 |
| 90-degree elbow | 2.0 | 2.5 | 3.0 | 4.0 | 5.0 | 7.0 | 8.0 | 10.0 |
| Tee, run | 0.6 | 0.8 | 0.9 | 1.2 | 1.5 | 2.0 | 2.5 | 3.0 |
| Tee, branch | 3.0 | 4.0 | 5.0 | 6.0 | 7.0 | 10.0 | 12.0 | 15.0 |
| Gate valve | 0.4 | 0.5 | 0.6 | 0.8 | 1.0 | 1.3 | 1.6 | 2.0 |
| Balancing valve | 0.8 | 1.1 | 1.5 | 1.9 | 2.2 | 3.0 | 3.7 | 4.5 |
| Plug-type cock | 0.8 | 1.1 | 1.5 | 1.9 | 2.2 | 3.0 | 3.7 | 4.5 |
| Check valve, swing | 5.6 | 8.4 | 11.2 | 14.0 | 16.8 | 22.4 | 28.0 | 33.6 |
| Globe valve | 15.0 | 20.0 | 25.0 | 35.0 | 45.0 | 55.0 | 65.0 | 80.0 |
| Angle valve | 8.0 | 12.0 | 15.0 | 18.0 | 22.0 | 28.0 | 34.0 | 40.0 |

For SI: 1 inch = 25.4 mm, 1 foot = 304.8 mm, 1 degree = 0.0175 rad.

**Figure 33.44** This chart examines the allowances involved in friction loss in valves and threaded fittings. *(Courtesy of International Code Council, Inc. and International Plumbing Code 2000).*

## PRESSURE LOSS IN FITTINGS AND VALVES EXPRESSED AS EQUIVALENT LENGTH OF TUBE[a] (feet)

| NOMINAL OR STANDARD SIZE (Inches) | FITTINGS ||||| VALVES |||
|---|---|---|---|---|---|---|---|---|
| | Standard Ell || 90-Degree Tee || Coupling | Ball | Gate | Butterfly | Check |
| | 90 Degree | 45 Degree | Side Branch | Straight Run | | | | | |
| 3/8 | 0.5 | — | 1.5 | — | — | — | — | — | 1.5 |
| 1/2 | 1 | 0.5 | 2 | — | — | — | — | — | 2 |
| 5/8 | 1.5 | 0.5 | 2 | — | — | — | — | — | 2.5 |
| 3/4 | 2 | 0.5 | 3 | — | — | — | — | — | 3 |
| 1 | 2.5 | 1 | 4.5 | — | — | 0.5 | — | — | 4.5 |
| 1 1/4 | 3 | 1 | 5.5 | 0.5 | 0.5 | 0.5 | — | — | 5.5 |
| 1 1/2 | 4 | 1.5 | 7 | 0.5 | 0.5 | 0.5 | — | — | 6.5 |
| 2 | 5.5 | 2 | 9 | 0.5 | 0.5 | 0.5 | 0.5 | 7.5 | 9 |
| 2 1/2 | 7 | 2.5 | 12 | 0.5 | 0.5 | — | 1 | 10 | 11.5 |
| 3 | 9 | 3.5 | 15 | 1 | 1 | — | 1.5 | 15.5 | 14.5 |
| 3 1/2 | 9 | 3.5 | 14 | 1 | 1 | — | 2 | — | 12.5 |
| 4 | 12.5 | 5 | 21 | 1 | 1 | — | 2 | 16 | 18.5 |
| 5 | 16 | 6 | 27 | 1.5 | 1.5 | — | 3 | 11.5 | 23.5 |
| 6 | 19 | 7 | 34 | 2 | 2 | — | 3.5 | 13.5 | 26.5 |
| 8 | 29 | 11 | 50 | 3 | 3 | — | 5 | 12.5 | 39 |

For SI: 1 inch = 25.4 mm, 1 foot = 304.8 mm, 1 degree = 0.0175 rad.

a. Allowances are for streamlined soldered fittings and recessed threaded fittings. For threaded fittings, double the allowances shown in the table. The equivalent lengths presented above are based on a C factor of 150 in the Hazen-Williams friction loss formula. The lengths shown are rounded to the nearest half-foot.

Figure 33.45  You can determine pressure losses as equivalent lengths from this table. (*Courtesy of International Code Council, Inc. and International Plumbing Code 2000*).

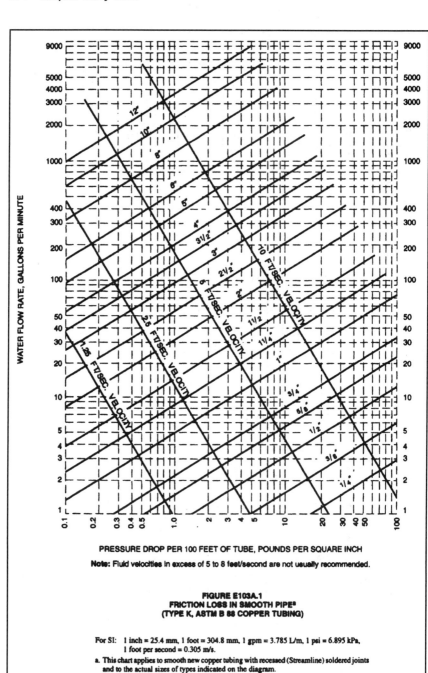

**Figure 33.46** This is one of several tables that determine friction loss. *(Courtesy of International Code Council, Inc. and International Plumbing Code 2000).*

# Mathematics for the Trade 375

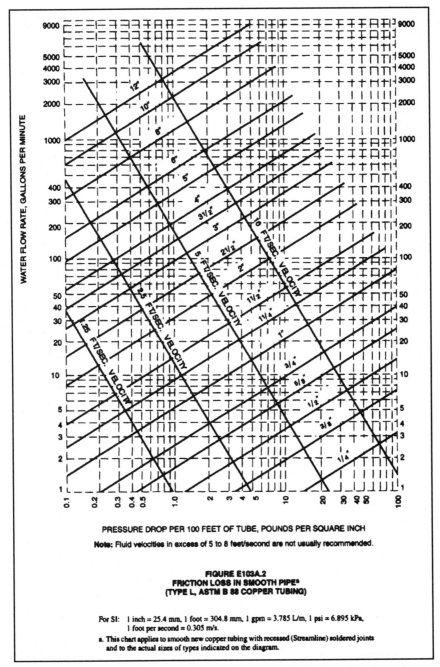

**Figure 33.47** This is one of several tables that determine friction loss. *(Courtesy of International Code Council, Inc. and International Plumbing Code 2000).*

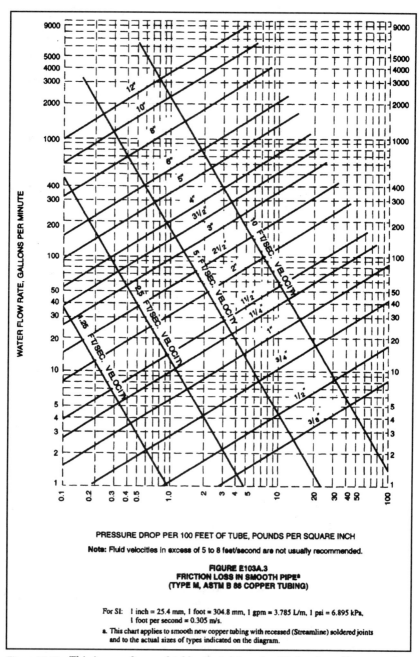

**Figure 33.48** This is one of several tables that determine friction loss. *(Courtesy of International Code Council, Inc. and International Plumbing Code 2000).*

Mathematics for the Trade 377

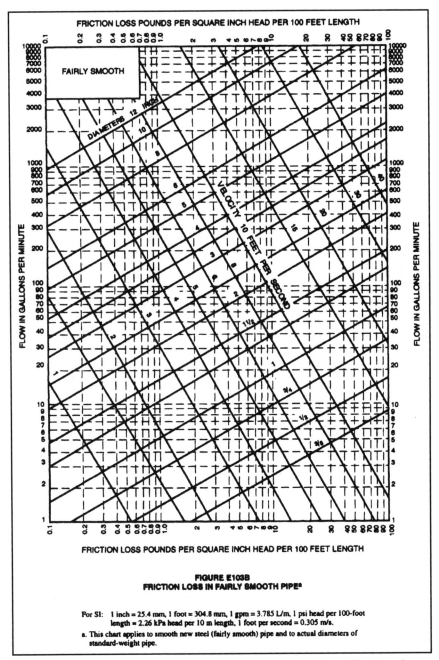

**Figure 33.49** This is one of several tables that determine friction loss. *(Courtesy of International Code Council, Inc. and International Plumbing Code 2000).*

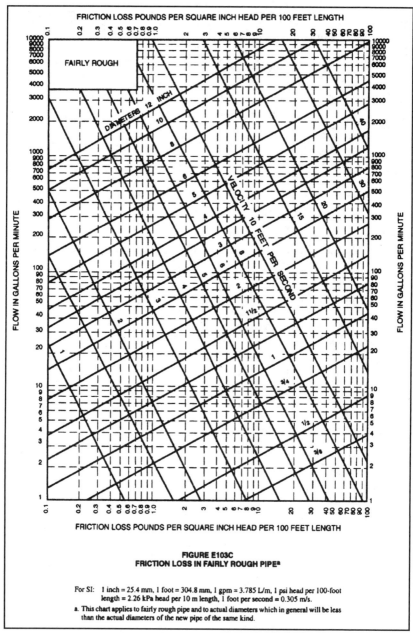

Figure 33.50 This is one of several tables that determine friction loss. *(Courtesy of International Code Council, Inc. and International Plumbing Code 2000).*

Mathematics for the Trade 379

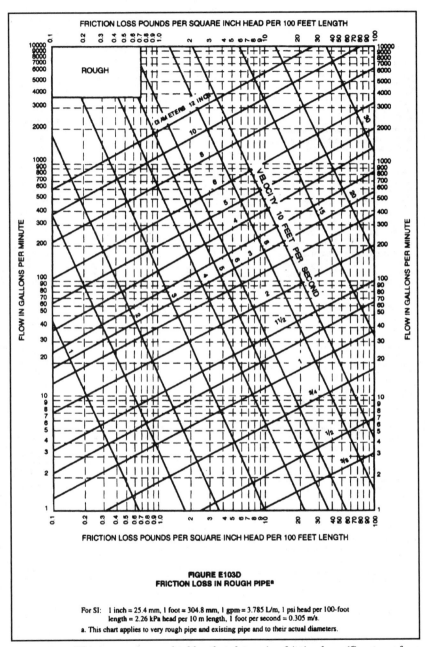

**FIGURE E103D
FRICTION LOSS IN ROUGH PIPE**ᵃ

For SI: 1 inch = 25.4 mm, 1 foot = 304.8 mm, 1 gpm = 3.785 L/m, 1 psi head per 100-foot length = 2.26 kPa head per 10 m length, 1 foot per second = 0.305 m/s.

a. This chart applies to very rough pipe and existing pipe and to their actual diameters.

**Figure 33.51** This is one of several tables that determine friction loss. *(Courtesy of International Code Council, Inc. and International Plumbing Code 2000).*

| Inch | mm |
|---|---|
| 1/2 | 15 |
| 3/4 | 20 |
| 1 | 25 |

**Water Supply Fixture Units (WSFU) and Minimum Fixture Branch Pipe Sizes[3]**

| Appliances, Appurtenances or Fixtures[2] | Minimum Fixture Branch Pipe Size[1,4] | Private | Public | Assembly[6] |
|---|---|---|---|---|
| Bathtub or Combination Bath/Shower (fill) | 1/2" | 4.0 | 4.0 | |
| 3/4" Bathtub Fill Valve | 3/4" | 10.0 | 10.0 | |
| Bidet | 1/2" | 1.0 | | |
| Clotheswasher | 1/2" | 4.0 | 4.0 | |
| Dental Unit, cuspidor | 1/2" | | 1.0 | |
| Dishwasher, domestic | 1/2" | 1.5 | 1.5 | |
| Drinking Fountain or Watercooler | 1/2" | 0.5 | 0.5 | 0.75 |
| Hose Bibb | 1/2" | 2.5 | 2.5 | |
| Hose Bibb, each additional[7] | 1/2" | 1.0 | 1.0 | |
| Lavatory | 1/2" | 1.0 | 1.0 | 1.0 |
| Lawn Sprinkler, each head[5] | | 1.0 | 1.0 | |
| Mobile Home, each (minimum) | | 12.0 | | |
| **Sinks** | | | | |
| Bar | 1/2" | 1.0 | 2.0 | |
| Clinic Faucet | 1/2" | | 3.0 | |
| Clinic Flushometer Valve with or without faucet | 1" | | 8.0 | |
| Kitchen, domestic | 1/2" | 1.5 | 1.5 | |
| Laundry | 1/2" | 1.5 | 1.5 | |
| Service or Mop Basin | 1/2" | 1.5 | 3.0 | |
| Washup, each set of faucets | 1/2" | | 2.0 | |
| Shower | 1/2" | 2.0 | 2.0 | |
| Urinal, 1.0 GPF | 3/4" | 3.0 | 4.0 | 5.0 |
| Urinal, greater than 1.0 GPF | 3/4" | 4.0 | 5.0 | 6.0 |
| Urinal, flush tank | 1/2" | 2.0 | 2.0 | 3.0 |
| Washfountain, circular spray | 3/4" | | 4.0 | |
| Water Closet, 1.6 GPF Gravity Tank | 1/2" | 2.5 | 2.5 | 3.5 |
| Water Closet, 1.6 GPF Flushometer Tank | 1/2" | 2.5 | 2.5 | 3.5 |
| Water Closet, 1.6 GPF Flushometer Valve | 1" | 5.0 | 5.0 | 6.0 |
| Water Closet, greater than 1.6 GPF Gravity Tank | 1/2" | 3.0 | 5.5 | 7.0 |
| Water Closet, greater than 1.6 GPF Flushometer Valve | 1" | 7.0 | 8.0 | 10.0 |

**Notes:**
1. Size of the cold branch outlet pipe, or both the hot and cold branch outlet pipes.
2. Appliances, Appurtenances or Fixtures not included in this Table may be sized by reference to fixtures having a similar flow rate and frequency of use.
3. The listed fixture unit values represent their total load on the cold water service. The separate cold water and hot water fixture unit value for fixtures having both cold and hot water connections shall each be taken as three-quarters (3/4) of the listed total value of the fixture.
4. The listed minimum supply branch pipe sizes for individual fixtures are the nominal (I.D.) pipe size.
5. For fixtures or supply connections likely to impose continuous flow demands, determine the required flow in gallons per minute (GPM) and add it separately to the demand (in GPM) for the distribution system or portions thereof.
6. Assembly [Public Use (See Table 4-1)].
7. Reduced fixture unit loading for additional hose bibbs as used is to be used only when sizing total building demand and for pipe sizing when more than one hose bibb is supplied by a segment of water distributing pipe. The fixture branch to each hose bibb shall be sized on the basis of 2.5 fixture units.

Figure 33.52 This shows the relationship between fixture units and fixture branch pipes for a water supply. *(Reprinted from the 2000 Uniform Plumbing Code (UPC) with the permission of the International Association of Plumbing and Mechanical Officials (IAPMO)).*

## Allowance in equivalent length of pipe for friction loss in valves and threaded fittings.*

### Equivalent Length of Pipe for Various Fittings

| Diameter of fitting Inches | 90° Standard Elbow Feet | 45° Standard Elbow Feet | Standard Tee 90° Feet | Coupling or Straight Run of Tee Feet | Gate Valve Feet | Globe Valve Feet | Angle Valve Feet |
|---|---|---|---|---|---|---|---|
| 3/8 | 1.0 | 0.6 | 1.5 | 0.3 | 0.2 | 8 | 4 |
| 1/2 | 2.0 | 1.2 | 3.0 | 0.6 | 0.4 | 15 | 8 |
| 3/4 | 2.5 | 1.5 | 4.0 | 0.8 | 0.5 | 20 | 12 |
| 1 | 3.0 | 1.8 | 5.0 | 0.9 | 0.6 | 25 | 15 |
| 1-1/4 | 4.0 | 2.4 | 6.0 | 1.2 | 0.8 | 35 | 18 |
| 1-1/2 | 5.0 | 3.0 | 7.0 | 1.5 | 1.0 | 45 | 22 |
| 2 | 7.0 | 4.0 | 10.0 | 2.0 | 1.3 | 55 | 28 |
| 2-1/2 | 8.0 | 5.0 | 12.0 | 2.5 | 1.6 | 65 | 34 |
| 3 | 10.0 | 6.0 | 15.0 | 3.0 | 2.0 | 80 | 40 |
| 4 | 14.0 | 8.0 | 21.0 | 4.0 | 2.7 | 125 | 55 |
| 5 | 17.0 | 10.0 | 25.0 | 5.0 | 3.3 | 140 | 70 |
| 6 | 20.0 | 12.0 | 30.0 | 6.0 | 4.0 | 165 | 80 |

**Figure 33.53** Pipe length in various fittings is described. *(Reprinted from the 2000 Uniform Plumbing Code (UPC) with the permission of the International Association of Plumbing and Mechanical Officials (IAPMO)).*

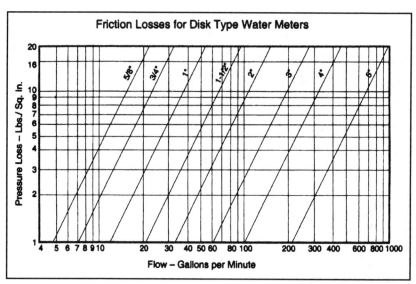

**Figure 33.54** English and metric unit information about friction loss. *(Reprinted from the 2000 Uniform Plumbing Code (UPC) with the permission of the International Association of Plumbing and Mechanical Officials (IAPMO)).*

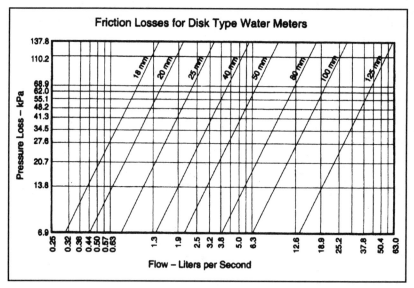

**Figure 33.55** English and metric unit information about friction loss. *(Reprinted from the 2000 Uniform Plumbing Code (UPC) with the permission of the International Association of Plumbing and Mechanical Officials (IAPMO)).*

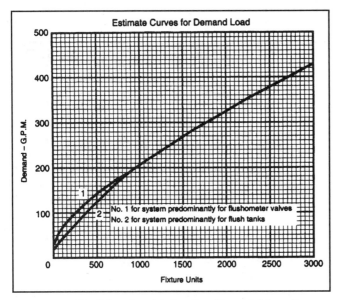

**Figure 33.56** English and metric lengths. *(Reprinted from the 2000 Uniform Plumbing Code (UPC) with the permission of the International Association of Plumbing and Mechanical Officials (IAPMO)).*

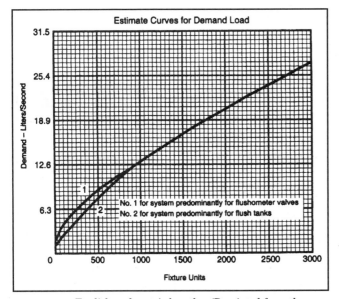

**Figure 33.57** English and metric lengths. *(Reprinted from the 2000 Uniform Plumbing Code (UPC) with the permission of the International Association of Plumbing and Mechanical Officials (IAPMO)).*

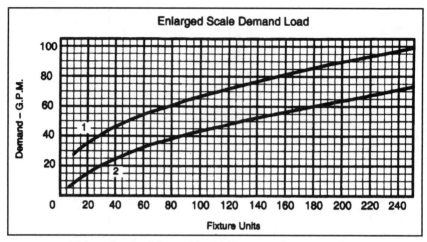

**Figure 33.58** Enlarged scale of demand loads. *(Reprinted from the 2000 Uniform Plumbing Code (UPC) with the permission of the International Association of Plumbing and Mechanical Officials (IAPMO)).*

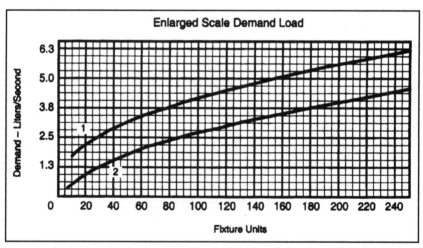

**Figure 33.59** Enlarged scale of demand loads. *(Reprinted from the 2000 Uniform Plumbing Code (UPC) with the permission of the International Association of Plumbing and Mechanical Officials (IAPMO)).*

Mathematics for the Trade 385

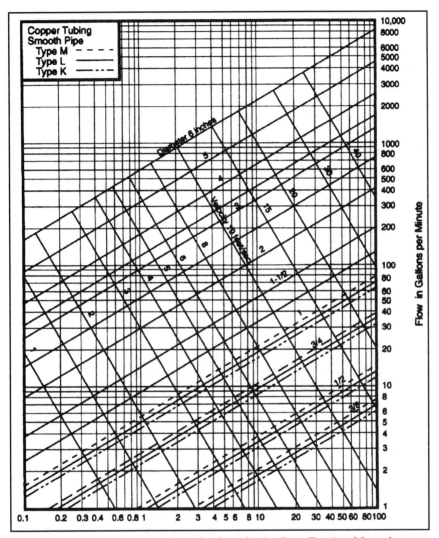

**Figure 33.60** This is one of several graphs about friction loss. *(Reprinted from the 2000 Uniform Plumbing Code (UPC) with the permission of the International Association of Plumbing and Mechanical Officials (IAPMO)).*

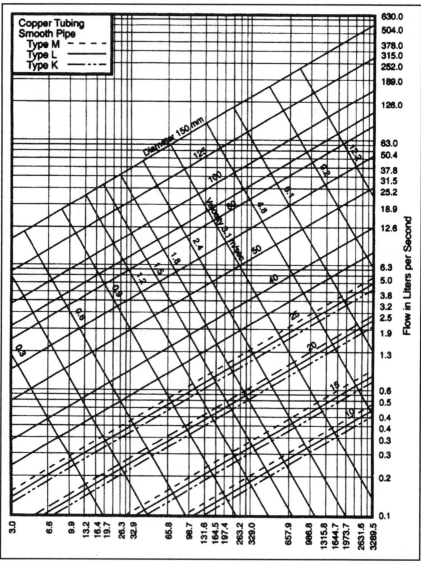

**Figure 33.61** This is one of several graphs about friction loss. *(Reprinted from the 2000 Uniform Plumbing Code (UPC) with the permission of the International Association of Plumbing and Mechanical Officials (IAPMO)).*

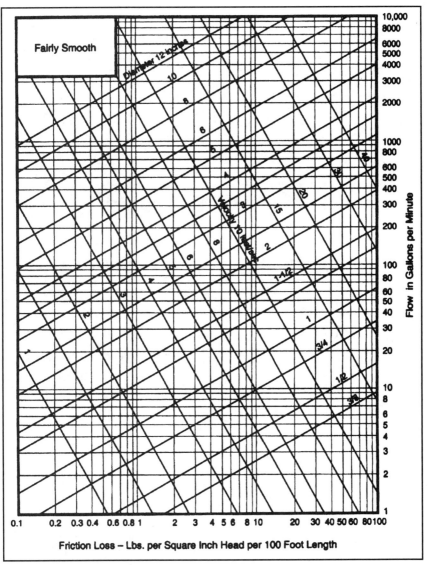

**Figure 33.62** This is one of several graphs about friction loss. *(Reprinted from the 2000 Uniform Plumbing Code (UPC) with the permission of the International Association of Plumbing and Mechanical Officials (IAPMO)).*

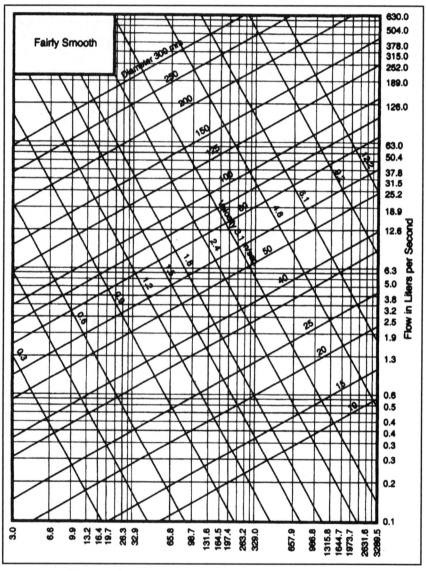

**Figure 33.63** This is one of several graphs about friction loss. *(Reprinted from the 2000 Uniform Plumbing Code (UPC) with the permission of the International Association of Plumbing and Mechanical Officials (IAPMO)).*

Mathematics for the Trade 389

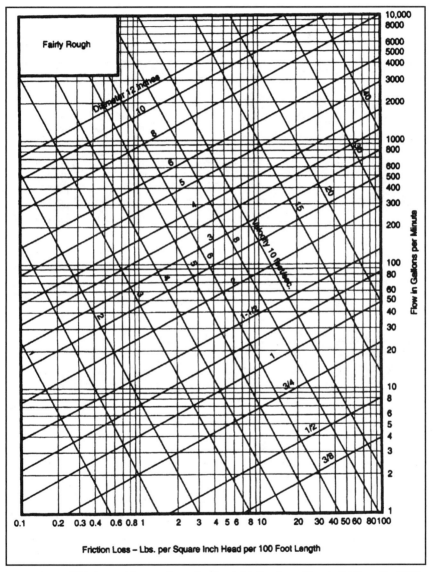

**Figure 33.64** This is one of several graphs about friction loss. *(Reprinted from the 2000 Uniform Plumbing Code (UPC) with the permission of the International Association of Plumbing and Mechanical Officials (IAPMO)).*

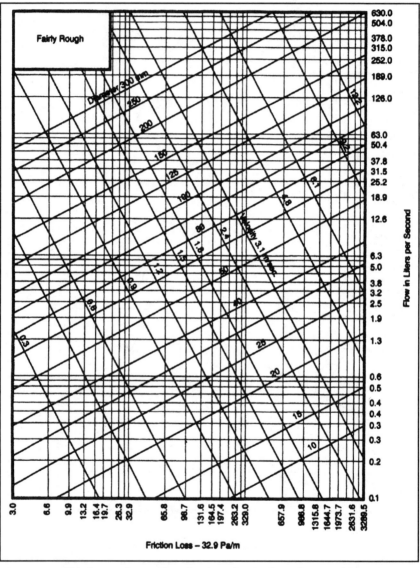

**Figure 33.65** This is one of several graphs about friction loss. *(Reprinted from the 2000 Uniform Plumbing Code (UPC) with the permission of the International Association of Plumbing and Mechanical Officials (IAPMO)).*

Mathematics for the Trade 391

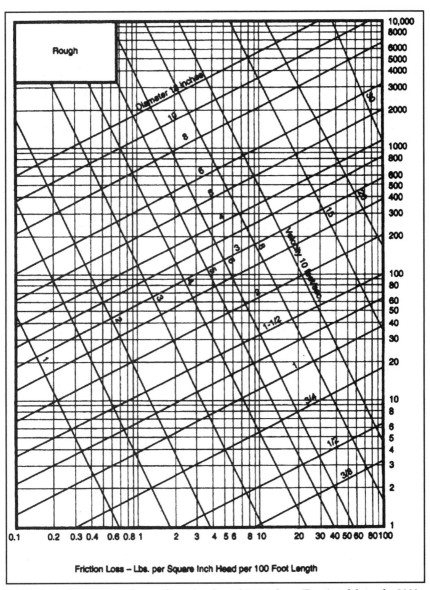

**Figure 33.66** This is one of several graphs about friction loss. *(Reprinted from the 2000 Uniform Plumbing Code (UPC) with the permission of the International Association of Plumbing and Mechanical Officials (IAPMO)).*

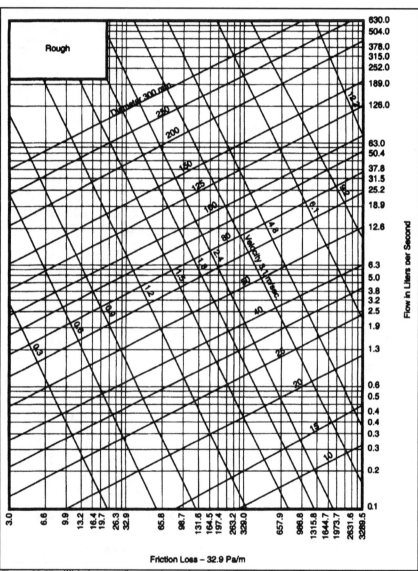

**Figure 33.67** This is one of several graphs about friction loss. *(Reprinted from the 2000 Uniform Plumbing Code (UPC) with the permission of the International Association of Plumbing and Mechanical Officials (IAPMO)).*

## Sizing of Grease Interceptors

Number of meals per peak hour[1] $\times$ Waste flow rate[2] $\times$ retention time[3] $\times$ storage factor[4] $=$ Interceptor size (liquid capacity)

1. **Meals Served at Peak Hour**
2. **Waste Flow Rate**
   a. With dishwashing machine ..................................6 gallon (22.7 L) flow
   b. Without dishwashing machine .........................5 gallon (18.9 L) flow
   c. Single service kitchen .........................................2 gallon (7.6 L) flow
   d. Food waste disposer .........................................1 gallon (3.8 L) flow
3. **Retention Times**
   Commercial kitchen waste
   Dishwasher...........................................................................2.5 hours
   Single service kitchen
   Single serving......................................................................1.5 hours
4. **Storage Factors**
   Fully equipped commercial kitchen .....................8 hour operation: 1
   ................................................................................16 hour operation: 2
   ................................................................................24 hour operation: 3
   Single Service Kitchen.................................................................1.5

**Figure 33.68** This chart shows information about grease interceptors and how to size them. *(Reprinted from the 2000 Uniform Plumbing Code (UPC) with the permission of the International Association of Plumbing and Mechanical Officials (IAPMO)).*

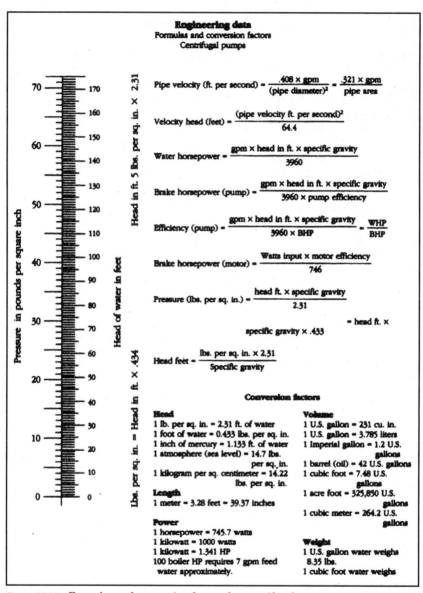

Figure 33.69  Formulas and conversion factors for centrifugal pumps.

| MULTIPLY | BY | TO OBTAIN |
|---|---|---|
| Gallons/minute | 8.0208 | Cubic feet/hour |
| Gallons water/minute | 6.0086 | Tons of water/24 hours |
| Inches | 2.540 | Centimeters |
| Inches of mercury | 0.03342 | Atmospheres |
| Inches of mercury | 1.133 | Feet of water |
| Inches of mercury | 0.4912 | Pounds/square inch |
| Inches of water | 0.002458 | Atmospheres |
| Inches of water | 0.07355 | Inches of mercury |
| Inches of water | 5.202 | Pounds/square feet |
| Inches of water | 0.03613 | Pounds/square inch |
| Liters | 1000 | Cubic centimeters |
| Liters | 61.02 | Cubic inches |
| Liters | 0.2642 | Gallons |
| Miles | 5280 | Feet |
| Miles/hour | 88 | Feet/minute |
| Miles/hour | 1.467 | Feet/second |
| Millimeters | 0.1 | Centimeters |
| Millimeters | 0.03937 | Inches |
| Million gallon/day | 1.54723 | Cubic feet/second |
| Pounds of water | 0.01602 | Cubic feet |
| Pounds of water | 27.68 | Cubic inches |
| Pounds of water | 0.1198 | Gallons |
| Pounds/cubic inch | 1728 | Pounds/cubic feet |
| Pounds/square foot | 0.01602 | Feet of water |
| Pounds/square inch | 0.06804 | Atmospheres |
| Pounds/square inch | 2.307 | Feet of water |
| Pounds/square inch | 2.036 | Inches of mercury |
| Quarts (dry) | 67.20 | Cubic inches |
| Quarts (liquid) | 57.75 | Cubic inches |
| Square feet | 144 | Square inches |
| Square miles | 640 | Acres |
| Square yards | 9 | Square feet |
| Temperature (°C) + 273 | 1 | Abs. temperature (°C) |
| Temperature (°C) + 17.28 | 1.8 | Temperature (°F) |
| Temperature (°F) + 460 | 1 | Abs. temperature (°F) |
| Temperature (°F) - 32 | 5/9 | Temperature (°C) |
| Tons (short) | 2000 | Pounds |
| Tons of water/24 hours | 83.333 | Pounds water/hour |
| Tons of water/24 hours | 0.16643 | Gallons/minute |
| Tons of water/24 hours | 1.3349 | Cubic feet/hour |

Figure 33.70 A useful set of tables to keep on hand. *(Reprinted from the 2000 Uniform Plumbing Code (UPC) with the permission of the International Association of Plumbing and Mechanical Officials (IAPMO)).*

## Chapter Thirty-three

### AREAS AND CIRCUMFERENCE OF CIRCLES

| Diameter | | Circumference | | Area | |
|---|---|---|---|---|---|
| Inches | mm | Inches | mm | Inches$^2$ | mm$^2$ |
| 1/8 | 6 | 0.40 | 10 | 0.01227 | 8.0 |
| 1/4 | 8 | 0.79 | 20 | 0.04909 | 31.7 |
| 3/8 | 10 | 1.18 | 30 | 0.11045 | 71.3 |
| 1/2 | 15 | 1.57 | 40 | 0.19635 | 126.7 |
| 3/4 | 20 | 2.36 | 60 | 0.44179 | 285.0 |
| 1 | 25 | 3.14 | 80 | 0.7854 | 506.7 |
| 1-1/4 | 32 | 3.93 | 100 | 1.2272 | 791.7 |
| 1-1/2 | 40 | 4.71 | 120 | 1.7671 | 1140.1 |
| 2 | 50 | 6.28 | 160 | 3.1416 | 2026.8 |
| 2-1/2 | 65 | 7.85 | 200 | 4.9087 | 3166.9 |
| 3 | 80 | 9.43 | 240 | 7.0686 | 4560.4 |
| 4 | 100 | 12.55 | 320 | 12.566 | 8107.1 |
| 5 | 125 | 15.71 | 400 | 19.635 | 12,667.7 |
| 6 | 150 | 18.85 | 480 | 28.274 | 18,241.3 |
| 7 | 175 | 21.99 | 560 | 38.485 | 24,828.9 |
| 8 | 200 | 25.13 | 640 | 50.265 | 32,428.9 |
| 9 | 225 | 28.27 | 720 | 63.617 | 41,043.1 |
| 10 | 250 | 31.42 | 800 | 78.540 | 50,670.9 |

**EQUAL PERIPHERIES**

$S = 0.7854\,D$
$D = 1.2732\,S$

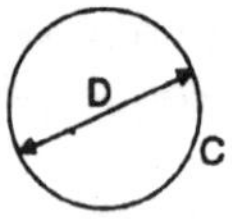

$S = 0.8862\,D$
$D = 1.1284\,S$
$S = 0.2821\,C$

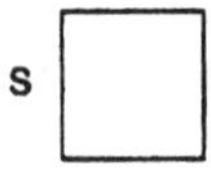

**EQUAL AREAS**

Area of square (S') = 1.2732 x area of circle

Area of square (S) = 0.6366 x area of circle

$C = \pi D = 2\pi R$

$C = 3.5446\,\sqrt{area}$

$D = 0.3183\,C = 2R$

$D = 1.1283\,\sqrt{area}$

Area $= \pi R^2 = 0.7854\,D^2$

Area $= 0.07958\,C^2 = \dfrac{\pi D^2}{4}$

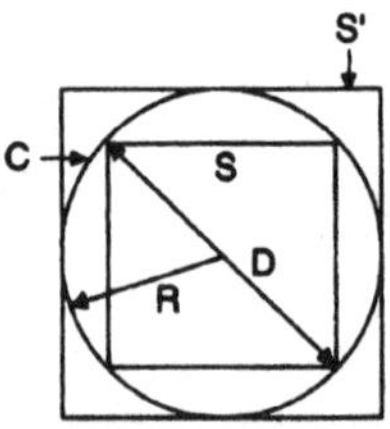

$\pi = 3.1416$

**Figure 33.71** More useful information. *(Reprinted from the 2000 Uniform Plumbing Code (UPC) with the permission of the International Association of Plumbing and Mechanical Officials (IAPMO)).*

## EQUAL PERIPHERIES

$S = 0.7854\ D$

$D = 1.2732\ S$

$S = 0.8862\ D$

$D = 1.1284\ S$

$S = 0.2821\ C$

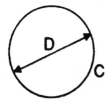

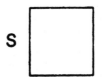

## EQUAL AREAS

Area of square (S') =
1.2732 x area of circle

Area of square (S) =
0.6366 x area of circle

$C = \pi D = 2\pi R$

$C = 3.5446\ \sqrt{area}$

$D = 0.3183\ C = 2R$

$D = 1.1283\ \sqrt{area}$

$Area = \pi R^2 = 0.7854\ D^2$

$Area = 0.07958\ C^2 = \dfrac{\pi D^2}{4}$

$\pi = 3.1416$

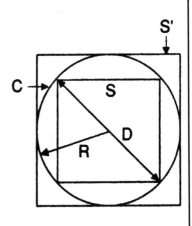

**Figure 33.72**  Mathematical formulas.

## Chapter Thirty-three

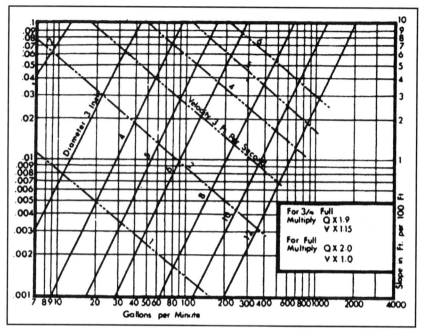

**Figure 33.73** Flow in partially filled pipes.

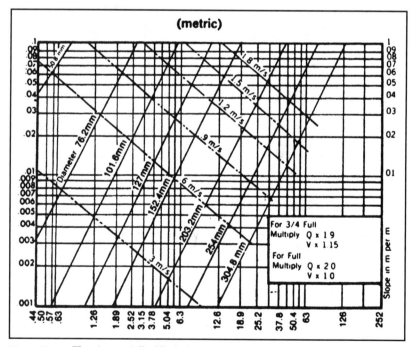

**Figure 33.74** Flow in partially filled pipes.

Circumference = Diameter × 3.1416
Circumference = Radius × 6.2832
Diameter = Radius × 2
Diameter = Square root of (area ÷ .7854)
Diameter = Square root of area × 1.1283
Diameter = Circumference × .31831
Radius = Diameter ÷ 2
Radius = Circumference × .15915
Radius = Square root of area × .56419
Area = Diameter × Diameter × .7854
Area = Half of the circumference × half of the diameter
Area = Square of the circumference × .0796
Arc length = Degrees × radius × .01745
Degrees of arc = Length ÷ (radius × .01745)
Radius of arc = Length ÷ (degrees × .01745)
Side of equal square = Diameter × .8862
Side of inscribed square = Diameter × .7071
Area of sector = Area of circle × degrees of arc ÷ 360

Figure 33.75  Formulas for a circle.

1. Circumference of a circle = π × diameter or 3.1416 × diameter
2. Diameter of a circle = Circumference × .31831
3. Area of a square = Length × width
4. Area of a rectangle = Length × width
5. Area of a parallelogram = Base × perpendicular height
6. Area of a triangle = ½ base × perpendicular height
7. Area of a circle = π × radius squared or diameter squared × .7854
8. Area of an ellipse = Length × width × .7854
9. Volume of a cube or rectangular prism = Length × width × height
10. Volue of a triangular prism = Area of triangle × length
11. Volume of a sphere = Diameter cubed × .5236 or (dia. × dia. × dia. × .5236)
12. Volume of a cone = π × radius square × ⅓ height
13. Volume of a cylinder = π × radius squared × height
14. Length of one side of a square × 1.128 = Diameter of an equal circle
15. Doubling the diameter of a pipe or cylinder increases its capacity 4 times
16. The pressure (in lbs. per sq. inch) of a column of water = Height of the column (in feet) × .434
17. The capacity of a pipe or tank (in U.S. gallons) = Diameter squared (in inches) × the length (in inches) × .0034
18. A gallon of water = 8½ lb. = 231 cu. inches
19. A cubic foot of water = 62½ lb. = 7½ gallons

Figure 33.76  Useful formulas.

> Area = Base × distance between the two parallel sides

**Figure 33.77** Parallelograms.

> Area = Length × width

**Figure 33.78** Rectangles.

> Area of surface = Diameter × diameter × 3.1416
> Side of inscribed cube = Radius × 1.547
> Volume = Diameter × diameter × diameter × .5236

**Figure 33.79** Spheres.

> Area = One-half of height times base

**Figure 33.80** Triangles.

> Area = One-half of the sum of the parallel sides × the height

**Figure 33.81** Trapezoids.

> Volume = Width × height × length

**Figure 33.82** Cubes.

> Area of surface = One half of circumference of base × slant height + area of base.
> Volume = Diameter × diameter × .7854 × one-third of the altitude.

**Figure 33.83** Cone calculation.

> Volume = Width × height × length

**Figure 33.84** Volume of a rectangular prism.

> Area = Length × width

**Figure 33.85** Finding the area of a square.

> Area = ½ perimeter of base × slant height + area of base
> Volume = Area of base × ⅓ of the altitude

**Figure 33.86** Finding the area and volume of a pyramid.

> These comprise the numerous figures having more than four sides, names according to the number of sides, thus:
>
> | Figure | Sides |
> |---|---|
> | Pentagon | 5 |
> | Hexagon | 6 |
> | Heptagon | 7 |
> | Octagon | 8 |
> | Nonagon | 9 |
> | Decagon | 10 |
>
> To find the area of a polygon: Multiply the sum of the sides (perimeter of the polygon) by the perpendicular dropped from its center to one of its sides, and half the product will be the area. This rule applies to all regular polygons.

**Figure 33.87** Polygons.

> Multiply Length × Width × Thickness
> Example: 50 ft. × 10 ft. × 8 in.
> 50' × 10' × 8/12' = 333.33 cu. feet
> To convert to cubic yards, divide by 27 cu. ft. per cu. yd.
> Example: 333.33 ÷ 27 = 12.35 cu. yd.

**Figure 33.88** Estimating volume.

> Area = Width of side × 2.598 × width of side

Figure 33.89  Hexagons.

> Area of surface = Diameter × 3.1416 × length + area of the two bases
> Area of base = Diameter × diameter × .7854
> Area of base = Volume ÷ length
> Length = Volume ÷ area of base
> Volume = Length × area of base
> Capacity in gallons = Volume in inches ÷ 231
> Capacity of gallons = Diameter × diameter × length × .0034
> Capacity in gallons = Volume in feet × 7.48

Figure 33.90  Cylinder formulas.

> Area = Short diameter × long diameter × .7854

Figure 33.91  Ellipse calculation.

> Deg. F. = Deg. C. × 1.8 + 32

Figure 33.92  Temperature conversion.

| To change | To | Multiply by |
|---|---|---|
| Feet of water | Pounds per square inch | 0.434 |
| Feet of water | Pounds per square foot | 62.5 |
| Feet of water | Inches of mercury | 0.8824 |
| Atmospheres | Pounds per square inch | 14.696 |
| Atmospheres | Inches of mercury | 29.92 |
| Atmospheres | Feet of water | 34 |
| Long tons | Pounds | 2240 |
| Short tons | Pounds | 2000 |
| Short tons | Long tons | 0.89295 |

Figure 33.93  Useful multipliers.

To figure the final temperature when two different temperatures of water are mixed together, use the following formula:

$$\frac{(A \times C) + (B \times D)}{A + B}$$

A = Weight of lower temperature water
B = Weight of higher temperature water
C = Lower temperature
D = Higher temperature

Figure 33.94  Temperature calculation.

Temperature can be expressed according to the Fahrenheit scale or the Celsius scale. To convert C to F or F to C, use the following formulas:

$$°F = 1.8 \times °C + 32$$

$$°C = 0.55555555 \times °F - 32$$

Figure 33.95  Temperature conversion.

Deg. C. = Deg. F. − 32 ÷ 1.8

Figure 33.96  Temperature conversion.

Temperature can be expressed according to the Fahrenheit scale or the Celsius scale. To convert C to F or F to C, use the following formulas:

$$°F = 1.8 \times °C + 32$$

$$°C = 0.55555555 \times °F - 32$$

Figure 33.97  Finding the area of a pipe.

The capacity of pipes is as the square of their diameters. Thus, doubling the diameter of a pipe increases its capacity four times.

Figure 33.98  A piping fact.

The formula for calculating expansion or contraction in plastic piping is:

$$L = Y \times \frac{T-F}{10} \times \frac{L}{100}$$

L = Expansion in inches
Y = Constant factor expressing inches of expansion per 100°F temperature change per 100 ft. of pipe
T = Maximum temperature (0°F)
F = Minimum temperature (0°F)
L = Length of pipe run in feet

**Figure 33.99** Expansion in plastic piping.

| Inch scale | Metric scale |
|---|---|
| ¼" | 1:50 |
| ⅛" | 1:100 |

**Figure 33.100** Scales used for building plans.

| Inch scale | Metric scale |
|---|---|
| ¹⁄₁₆" | 1:200 |

**Figure 33.101** Scale used for site plans.

(Surface area ÷ R value) × (temperature inside − temperature outside)

Surface area of a material (in square feet) divided by its "R" value and multiplied by the difference in Fahrenheit degrees between inside and outside temperature equals heat loss in BTU's per hour.

**Figure 33.102** Calculating heat loss per hour with R-value.

$$L = \frac{144}{D \times 3.1414} \times R \div 12$$

D = O.D. of pipe
L = length of pipe needed in ft.
R = sq. ft. of radiation needed

**Figure 33.103** Formulas for pipe radiation.

- 3 feet of 1-in. pipe equal 1 square foot of radiation.
- 2⅓ linear feet of 1¼ in. pipe equal 1 square foot of radiation.
- Hot water radiation gives off 150 BTU per square foot of radiation per hour.
- Steam radiation gives off 240 BTU per square foot of radiation per hour.
- On greenhouse heating, figure ⅔ square foot of radiation per square foot of glass.
- One square foot of direct radiation condenses .25 pound of water per hour.

**Figure 33.104** Radiant heat facts.

The approximate weight of a piece of pipe can be determined by the following formulas:

Cast Iron Pipe: weight = $(A^2 - B^2) \times C \times .2042$

Steel Pipe: weight = $(A^2 - B^2) \times C \times .2199$

Copper Pipe: weight = $(A^2 - B^2) \times C \times .2537$

A = outside diameter of the pipe in inches

B = inside diameter of the pipe in inches

C = length of the pipe in inches

**Figure 33.105** Finding the weight of piping.

# Appendix

# Reference and Conversion Tables and Data

This appendix contains a wealth of useful information. It is all presented in a visual format. You can use the material in this appendix to do everything from metric conversions to finding the weight of cast iron pipe. Many plumbing questions are answered in the illustrations found in this appendix.

| Abbreviation | Meaning |
|---|---|
| A or a | Area |
| A.W.G. | American wire gauge |
| bbl. | Barrels |
| B or b | Breadth |
| bhp | Brake horse power |
| B.M. | Board measure |
| Btu | British thermal units |
| B.W.G. | Birmingham wire gauge |
| B & S | Brown and Sharpe wire gauge (American wire |
| C of g | Center gravity |
| cond. | Condensing |
| cu. | Cubic |
| cyl. | Cylinder |
| D or d | Depth or diameter |
| evap. | Evaporation |
| F | Coefficient of friction; Fahrenheit |
| F or f | Force or factor of safety |
| ft. lbs. | Foot pounds |
| gals. | Gallons |
| H or h | Height or head of water |
| HP | Horsepower |
| IHP | Indiated horsepower |
| L or l | Length |
| lbs. | Pounds |
| lbs. per sq. in. | Pounds per square inch |
| o.d. | Outside diameter (pipes) |
| oz. | Ounces |
| pt. | Pint |
| P or p | Pressure or load |
| psi | Pounds per square inch |
| R or r | Radius |
| rpm | Revolutions per minute |

Figure A.1 Commonly used abbreviations.

| | |
|---|---|
| A or a | Area |
| A.W.G. | American wire gauge |
| bbl. | Barrels |
| B or b | Breadth |
| bhp | Brake horse power |
| B.M. | Board measure |
| Btu | British thermal units |
| B.W.G. | Birmingham wire gauge |
| B & S | Brown and Sharpe wire gauge (American wire gauge) |
| C of g | Center gravity |
| cond. | Condensing |
| cu. | Cubic |
| cyl. | Cylinder |
| D or d | Depth or diameter |
| evap. | Evaporation |
| F | Coefficient of friction; Fahrenheit |
| F or f | Force or factor of safety |
| ft. lbs. | Foot pounds |
| gals. | Gallons |
| H or h | Height or head of water |
| HP | Horsepower |
| IHP | Indiated horsepower |
| L or l | Length |
| lbs. | Pounds |
| lbs. per sq. in. | Pounds per square inch |
| o.d. | Outside diameter (pipes) |
| oz. | Ounces |
| pt. | Pint |
| P or p | Pressure or load |
| psi | Pounds per square inch |
| R or r | Radius |
| rpm | Revolutions per minute |
| sq. ft. | Square foot |
| sq. in. | Square inch |

Figure A.2  Abbreviations.

| | |
|---|---|
| A or a | Area, acre |
| AWG | American Wire Gauge |
| B or b | Breadth |
| bbl | Barrels |
| bhp | Brake horsepower |
| BM | Board measure |
| Btu | British thermal units |
| BWG | Birmingham Wire Gauge |
| B & S | Brown and Sharpe Wire Gauge (American Wire Gauge) |
| C of g | Center of gravity |
| cond | Condensing |
| cu | Cubic |
| cyl | Cylinder |
| D or d | Depth, diameter |
| dr | Dram |
| evap | Evaporation |
| F | Coefficient of friction; Fahrenheit |
| F or f | Force, factor of safety |
| ft (or ') | Foot |
| ft lb | Foot pound |
| fur | Furlong |
| gal | Gallon |
| gi | Gill |
| ha | Hectare |
| H or h | Height, head of water |
| HP | horsepower |
| IHP | Indicated horsepower |
| in (or ") | Inch |
| L or l | Length |
| lb | Pound |
| lb/sq in. | Pounds per square inch |
| mi | Mile |
| o.d. | Outside diameter (pipes) |
| oz | Ounces |
| pt | Pint |
| P or p | Pressure, load |
| psi | Pounds per square inch |
| R or r | Radius |
| rpm | Revolutions per minute |
| sq ft | Square foot |
| sq in. | Square inch |
| sq yd | Square yard |
| T or t | Thickness, temperature |
| temp | Temperature |
| V or v | Velocity |
| vol | Volume |
| W or w | Weight |
| W. I. | Wrought iron |

Figure A.3   Abbreviations.

| Material | Chemical symbol |
|---|---|
| Aluminum | AL |
| Antimony | Sb |
| Brass | .. |
| Bronze | .. |
| Chromium | Cr |
| Copper | Cu |
| Gold | Au |
| Iron (cast) | Fe |
| Iron (wrought) | Fe |
| Lead | Pb |
| Manganese | Mn |
| Mercury | Hg |
| Molybdenum | Mo |
| Monel | .. |
| Platinum | Pt |
| Steel (mild) | Fe |
| Steel (stainless) | .. |
| Tin | Sn |
| Titanium | Ti |
| Zinc | Zn |

Figure A.4  Symbols of various materials.

|  | Drawn (hard copper) (feet) |  | Annealed (soft copper) (feet) |
| --- | --- | --- | --- |
| **Type K Tube** | | | |
| Straight Lengths: | | Straight Lengths: | |
| Up to 8-in. diameter | 20 | Up to 8-in. diameter | 20 |
| 10-in. diameter | 18 | 10-in. diameter | 18 |
| 12-in. diameter | 12 | 12-in. diameter | 12 |
| | | Coils: | |
| | | Up to 1-in. diameter | 60 |
| | | | 100 |
| | | 1¼-in. diameter | 60 |
| | | | 100 |
| | | 2-in. diameter | 40 |
| | | | 45 |
| **Type L Tube** | | | |
| Straight Lengths: | | Straight Lengths: | |
| Up to 10-in. diameter | 20 | Up to 10-in. diameter | 20 |
| 12-in. diameter | 18 | 12-inch diameter | 18 |
| | | Coils: | |
| | | Up to 1-in. diameter | 60 |
| | | | 100 |
| | | 1¼- and 1½-in. diameter | 60 |
| | | | 100 |
| | | 2-in. diameter | 40 |
| | | | 45 |
| **DWV Tube** | | | |
| Straight Lengths: | | Not available | |
| All diameters | 20 | | |
| **Type M Tube** | | | |
| Straight Lengths: | | Not available | |
| All diameters | 20 | | |

**Figure A.5** Available lengths of copper plumbing tube.

|  | Diameter (inches) | Service Weight (lb) | Extra Heavy Weight (lb) |
|---|---|---|---|
| Double Hub, 5-ft Lengths | 2 | 21 | 26 |
|  | 3 | 31 | 47 |
|  | 4 | 42 | 63 |
|  | 5 | 54 | 78 |
|  | 6 | 68 | 100 |
|  | 8 | 105 | 157 |
|  | 10 | 150 | 225 |
| Double Hub, 30-ft Length | 2 | 11 | 14 |
|  | 3 | 17 | 26 |
|  | 4 | 23 | 33 |
| Single Hub, 5-ft Lengths | 2 | 20 | 25 |
|  | 3 | 30 | 45 |
|  | 4 | 40 | 60 |
|  | 5 | 52 | 75 |
|  | 6 | 65 | 95 |
|  | 8 | 100 | 150 |
|  | 10 | 145 | 215 |
| Single Hub, 10-ft Lengths | 2 | 38 | 43 |
|  | 3 | 56 | 83 |
|  | 4 | 75 | 108 |
|  | 5 | 98 | 133 |
|  | 6 | 124 | 160 |
|  | 8 | 185 | 265 |
|  | 10 | 270 | 400 |
| No-Hub Pipe, 10-ft Lengths | 1½ | 27 |  |
|  | 2 | 38 |  |
|  | 3 | 54 |  |
|  | 4 | 74 |  |
|  | 5 | 95 |  |
|  | 6 | 118 |  |
|  | 8 | 180 |  |

**Figure A.6** Weight of cast iron pipe.

| Single-Family Dwellings; Number of Bedrooms | Multiple Dwelling Units or Apartments; One Bedroom Each | Other Uses; Maximum Fixture-Units Served | Minimum Septic Tank Capacity in Gallons |
|---|---|---|---|
| 1–3 |  | 20 | 1000 |
| 4 | 2 | 25 | 1200 |
| 5–6 | 3 | 33 | 1500 |
| 7–8 | 4 | 45 | 2000 |
|  | 5 | 55 | 2250 |
|  | 6 | 60 | 2500 |
|  | 7 | 70 | 2750 |
|  | 8 | 80 | 3000 |
|  | 9 | 90 | 3250 |
|  | 10 | 100 | 3500 |

**Figure A.7** Common septic tank capacities.

1 ft$^3$ of water contains 7½ gal, 1728 in.$^3$, and weighs 62½ lb.

1 gal of water weighs 8⅓ lb and contains 231 in.$^3$

Water expands 1/23 of its volume when heated from 40° to 212°.

The height of a column of water, equal to a pressure of 1 lb/in.$^2$, is 2.31 ft.

To find the pressure in lb/in.$^2$ of a column of water, multiply the height of the column in feet by 0.434.

The average pressure of the atmosphere is estimated at 14.7 lb/in.$^2$ so that with a perfect vacuum it will sustain a column of water 34 ft high.

The friction of water in pipes varies as the square of the velocity.

To evaporate 1 ft$^3$ of water requires the consumption of 7½ lb of ordinary coal or about 1 lb of coal to 1 gal of water.

1 in.$^3$ of water evaporated at atmospheric pressure is converted into approximately 1 ft$^3$ of steam.

**Figure A.8** Facts about water.

| Fixture | Flow Rate (gpm)[a] |
|---|---|
| Ordinary basin faucet | 2.0 |
| Self-closing basin faucet | 2.5 |
| Sink faucet, 3/8 in. | 4.5 |
| Sink faucet, 1/2 in. | 4.5 |
| Bathtub faucet | 6.0 |
| Laundry tub cock, 1/2 in. | 5.0 |
| Shower | 5.0 |
| Ballcock for water closet | 3.0 |
| Flushometer valve for water closet | 15-35 |
| Flushometer valve for urinal | 15.0 |
| Drinking fountain | .75 |
| Sillcock (wall hydrant) | 5.0 |

[a]Figures do not represent the use of water-conservation devices.

Figure A.9  Rates of water flow.


| Activity | Normal Use (gallons) | Conservative Use (gallons) |
|---|---|---|
| Shower | 25 (water running) | 4 (wet down, soap up, rinse off) |
| Tub bath | 36 (full) | 10–12 (minimal water level) |
| Dishwashing | 50 (tap running) | 5 (wash and rinse in sink) |
| Toilet flushing | 5–7 (depends on tank size) | 1½–3 (water-saver toilets or tank displacement bottles) |
| Automatic dishwasher | 16 (full cycle) | 7 (short cycle) |
| Washing machine | 60 (full cycle, top water level) | 27 (short cycle, minimal water level) |
| Washing hands | 2 (tap running) | 1 (full basin) |
| Brushing teeth | 1 (tap running) | 1/2 (wet and rinse briefly) |

Figure A.10  Conserving water.

| Material | Pounds per Cubic Inch | Pounds per Cubic Foot |
|---|---|---|
| Aluminum | 0.093 | 160 |
| Antimony | 0.2422 | 418 |
| Brass | 0.303 | 524 |
| Bronze | 0.320 | 552 |
| Chromium | 0.2348 | 406 |
| Copper | 0.323 | 558 |
| Gold | 0.6975 | 1205 |
| Iron (cast) | 0.260 | 450 |
| Iron (wrought) | 0.2834 | 490 |
| Lead | 0.4105 | 710 |
| Manganese | 0.2679 | 463 |
| Mercury | 0.491 | 849 |
| Molybdenum | 0.309 | 534 |
| Monel | 0.318 | 550 |
| Platinum | 0.818 | 1413 |
| Steel (mild) | 0.2816 | 490 |
| Steel (stainless) | 0.277 | 484 |
| Tin | 0.265 | 459 |
| Titanium | 0.1278 | 221 |
| Zinc | 0.258 | 446 |

**Figure A.11** Weights of various materials.

| Metal | Degrees Fahrenheit |
|---|---|
| Aluminum | 1200 |
| Antimony | 1150 |
| Bismuth | 500 |
| Brass | 1700/1850 |
| Copper | 1940 |
| Cadmium | 610 |
| Iron (cast) | 2300 |
| Iron (wrought) | 2900 |
| Lead | 620 |
| Mercury | 139 |
| Steel | 2500 |
| Tin | 446 |
| Zinc, cast | 785 |

**Figure A.12** Melting points of commercial metals.

| Pipe size (in inches) | PSI | Length of pipe is 50 feet |
|---|---|---|
| ¾ | 20 | 16 |
| ¾ | 40 | 24 |
| ¾ | 60 | 29 |
| ¾ | 80 | 34 |
| 1 | 20 | 31 |
| 1 | 40 | 44 |
| 1 | 60 | 55 |
| 1 | 80 | 65 |
| 1¼ | 20 | 84 |
| 1¼ | 40 | 121 |
| 1¼ | 60 | 151 |
| 1¼ | 80 | 177 |
| 1½ | 20 | 94 |
| 1½ | 40 | 137 |
| 1½ | 60 | 170 |
| 1½ | 80 | 200 |

Figure A.13  Discharge of pipes in gallons per minute.

| Pipe size (in inches) | PSI | Length of pipe is 100 feet |
|---|---|---|
| ¾ | 20 | 11 |
| ¾ | 40 | 16 |
| ¾ | 60 | 20 |
| ¾ | 80 | 24 |
| 1 | 20 | 21 |
| 1 | 40 | 31 |
| 1 | 60 | 38 |
| 1 | 80 | 44 |
| 1¼ | 20 | 58 |
| 1¼ | 40 | 84 |
| 1¼ | 60 | 104 |
| 1¼ | 80 | 121 |
| 1½ | 20 | 65 |
| 1½ | 40 | 94 |
| 1½ | 60 | 117 |
| 1½ | 80 | 137 |

Figure A.14  Discharge of pipes in gallons per minute.

| Pipe diameter (in inches) | Approximate capacity (in U.S. gallons) per foot of pipe |
|---|---|
| ¾ | .0230 |
| 1 | .0408 |
| 1¼ | .0638 |
| 1½ | .0918 |
| 2 | .1632 |
| 3 | .3673 |
| 4 | .6528 |
| 6 | 1.469 |
| 8 | 2.611 |
| 10 | 4.018 |

Figure A.15  Pipe capacaties.

| Pipe material | Coefficient in/in/°F | (°C) |
|---|---|---|
| **Metallic pipe** | | |
| Carbon steel | 0.000005 | (14.0) |
| Stainless steel | 0.000115 | (69) |
| Cast iron | 0.0000056 | (1.0) |
| Copper | 0.000010 | (1.8) |
| Aluminum | 0.0000980 | (1.7) |
| Brass (yellow) | 0.000001 | (1.8) |
| Brass (red) | 0.000009 | (1.4) |
| **Plastic pipe** | | |
| ABS | 0.00005 | (8) |
| PVC | 0.000060 | (33) |
| PB | 0.000150 | (72) |
| PE | 0.000080 | (14.4) |
| CPVC | 0.000035 | (6.3) |
| Styrene | 0.000060 | (33) |
| PVDF | 0.000085 | (14.5) |
| PP | 0.000065 | (77) |
| Saran | 0.000038 | (6.5) |
| CAB | 0.000080 | (14.4) |
| FRP (average) | 0.000011 | (1.9) |
| PVDF | 0.000096 | (15.1) |
| CAB | 0.000085 | (14.5) |
| HDPE | 0.00011 | (68) |
| **Glass** | | |
| Borosilicate | 0.0000018 | (0.33) |

Figure A.16  Thermal expansion of piping materials.

| Length (ft) | Temperature Change (°F) | | | | | | |
|---|---|---|---|---|---|---|---|
| | 40 | 50 | 60 | 70 | 80 | 90 | 100 |
| 20 | 0.278 | 0.348 | 0.418 | 0.487 | 0.557 | 0.626 | 0.696 |
| 40 | 0.557 | 0.696 | 0.835 | 0.974 | 1.114 | 1.235 | 1.392 |
| 60 | 0.835 | 1.044 | 1.253 | 1.462 | 1.670 | 1.879 | 2.088 |
| 80 | 1.134 | 1.392 | 1.670 | 1.879 | 2.227 | 2.506 | 2.784 |
| 100 | 1.192 | 1.740 | 2.088 | 2.436 | 2.784 | 3.132 | 3.480 |

**Figure A.17** Thermal expansion of PVC-DWV pipe.


| Pipe size (in inches) | Maximum outside diameter | Threads per inch |
|---|---|---|
| ¼ | 1.375 | 8 |
| 1 | 1.375 | 8 |
| 1¼ | 1.6718 | 9 |
| 1½ | 1.990 | 9 |
| 2 | 2.5156 | 8 |
| 3 | 3.6239 | 6 |
| 4 | 5.0109 | 4 |
| 5 | 6.260 | 4 |
| 6 | 7.025 | 4 |

**Figure A.18** Threads per inch for national standard threads.


| Pipe size (in inches) | Maximum outside diameter | Threads per inch |
|---|---|---|
| ¼ | 1.375 | 8 |
| 1 | 1.375 | 8 |
| 1¼ | 1.6718 | 9 |
| 1½ | 1.990 | 9 |
| 2 | 2.5156 | 8 |
| 3 | 3.6239 | 6 |
| 4 | 5.0109 | 4 |
| 5 | 6.260 | 4 |
| 6 | 7.025 | 4 |

**Figure A.19** Threads per inch for American standard straight pipe.

| IPS, in | Weight per foot, lb | Length in feet containing 1 ft³ of water | Gallons in 1 linear ft |
|---|---|---|---|
| ¼ | 0.42 | | 0.005 |
| ⅜ | 0.57 | 754 | 0.0099 |
| ½ | 0.85 | 473 | 0.016 |
| ¾ | 1.13 | 270 | 0.027 |
| 1 | 1.67 | 166 | 0.05 |
| 1¼ | 2.27 | 96 | 0.07 |
| 1½ | 2.71 | 70 | 0.1 |
| 2 | 3.65 | 42 | 0.17 |
| 2½ | 5.8 | 30 | 0.24 |
| 3 | 7.5 | 20 | 0.38 |
| 4 | 10.8 | 11 | 0.66 |
| 5 | 14.6 | 7 | 1.03 |
| 6 | 19.0 | 5 | 1.5 |
| 8 | 25.5 | 3 | 2.6 |
| 10 | 40.5 | 1.8 | 4.1 |
| 12 | 53.5 | 1.2 | 5.9 |

Figure A.20  Weight of steel pipe and contained water.

| Nominal rod diameter, in | Root area of thread, in² | Maximum safe load at rod temperature of 650°F, lb |
|---|---|---|
| ¼ | 0.027 | 240 |
| ⁵⁄₁₆ | 0.046 | 410 |
| ⅜ | 0.068 | 610 |
| ½ | 0.126 | 1,130 |
| ⅝ | 0.202 | 1,810 |
| ¾ | 0.302 | 2,710 |
| ⅞ | 0.419 | 3,770 |
| 1 | 0.552 | 4,960 |
| 1⅛ | 0.693 | 6,230 |
| 1¼ | 0.889 | 8,000 |
| 1⅜ | 1.053 | 9,470 |
| 1½ | 1.293 | 11,630 |
| 1⅝ | 1.515 | 13,630 |
| 1¾ | 1.744 | 15,690 |
| 1⅞ | 2.048 | 18,430 |
| 2 | 2.292 | 20,690 |
| 2¼ | 3.021 | 27,200 |
| 2½ | 3.716 | 33,500 |
| 2¾ | 4.619 | 41,600 |
| 3 | 5.621 | 50,600 |
| 3¼ | 6.720 | 60,500 |
| 3½ | 7.918 | 71,260 |

Figure A.21  Load rating of threaded rods.

| Pipe size, in | Rod size, in |
|---|---|
| 2 and smaller | 3/8 |
| 2½ to 3½ | ½ |
| 4 and 5 | 5/8 |
| 6 | ¾ |
| 8 to 12 | 7/8 |
| 14 and 16 | 1 |
| 18 | 1⅛ |
| 20 | 1¼ |
| 24 | 1½ |

**Figure A.22** Recommended rod size for individual pipes.

| Pipe | Weight factor* |
|---|---|
| Aluminum | 0.35 |
| Brass | 1.12 |
| Cast iron | 0.91 |
| Copper | 1.14 |
| Stainless steel | 1.0 |
| Carbon steel | 1.0 |
| Wrought iron | 0.98 |

*Average plastic pipe weights one-fifth as much as carbon steel pipe.

**Figure A.23** Relative weight factor for metal pipe.

| To change | To | Multiply by |
|---|---|---|
| Inches | Millimeters | 25.4 |
| Feet | Meters | .3048 |
| Miles | Kilometers | 1.6093 |
| Square inches | Square centimeters | 6.4515 |
| Square feet | Square meters | .09290 |
| Acres | Hectares | .4047 |
| Acres | Square kilometers | .00405 |
| Cubic inches | Cubic centimeters | 16.3872 |
| Cubic feet | Cubic meters | .02832 |
| Cubic yards | Cubic meters | .76452 |
| Cubic inches | Liters | .01639 |
| U.S. gallons | Liters | 3.7854 |
| Ounces (avoirdupois) | Grams | 28.35 |
| Pounds | Kilograms | .4536 |
| Lbs. per sq. in. (P.S.I.) | Kg.'s per sq. cm. | .0703 |
| Lbs. per cu. ft. | Kg.'s per cu. meter | 16.0189 |
| Tons (2000 lbs.) | Metric tons (1000 kg.) | .9072 |
| Horsepower | Kilowatts | .746 |

**Figure A.24** English to metric conversion.

| Quantity | Equals |
|---|---|
| 100 sq. millimeters | 1 sq. centimeter |
| 100 sq. centimeters | 1 sq. decimeter |
| 100 sq. decimeters | 1 sq. meter |

Figure A.25  Metric square measure.

| Unit | Equals |
|---|---|
| 1 cubic meter | 35.314 cubic feet |
| | 1.308 cubic yards |
| | 264.2 U.S. gallons (231 cubic inches) |
| 1 cubic decimeter | 61.0230 cubic inches |
| | .0353 cubic feet |
| 1 cubic centimeter | .061 cubuic inch |
| 1 liter | 1 cubic decimeter |
| | 61.0230 cubic inches |
| | 0.0353 cubic foot |
| | 1.0567 quarts (U.S.) |
| | 0.2642 gallon (U.S.) |
| | 2.2020 lb. of water at 62°F. |
| 1 cubic yard | .7645 cubic meter |
| 1 cubic foot | .02832 cubic meter |
| | 28.317 cubic decimeters |
| | 28.317 liters |
| 1 cubic inch | 16.383 cubic centimeters |
| 1 gallon (British) | 4.543 liters |
| 1 gallon (U.S.) | 3.785 liters |
| 1 gram | 15.432 grains |
| 1 kilogram | 2.2046 pounds |
| 1 metric ton | .9842 ton of 2240 pounds |
| | 19.68 cwts. |
| | 2204.6 pounds |
| 1 grain | .0648 gram |
| 1 ounce avoirdupois | 28.35 grams |
| 1 pound | .4536 kilograms |
| 1 ton of 2240 lb. | 1.1016 metric tons |
| | 1016 kilograms |

Figure A.26  Measures of volume and capacity.

| To change | To | Multiply by |
|---|---|---|
| Atmospheres | Pounds per square inch | 14.696 |
| Atmospheres | Inches of mercury | 29.92 |
| Atmospheres | Feet of water | 34 |
| Long tons | Pounds | 2240 |
| Short tons | Pounds | 2000 |
| Short tons | Long tons | 0.89295 |

Figure 33.27  Measurement conversion factors.

| To find | Multiply | By |
|---|---|---|
| Microns | Mils | 25.4 |
| Centimeters | Inches | 2.54 |
| Meters | Feet | 0.3048 |
| Meters | Yards | 0.19144 |
| Kilometers | Miles | 1.609344 |
| Grams | Ounces | 28.349523 |
| Kilograms | Pounds | 0.4539237 |
| Liters | Gallons (U.S.) | 3.7854118 |
| Liters | Gallons (Imperial) | 4.546090 |
| Milliliters (cc) | Fluid ounces | 29.573530 |
| Milliliters (cc) | Cubic inches | 16.387064 |
| Square centimeters | Square inches | 6.4516 |
| Square meters | Square feet | 0.09290304 |
| Square meters | Square yards | 0.83612736 |
| Cubic meters | Cubic feet | $2.8316847 \times 10^{-2}$ |
| Cubic meters | Cubic yards | 0.76455486 |
| Joules | BTU | 1054.3504 |
| Joules | Foot-pounds | 1.35582 |
| Kilowatts | BTU per minute | 0.01757251 |
| Kilowatts | Foot-pounds per minute | $2.2597 \times 10^{-5}$ |
| Kilowatts | Horsepower | 0.7457 |
| Radians | Degrees | 0.017453293 |
| Watts | BTU per minute | 17.5725 |

Figure A.28  Conversion factors in converting from customary units to metric units.

| To change | To | Multiply by |
|---|---|---|
| Inches | Feet | 0.0833 |
| Inches | Millimeters | 25.4 |
| Feet | Inches | 12 |
| Feet | Yards | 0.3333 |
| Yards | Feet | 3 |
| Square inches | Square feet | 0.00694 |
| Square feet | Square inches | 144 |
| Square feet | Square yards | 0.11111 |
| Square yards | Square feet | 9 |
| Cubic inches | Cubic feet | 0.00058 |
| Cubic feet | Cubic inches | 1728 |
| Cubic feet | Cubic yards | 0.03703 |
| Gallons | Cubic inches | 231 |
| Gallons | Cubic feet | 0.1337 |
| Gallons | Pounds of water | 8.33 |
| Pounds of water | Gallons | 0.12004 |
| Ounces | Pounds | 0.0625 |
| Pounds | Ounces | 16 |
| Inches of water | Pounds per square inch | 0.0361 |
| Inches of water | Inches of mercury | 0.0735 |
| Inches of water | Ounces per square inch | 0.578 |
| Inches of water | Pounds per square foot | 5.2 |
| Inches of mercury | Inches of water | 13.6 |
| Inches of mercury | Feet of water | 1.1333 |
| Inches of mercury | Pounds per square inch | 0.4914 |
| Ounces per square inch | Inches of mercury | 0.127 |
| Ounces per square inch | Inches of water | 1.733 |
| Pounds per square inch | Inches of water | 27.72 |
| Pounds per square inch | Feet of water | 2.310 |
| Pounds per square inch | Inches of mercury | 2.04 |
| Pounds per square inch | Atmospheres | 0.0681 |
| Feet of water | Pounds per square inch | 0.434 |
| Feet of water | Pounds per square foot | 62.5 |
| Feet of water | Inches of mercury | 0.8824 |

Figure A.29  Measurement conversion factors.

| Pipe size | Projected flow rate (gallons per minute) |
|---|---|
| ½ inch | 2 to 5 |
| ¾ inch | 5 to 10 |
| 1 inch | 10 to 20 |
| 1¼ inch | 20 to 30 |
| 1½ inch | 30 to 40 |

Figure A.30 Projected flow rates for various pipe sizes.

| Material | Weight in pounds per cubic inch | Weight in pounds per cubic foot |
|---|---|---|
| Aluminum | .093 | 160 |
| Antimony | .2422 | 418 |
| Brass | .303 | 524 |
| Bronze | .320 | 552 |
| Chromium | .2348 | 406 |
| Copper | .323 | 558 |
| Gold | .6975 | 1205 |
| Iron (cast) | .260 | 450 |
| Iron (wrought) | .2834 | 490 |
| Lead | .4105 | 710 |
| Manganese | .2679 | 463 |
| Mercury | .491 | 849 |
| Molybdenum | .309 | 534 |
| Monel | .318 | 550 |
| Platinum | .818 | 1413 |
| Steel (mild) | .2816 | 490 |
| Steel (stainless) | .277 | 484 |
| Tin | .265 | 459 |
| Titanium | .1278 | 221 |
| Zinc | .258 | 446 |

Figure A.31 Weight of various materials.

| GPM | Liters/Minute |
|---|---|
| 1 | 3.75 |
| 2 | 6.50 |
| 3 | 11.25 |
| 4 | 15.00 |
| 5 | 18.75 |
| 6 | 22.50 |
| 7 | 26.25 |
| 8 | 30.00 |
| 9 | 33.75 |
| 10 | 37.50 |

**Figure A.32** Flow rate conversion from gallons per minute to approximate liters per minute.

| Pounds per square inch | Feet head |
|---|---|
| 1 | 2.31 |
| 2 | 4.62 |
| 3 | 6.93 |
| 4 | 9.24 |
| 5 | 11.54 |
| 6 | 13.85 |
| 7 | 16.16 |
| 8 | 18.47 |
| 9 | 20.78 |
| 10 | 23.09 |
| 15 | 34.63 |
| 20 | 46.18 |
| 25 | 57.72 |
| 30 | 69.27 |
| 40 | 92.36 |
| 50 | 115.45 |
| 60 | 138.54 |
| 70 | 161.63 |
| 80 | 184.72 |
| 90 | 207.81 |
| 100 | 230.90 |
| 110 | 253.98 |
| 120 | 277.07 |
| 130 | 300.16 |
| 140 | 323.25 |
| 150 | 346.34 |
| 160 | 369.43 |
| 170 | 392.52 |
| 180 | 415.61 |
| 200 | 461.78 |
| 250 | 577.24 |
| 300 | 692.69 |
| 350 | 808.13 |
| 400 | 922.58 |
| 500 | 1154.48 |
| 600 | 1385.39 |
| 700 | 1616.30 |
| 800 | 1847.20 |
| 900 | 2078.10 |
| 1000 | 2309.00 |

**Figure A.33** Water pressure in pounds with equivalent feet heads.

| Feet head | Pounds per square inch | Feet head | Pounds per square inch |
|---|---|---|---|
| 1 | 0.43 | 50 | 21.65 |
| 2 | 0.87 | 60 | 25.99 |
| 3 | 1.30 | 70 | 30.32 |
| 4 | 1.73 | 80 | 34.65 |
| 5 | 2.17 | 90 | 38.98 |
| 6 | 2.60 | 100 | 43.34 |
| 7 | 3.03 | 110 | 47.64 |
| 8 | 3.46 | 120 | 51.97 |
| 9 | 3.90 | 130 | 56.30 |
| 10 | 4.33 | 140 | 60.63 |
| 15 | 6.50 | 150 | 64.96 |
| 20 | 8.66 | 160 | 69.29 |
| 25 | 10.83 | 170 | 73.63 |
| 30 | 12.99 | 180 | 77.96 |
| 40 | 17.32 | 200 | 86.62 |

Figure A.34  Water feet head to pounds per square inch.

| Vacuum in inches of mercury | Boiling point |
|---|---|
| 29 | 76.62 |
| 28 | 99.93 |
| 27 | 114.22 |
| 26 | 124.77 |
| 25 | 133.22 |
| 24 | 140.31 |
| 23 | 146.45 |

Figure A.35  Boiling points of water at various pressures.

| To change | to | Multiply by |
|---|---|---|
| Inches | Feet | 0.0833 |
| Inches | Millimeters | 25.4 |
| Feet | Inches | 12 |
| Feet | Yards | 0.3333 |
| Yards | Feet | 3 |
| Square inches | Square feet | 0.00694 |
| Square feet | Square inches | 144 |
| Square feet | Square yards | 0.11111 |
| Square yards | Square feet | 9 |
| Cubic inches | Cubic feet | 0.00058 |
| Cubic feet | Cubic inches | 1728 |
| Cubic feet | Cubic yards | 0.03703 |
| Gallons | Cubic inches | 231 |
| Gallons | Cubic feet | 0.1337 |
| Gallons | Pounds of water | 8.33 |
| Pounds of water | Gallons | 0.12004 |
| Ounces | Pounds | 0.0625 |
| Pounds | Ounces | 16 |
| Inches of water | Pounds per square inch | 0.0361 |
| Inches of water | Inches of mercury | 0.0735 |
| Inches of water | Ounces per square inch | 0.578 |
| Inches of water | Pounds per square foot | 5.2 |
| Inches of mercury | Inches of water | 13.6 |
| Inches of mercury | Feet of water | 1.1333 |
| Inches of mercury | Pounds per square inch | 0.4914 |
| Ounces per square inch | Inches of mercury | 0.127 |
| Ounces per square inch | Inches of water | 1.733 |
| Pounds per square inch | Inches of water | 27.72 |
| Pounds per square inch | Feet of water | 2.310 |
| Pounds per square inch | Inches of mercury | 2.04 |
| Pounds per square inch | Atmospheres | 0.0681 |
| Feet of water | Pounds per square inch | 0.434 |
| Feet of water | Pounds per square foot | 62.5 |
| Feet of water | Inches of mercury | 0.8824 |
| Atmospheres | Pounds per square inch | 14.696 |
| Atmospheres | Inches of mercury | 29.92 |
| Atmospheres | Feet of water | 34 |
| Long tons | Pounds | 2240 |
| Short tons | Pounds | 2000 |
| Short tons | Long tons | 0.89295 |

Figure A.36  Measurement conversion factors.

| To change | to | Multiply by |
|---|---|---|
| Inches | Feet | 0.0833 |
| Inches | Millimeters | 25.4 |
| Feet | Inches | 12 |
| Feet | Yards | 0.3333 |
| Yards | Feet | 3 |
| Square inches | Square feet | 0.00694 |
| Square feet | Square inches | 144 |
| Square feet | Square yards | 0.11111 |
| Square yards | Square feet | 9 |
| Cubic inches | Cubic feet | 0.00058 |
| Cubic feet | Cubic inches | 1728 |
| Cubic feet | Cubic yards | 0.03703 |
| Cubic yards | Cubic feet | 27 |
| Cubic inches | Gallons | 0.00433 |
| Cubic feet | Gallons | 7.48 |
| Gallons | Cubic inches | 231 |
| Gallons | Cubic feet | 0.1337 |
| Gallons | Pounds of water | 8.33 |
| Pounds of water | Gallons | 0.12004 |
| Ounces | Pounds | 0.0625 |
| Pounds | Ounces | 16 |
| Inches of water | Pounds per square inch | 0.0361 |
| Inches of water | Inches of mercury | 0.0735 |
| Inches of water | Ounces per square inch | 0.578 |
| Inches of water | Pounds per square foot | 5.2 |
| Inches of mercury | Inches of water | 13.6 |
| Inches of mercury | Feet of water | 1.1333 |
| Inches of mercury | Feet of water | 0.4914 |
| Ounces per square inch | Pounds per square inch | 0.127 |
| Ounces per square inch | Inches of mercury | 1.733 |
| Pounds per square inch | Inches of water | 27.72 |
| Pounds per square inch | Feet of water | 2.310 |
| Pounds per square inch | Inches of mercury | 2.04 |
| Pounds per square inch | Atmospheres | 0.0681 |
| Feet of water | Pounds per square inch | 0.434 |
| Feet of water | Pounds per square foot | 62.5 |
| Feet of water | Inches of mercury | 0.8824 |
| Atmospheres | Pounds per square inch | 14.696 |
| Atmospheres | Inches of mercury | 29.92 |
| Atmospheres | Feet of water | 34 |
| Long tons | Pounds | 2240 |
| Short tons | Pounds | 2000 |
| Short tons | Long tons | 0.89295 |

Figure A.37  Useful multipliers.

> Outside design temperature = average of lowest recorded temperature in each month from October to March
>
> Inside design temperature = 70°F or as specified by owner
>
> A degree day is one day multiplied by the number of Fahrenheit degrees the mean temperature is below 65°F. The number of degree days in a year is a good guideline for designing heating and insulation systems.

Figure A.38  Design temperatures.

| Temperature (°F) | Steel | Cast iron | Brass and copper |
|---|---|---|---|
| 0   | 0    | 0    | 0    |
| 20  | 0.15 | 0.10 | 0.25 |
| 40  | 0.30 | 0.25 | 0.45 |
| 60  | 0.45 | 0.40 | 0.65 |
| 80  | 0.60 | 0.55 | 0.90 |
| 100 | 0.75 | 0.70 | 1.15 |
| 120 | 0.90 | 0.85 | 1.40 |
| 140 | 1.10 | 1.00 | 1.65 |
| 160 | 1.25 | 1.15 | 1.90 |
| 180 | 1.45 | 1.30 | 2.15 |
| 200 | 1.60 | 1.50 | 2.40 |
| 220 | 1.80 | 1.65 | 2.65 |
| 240 | 2.00 | 1.80 | 2.90 |
| 260 | 2.15 | 1.95 | 3.15 |
| 280 | 2.35 | 2.15 | 3.45 |
| 300 | 2.50 | 2.35 | 3.75 |
| 320 | 2.70 | 2.50 | 4.05 |
| 340 | 2.90 | 2.70 | 4.35 |
| 360 | 3.05 | 2.90 | 4.65 |
| 380 | 3.25 | 3.10 | 4.95 |
| 400 | 3.45 | 3.30 | 5.25 |
| 420 | 3.70 | 3.50 | 5.60 |
| 440 | 3.95 | 3.75 | 5.95 |
| 460 | 4.20 | 4.00 | 6.30 |
| 480 | 4.45 | 4.25 | 6.65 |
| 500 | 4.70 | 4.45 | 7.05 |

Figure A.39  Steam pipe expansion (inches increase per 100 inches).

| Unit | Symbol |
|---|---|
| *Length* | |
| Millimeter | mm |
| Centimeter | cm |
| Meter | m |
| Kilometer | km |
| *Area* | |
| Square millimeter | $mm^2$ |
| Square centimeter | $cm^2$ |
| Square decimeter | $dm^2$ |
| Square meter | $m^2$ |
| Square kilometer | $km^2$ |
| *Volume* | |
| Cubic centimeter | $cm^3$ |
| Cubic decimeter | $dm^3$ |
| Cubic meter | $m^3$ |
| *Mass* | |
| Milligram | mg |
| Gram | g |
| Kilogram | kg |
| Tonne | t |
| *Temperature* | |
| Degrees Celsius | °C |
| Kelvin | K |
| *Time* | |
| Second | s |
| *Plane angle* | |
| Radius | rad |
| *Force* | |
| Newton | N |
| *Energy, work, quantity of heat* | |
| Joule | J |
| Kilojoule | kJ |
| Megajoule | MJ |
| *Power, heat flow rate* | |
| Watt | W |
| Kilowatt | kW |
| *Pressure* | |
| Pascal | Pa |
| Kilopascal | kPa |
| Megapascal | MPa |
| *Velocity, speed* | |
| Meter per second | m/s |
| Kilometer per hour | km/h |

**Figure A.40** Metric abbreviations.

| Appliance | Size (inches) |
|---|---|
| Clothes washer | 2 |
| Bathtub with or without shower | 1½ |
| Bidet | 1½ |
| Dental unit or cuspidor | 1¼ |
| Drinking fountain | 1¼ |
| Dishwasher, domestic | 1½ |
| Dishwasher, commercial | 2 |
| Floor drain | 2, 3, or 4 |
| Lavatory | 1¼ |
| Laundry tray | 1½ |
| Shower stall, domestic | 2 |
| Sinks: | |
|   Combination, sink and tray (with disposal unit) | 1½ |
|   Combination, sink and tray (with one trap) | 1½ |
|   Domestic, with or without disposal unit | 1½ |
|   Surgeon's | 1½ |
|   Laboratory | 1½ |
|   Flushrim or bedpan washer | 3 |
|   Service sink | 2 or 3 |
|   Pot or scullery sink | 2 |
|   Soda fountain | 1½ |
|   Commercial, flat rim, bar, or counter sink | 1½ |
|   Wash sinks circular or multiple | 1½ |
| Urinals: | |
|   Pedestal | 3 |
|   Wall-hung | 1½ or 2 |
|   Trough (per 6-ft section) | 1½ |
|   Stall | 2 |
|   Water closet | 3 |

**Figure A.41** Common trap sizes.

| | |
|---|---|
| 1 ft$^3$ | 62.4 lbs |
| 1 ft$^3$ | 7.48 gal |
| 1 gal | 8.33 lbs |
| 1 gal | 0.1337 ft$^3$ |

**Figure A.42** Water weight.

| | |
|---|---|
| 1 ft$^3$ | 62.4 lbs |
| 1 ft$^3$ | 7.48 gal |
| 1 gal | 8.33 lbs |
| 1 gal | 0.1337 ft$^3$ |

**Figure A.43** Water volume to weight.

| Category | Estimated water usage per day |
|---|---|
| Barber shop | 100 gal per chair |
| Beauty shop | 125 gal per chair |
| Boarding school, elementary | 75 gal per student |
| Boarding school, secondary | 100 gal per student |
| Clubs, civic | 3 gal per person |
| Clubs, country | 25 gal per person |
| College, day students | 25 gal per student |
| College, junior | 100 gal per student |
| College, senior | 100 gal per student |
| Dentist's office | 750 gal per chair |
| Department store | 40 gal per employee |
| Drugstore | 500 gal per store |
| Drugstore with fountain | 2000 gal per store |
| Elementary school | 16 gal per student |
| Hospital | 400 gal per patient |
| Industrial plant | 30 gal per employee + process water |
| Junior and senior high school | 25 gal per student |
| Laundry | 2000–20,000 gal |
| Launderette | 1000 gal per unit |
| Meat market | 5 gal per 100 ft$^2$ of floor area |
| Motel or hotel | 125 gal per room |
| Nursing home | 150 gal per patient |
| Office building | 25 gal per employee |
| Physician's office | 200 gal per examining room |
| Prison | 60 gal per inmate |
| Restaurant | 20–120 gal per seat |
| Rooming house | 100 gal per tenant |
| Service station | 600–1500 gal per stall |
| Summer camp | 60 gal per person |
| Theater | 3 gal per seat |

Figure A.44  Estimating guidelines for daily water usage.

| Unit | Pounds per square inch | Feet of water | Meters of water | Inches of mercury | Atmospheres |
|---|---|---|---|---|---|
| 1 pound per square inch | 1.0 | 2.31 | 0.704 | 2.04 | 0.0681 |
| 1 foot of water | 0.433 | 1.0 | 0.305 | 0.882 | 0.02947 |
| 1 meter of water | 1.421 | 3.28 | 1.00 | 2.89 | 0.0967 |
| 1 inch of mercury | 0.491 | 1.134 | 0.3456 | 1.00 | 0.0334 |
| 1 atmosphere (sea level) | 14.70 | 33.93 | 10.34 | 29.92 | 1.0000 |

Figure A.45  Conversion of water values.

| To convert | Multiply by | To obtain |
|---|---|---|
| | **A** | |
| acres | $4.35 \times 10^4$ | sq. ft. |
| acres | $4.047 \times 10^3$ | sq. meters |
| acre-feet | $4.356 \times 10^4$ | cu. feet |
| acre-feet | $3.259 \times 10^5$ | gallons |
| atmospheres | $2.992 \times 10^1$ | in. of mercury (at 0°C.) |
| atmospheres | 1.0333 | kgs./sq. cm. |
| atmospheres | $1.0333 \times 10^4$ | kgs./sq. meter |
| atmospheres | $1.47 \times 10^1$ | pounds/sq. in. |
| | **B** | |
| barrels (u.s., liquid) | $3.15 \times 10^1$ | gallons |
| barrels (oil) | $4.2 \times 10^1$ | gallons (oil) |
| bars | $9.869 \times 10^{-1}$ | atmospheres |
| btu | $7.7816 \times 10^2$ | foot-pounds |
| btu | $3.927 \times 10^{-4}$ | horsepower-hours |
| btu | $2.52 \times 10^{-1}$ | kilogram-calories |
| btu | $2.928 \times 10^{-4}$ | kilowatt-hours |
| btu/hr. | $2.162 \times 10^{-1}$ | ft. pounds/sec. |
| btu/hr. | $3.929 \times 10^{-4}$ | horsepower |
| btu/hr. | $2.931 \times 10^{-1}$ | watts |
| btu/min. | $1.296 \times 10^1$ | ft.-pounds/sec. |
| btu/min. | $1.757 \times 10^{-2}$ | kilowatts |
| | **C** | |
| centigrade (degrees) | (°C × 9/5) + 32 | fahrenheit (degrees) |
| centigrade (degrees) | °C + 273.18 | kelvin (degrees) |
| centigrams | $1. \times 10^{-2}$ | grams |
| centimeters | $3.281 \times 10^{-2}$ | feet |
| centimeters | $3.937 \times 10^{-1}$ | inches |
| centimeters | $1. \times 10^{-5}$ | kilometers |
| centimeters | $1. \times 10^{-2}$ | meters |
| centimeters | $1. \times 10^1$ | millimeters |
| centimeters | $3.937 \times 10^2$ | mils |
| centimeters of mercury | $1.316 \times 10^{-2}$ | atmospheres |
| centimeters of mercury | $4.461 \times 10^{-1}$ | ft. of water |
| centimeters of mercury | $1.934 \times 10^{-1}$ | pounds/sq. in. |
| centimeters/sec. | 1.969 | feet/min. |
| centimeters/sec. | $3.281 \times 10^{-2}$ | feet/sec. |
| centimeters/sec. | $6.0 \times 10^{-1}$ | meters/min. |
| centimeters/sec./sec. | $3.281 \times 10^{-2}$ | ft./sec./sec. |
| cubic centimeters | $3.531 \times 10^{-5}$ | cubic ft. |
| cubic centimeters | $6.102 \times 10^{-2}$ | cubic in. |
| cubic centimeters | $1.0 \times 10^{-6}$ | cubic meters |
| cubic centimeters | $2.642 \times 10^{-4}$ | gallons (u.s. liquid) |
| cubic centimeters | $2.113 \times 10^{-3}$ | pints (u.s. liquid) |
| cubic centimeters | $1.057 \times 10^{-3}$ | quarts (u.s. liquid) |

**Figure A.46a** Measurement conversions. *(Continued)*

**Figure A.46b** *(Continued)* Measurement conversions.

| To convert | Multiply by | To obtain |
|---|---|---|
| cubic feet | $2.8320 \times 10^4$ | cu. cms. |
| cubic feet | $1.728 \times 10^3$ | cu. inches |
| cubic feet | $2.832 \times 10^{-2}$ | cu. meters |
| cubic feet | $7.48052$ | gallons (u.s. liquid) |
| cubic feet | $5.984 \times 10^1$ | pints (u.s. liquid) |
| cubic feet | $2.992 \times 10^1$ | quarts (u.s. liquid) |
| cubic feet/min. | $4.72 \times 10^1$ | cu. cms./sec. |
| cubic feet/min. | $1.247 \times 10^{-1}$ | gallons/sec. |
| cubic feet/min. | $4.720 \times 10^{-1}$ | liters/sec. |
| cubic feet/min. | $6.243 \times 10^1$ | pounds water/min. |
| cubic feet/sec. | $6.46317 \times 10^{-1}$ | million gals./day |
| cubic feet/sec. | $4.48831 \times 10^2$ | gallons/min. |
| cubic inches | $5.787 \times 10^{-4}$ | cu. ft. |
| cubic inches | $1.639 \times 10^{-5}$ | cu. meters |
| cubic inches | $2.143 \times 10^{-5}$ | cu. yards |
| cubic inches | $4.329 \times 10^{-3}$ | gallons |
| **D** | | |
| degrees (angle) | $1.745 \times 10^{-2}$ | radians |
| degrees (angle) | $3.6 \times 10^3$ | seconds |
| degrees/sec. | $2.778 \times 10^{-3}$ | revolutions/sec. |
| dynes/sq. cm. | $4.015 \times 10^{-4}$ | in. of water (at 4°C.) |
| dynes | $1.020 \times 10^{-6}$ | kilograms |
| dynes | $2.248 \times 10^{-6}$ | pounds |
| **F** | | |
| fathoms | $1.8288$ | meters |
| fathoms | $6.0$ | feet |
| feet | $3.048 \times 10^1$ | centimeters |
| feet | $3.048 \times 10^{-1}$ | meters |
| feet of water | $2.95 \times 10^{-2}$ | atmospheres |
| feet of water | $3.048 \times 10^{-2}$ | kgs./sq. cm. |
| feet of water | $6.243 \times 10^1$ | pounds/sq. ft. |
| feet/min. | $5.080 \times 10^{-1}$ | cms./sec. |
| feet/min. | $1.667 \times 10^{-2}$ | feet/sec. |
| feet/min. | $3.048 \times 10^{-1}$ | meters/min. |
| feet/min. | $1.136 \times 10^{-2}$ | miles/hr. |
| feet/sec. | $1.829 \times 10^1$ | meters/min. |
| feet/100 feet | $1.0$ | per cent grade |
| foot-pounds | $1.286 \times 10^{-3}$ | btu |
| foot-pounds | $1.356 \times 10^7$ | ergs |
| foot-pounds | $3.766 \times 10^{-7}$ | kilowatt-hrs. |
| foot-pounds/min. | $1.286 \times 10^{-3}$ | btu/min. |
| foot-pounds/min. | $3.030 \times 10^{-5}$ | horsepower |
| foot-pounds/min. | $3.241 \times 10^{-4}$ | kg.-calories/min. |
| foot-pounds/sec. | $4.6263$ | btu/hr. |
| foot-pounds/sec. | $7.717 \times 10^{-2}$ | btu/min. |
| foot-pounds/sec. | $1.818 \times 10^{-3}$ | horsepower |
| foot-pounds/sec. | $1.356 \times 10^{-3}$ | kilowatts |
| furlongs | $1.25 \times 10^{-1}$ | miles (u.s.) |

Figure A.46c *(Continued)* Measurement conversions.

| To convert | Multiply by | To obtain |
|---|---|---|
| | **G** | |
| gallons | $3.785 \times 10^3$ | cu. cms. |
| gallons | $1.337 \times 10^{-1}$ | cu. feet |
| gallons | $2.31 \times 10^2$ | cu. inches |
| gallons | $3.785 \times 10^{-3}$ | cu. meters |
| gallons | $4.951 \times 10^{-3}$ | cu. yards |
| gallons | 3.785 | liters |
| gallons (liq. br. imp.) | 1.20095 | gallons (u.s. liquid) |
| gallons (u.s.) | $8.3267 \times 10^{-1}$ | gallons (imp.) |
| gallons of water | 8.337 | pounds of water |
| gallons/min. | $2.228 \times 10^{-3}$ | cu. feet/sec. |
| gallons/min. | $6.308 \times 10^{-2}$ | liters/sec. |
| gallons/min. | 8.0208 | cu. feet/hr. |
| | **H** | |
| horsepower | $4.244 \times 10^1$ | btu/min. |
| horsepower | $3.3 \times 10^4$ | foot-lbs./min. |
| horsepower | $5.50 \times 10^2$ | foot-lbs./sec. |
| horsepower (metric) | $9.863 \times 10^{-1}$ | horsepower |
| horsepower | 1.014 | horsepower (metric) |
| horsepower | $7.457 \times 10^{-1}$ | kilowatts |
| horsepower | $7.457 \times 10^2$ | watts |
| horsepower (boiler) | $3.352 \times 10^4$ | btu/hr. |
| horsepower (boiler) | 9.803 | kilowatts |
| horsepower-hours | $2.547 \times 10^3$ | btu |
| horsepower-hours | $1.98 \times 10^6$ | foot-lbs. |
| horsepower-hours | $6.4119 \times 10^5$ | gram-calories |
| hours | $5.952 \times 10^{-3}$ | weeks |
| | **I** | |
| inches | 2.540 | centimeters |
| inches | $2.540 \times 10^{-2}$ | meters |
| inches | $1.578 \times 10^{-5}$ | miles |
| inches | $2.54 \times 10^1$ | millimeters |
| inches | $1.0 \times 10^3$ | mils |
| inches | $2.778 \times 10^{-2}$ | yards |
| inches of mercury | $3.342 \times 10^{-2}$ | atmospheres |
| inches of mercury | 1.133 | feet of water |
| inches of mercury | $3.453 \times 10^{-2}$ | kgs./sq. cm. |
| inches of mercury | $3.453 \times 10^2$ | kgs./sq. meter |
| inches of mercury | $7.073 \times 10^1$ | pounds/sq. ft. |
| inches of mercury | $4.912 \times 10^{-1}$ | pounds/sq. in. |
| in. of water (at 4°C.) | $7.355 \times 10^{-2}$ | inches of mercury |
| in. of water (at 4°C.) | $2.54 \times 10^{-3}$ | kgs./sq. cm. |
| in. of water (at 4°C.) | 5.204 | pounds/sq. ft. |
| in. of water (at 4°C.) | $3.613 \times 10^{-2}$ | pounds/sq. in. |

**Figure A.46d** *(Continued)* Measurement conversions.

| To convert | Multiply by | To obtain |
|---|---|---|
| | **J** | |
| joules | $9.486 \times 10^{-4}$ | btu |
| joules/cm. | $1.0 \times 10^{7}$ | dynes |
| joules/cm. | $1.0 \times 10^{2}$ | joules/meter (newtons) |
| joules/cm. | $2.248 \times 10^{1}$ | pounds |
| | **K** | |
| kilograms | $9.80665 \times 10^{5}$ | dynes |
| kilograms | $1.0 \times 10^{3}$ | grams |
| kilograms | $2.2046$ | pounds |
| kilograms | $9.842 \times 10^{-4}$ | tons (long) |
| kilograms | $1.102 \times 10^{-3}$ | tons (short) |
| kilograms/sq. cm. | $9.678 \times 10^{-1}$ | atmospheres |
| kilograms/sq. cm. | $3.281 \times 10^{1}$ | feet of water |
| kilograms/sq. cm. | $2.896 \times 10^{1}$ | inches of mercury |
| kilograms/sq. cm. | $1.422 \times 10^{1}$ | pounds/sq. in. |
| kilometers | $1.0 \times 10^{5}$ | centimeters |
| kilometers | $3.281 \times 10^{3}$ | feet |
| kilometers | $3.937 \times 10^{4}$ | inches |
| kilometers | $1.0 \times 10^{3}$ | meters |
| kilometers | $6.214 \times 10^{-1}$ | miles (statute) |
| kilometers | $5.396 \times 10^{-1}$ | miles (nautical) |
| kilometers | $1.0 \times 10^{6}$ | millimeters |
| kilowatts | $5.692 \times 10^{1}$ | btu/min. |
| kilowatts | $4.426 \times 10^{4}$ | foot-lbs./min. |
| kilowatts | $7.376 \times 10^{2}$ | foot-lbs./sec. |
| kilowatts | $1.341$ | horsepower |
| kilowatts | $1.434 \times 10^{1}$ | kg.-calories/min. |
| kilowatts | $1.0 \times 10^{3}$ | watts |
| kilowatt-hrs. | $3.413 \times 10^{3}$ | btu |
| kilowatt-hrs. | $2.655 \times 10^{6}$ | foot-lbs. |
| kilowatt-hrs. | $8.5985 \times 10^{3}$ | gram calories |
| kilowatt-hrs. | $1.341$ | horsepower-hours |
| kilowatt-hrs. | $3.6 \times 10^{6}$ | joules |
| kilowatt-hrs. | $8.605 \times 10^{2}$ | kg.-calories |
| kilowatt-hrs. | $8.5985 \times 10^{3}$ | kg.-meters |
| kilowatt-hrs. | $2.275 \times 10^{1}$ | pounds of water raised from 62° to 212°F. |
| | **L** | |
| links (engineers) | $1.2 \times 10^{1}$ | inches |
| links (surveyors) | $7.92$ | inches |
| liters | $1.0 \times 10^{3}$ | cu. cm. |
| liters | $6.102 \times 10^{1}$ | cu. inches |
| liters | $1.0 \times 10^{-3}$ | cu. meters |
| liters | $2.642 \times 10^{-1}$ | gallons (u.s. liquid) |
| liters | $2.113$ | pints (u.s. liquid) |
| liters | $1.057$ | quarts (u.s. liquid) |

**Figure A.46e** *(Continued)* Measurement conversions.

| To convert | Multiply by | To obtain |
|---|---|---|
| | **M** | |
| meters | $1.0 \times 10^2$ | centimeters |
| meters | 3.281 | feet |
| meters | $3.937 \times 10^1$ | inches |
| meters | $1.0 \times 10^{-3}$ | kilometers |
| meters | $5.396 \times 10^{-4}$ | miles (nautical) |
| meters | $6.214 \times 10^{-4}$ | miles (statute) |
| meters | $1.0 \times 10^3$ | millimeters |
| meters/min. | 1.667 | cms./sec. |
| meters/min. | 3.281 | feet/min. |
| meters/min. | $5.468 \times 10^{-2}$ | feet/sec. |
| meters/min. | $6.0 \times 10^{-2}$ | kms./hr. |
| meters/min. | $3.238 \times 10^{-2}$ | knots |
| meters/min. | $3.728 \times 10^{-2}$ | miles/hr. |
| meters/sec. | $1.968 \times 10^2$ | feet/min. |
| meters/sec. | 3.281 | feet/sec. |
| meters/sec. | 3.6 | kilometers/hr. |
| meters/sec. | $6.0 \times 10^{-2}$ | kilometers/min. |
| meters/sec. | 2.237 | miles/hr. |
| meters/sec. | $3.728 \times 10^{-2}$ | miles/min. |
| miles (neutical) | $6.076 \times 10^3$ | feet |
| miles (statute) | $5.280 \times 10^3$ | feet |
| miles/hr. | $8.8 \times 10^1$ | ft./min. |
| millimeters | $1.0 \times 10^{-1}$ | centimeters |
| millimeters | $3.281 \times 10^{-3}$ | feet |
| millimeters | $3.937 \times 10^{-2}$ | inches |
| millimeters | $1.0 \times 10^{-1}$ | meters |
| minutes (time) | $9.9206 \times 10^{-5}$ | weeks |
| | **O** | |
| ounces | $2.8349 \times 10^1$ | grams |
| ounces | $6.25 \times 10^{-2}$ | pounds |
| ounces (fluid) | 1.805 | cu. inches |
| ounces (fluid) | $2.957 \times 10^{-2}$ | liters |
| | **P** | |
| parts/million | $5.84 \times 10^{-2}$ | grains/u.s. gal. |
| parts/million | $7.016 \times 10^{-2}$ | grains/imp. gal. |
| parts/million | 8.345 | pounds/million gal. |
| pints (liquid) | $4.732 \times 10^2$ | cubic cms. |
| pints (liquid) | $1.671 \times 10^{-2}$ | cubic ft. |
| pints (liquid) | $2.887 \times 10^1$ | cubic inches |
| pints (liqui) | $4.732 \times 10^{-4}$ | cubic meters |
| pints (liquid) | $1.25 \times 10^{-1}$ | gallons |
| pints (liquid) | $4.732 \times 10^{-1}$ | liters |
| pints (liquid) | $5.0 \times 10^{-1}$ | quarts (liquid) |

**Figure A.46f** *(Continued)* Measurement conversions.

| To convert | Multiply by | To obtain |
|---|---|---|
| pounds | $2.56 \times 10^2$ | drams |
| pounds | $4.448 \times 10^5$ | dynes |
| pounds | $7.0 \times 10^1$ | grains |
| pounds | $4.5359 \times 10^2$ | grams |
| pounds | $4.536 \times 10^{-1}$ | kilograms |
| pounds | $1.6 \times 10^1$ | ounces |
| pounds | $3.217 \times 10^1$ | pounds |
| pounds | 1.21528 | pounds (troy) |
| pounds of water | $1.602 \times 10^{-2}$ | cu. ft. |
| pounds of water | $2.768 \times 10^1$ | cu. inches |
| pounds of water | $1.198 \times 10^{-1}$ | gallons |
| pounds of water/min. | $2.670 \times 10^{-4}$ | cu. ft./sec. |
| pound-feet | $1.356 \times 10^7$ | cm.-dynes |
| pound-feet | $1.3825 \times 10^4$ | cm.-grams |
| pound-feet | $1.383 \times 10^{-1}$ | meter-kgs. |
| pounds/cu. ft. | $1.602 \times 10^{-2}$ | grams/cu. cm. |
| pounds/cu. ft. | $5.787 \times 10^{-4}$ | pounds/cu. inches |
| pounds/sq. in. | $6.804 \times 10^{-2}$ | atmospheres |
| pounds/sq. in. | 2.307 | feet of water |
| pounds/sq. in. | 2.036 | inches of mercury |
| pounds/sq. in. | $7.031 \times 10^2$ | kgs./sq. meter |
| pounds/sq. in. | $1.44 \times 10^2$ | pounds/sq. ft. |
| **Q** | | |
| quarts (dry) | $6.72 \times 10^1$ | cu. inches |
| quarts (liquid) | $9.464 \times 10^2$ | cu. cms. |
| quarts (liquid) | $3.342 \times 10^{-2}$ | cu. ft. |
| quarts (liquid) | $5.775 \times 10^1$ | cu. inches |
| quarts (liquid) | $2.5 \times 10^{-1}$ | gallons |
| **R** | | |
| revolutions | $3.60 \times 10^2$ | degrees |
| revolutions | 4.0 | quadrants |
| rods (surveyors' meas.) | 5.5 | yards |
| rods | $1.65 \times 10^1$ | feet |
| rods | $1.98 \times 10^2$ | inches |
| rods | $3.125 \times 10^{-3}$ | miles |

Figure A.46g *(Continued)* Measurement conversions.

| To convert | Multiply by | To obtain |
|---|---|---|
| **S** | | |
| slugs | $3.217 \times 10^1$ | pounds |
| square centimeters | $1.076 \times 10^{-3}$ | sq. feet |
| square centimeters | $1.550 \times 10^{-1}$ | sq. inches |
| square centimeters | $1.0 \times 10^{-4}$ | sq. meters |
| square centimeters | $1.0 \times 10^2$ | sq. millimeters |
| square feet | $2.296 \times 10^{-5}$ | acres |
| square feet | $9.29 \times 10^2$ | sq. cms. |
| square feet | $1.44 \times 10^2$ | sq. inches |
| square feet | $9.29 \times 10^{-2}$ | sq. meters |
| square feet | $3.587 \times 10^{-3}$ | sq. miles |
| square inches | $6.944 \times 10^{-3}$ | sq. ft. |
| square inches | $6.452 \times 10^2$ | sq. millimeters |
| square miles | $6.40 \times 10^2$ | acres |
| square miles | $2.788 \times 10^7$ | sq. ft. |
| square yards | $2.066 \times 10^{-4}$ | acres |
| square yards | $8.361 \times 10^3$ | sq. cms. |
| square yards | 9.0 | sq. ft. |
| square yards | $1.296 \times 10^3$ | sq. inches |
| **T** | | |
| temperature (°C.) +273 | 1.0 | absolute temperature (°K.) |
| temperature (°C.) +17.78 | 1.8 | temperature (°F.) |
| temperature (°F.) +460 | 1.0 | absolute temperature (°R.) |
| temperature (°F.) −32 | 8/9 | temperature (°C.) |
| tons (long) | $2.24 \times 10^2$ | pounds |
| tons (long) | 1.12 | tons (short) |
| tons (metric) | $2.205 \times 10^5$ | pounds |
| tons (short) | $2.0 \times 10^3$ | pounds |
| **W** | | |
| watts | 3.4129 | btu/hr. |
| watts | $5.688 \times 10^{-2}$ | btu/min. |
| watts | $4.427 \times 10^1$ | ft.-lbs/min. |
| watts | $7.378 \times 10^{-1}$ | ft.-lbs./sec |
| watts | $1.341 \times 10^{-3}$ | horsepower |
| watts | $1.36 \times 10^{-3}$ | horsepower (metric) |
| watts | $1.0 \times 10^{-3}$ | kilowatts |
| watt-hours | 3.413 | btu |
| watt-hours | $2.656 \times 10^3$ | foot-lbs. |
| watt-hours | $1.341 \times 10^{-3}$ | horsepower-hours |
| watt (international) | 1.000165 | watt (absolute) |
| weeks | $1.68 \times 10^2$ | hours |
| weeks | $1.008 \times 10^4$ | minutes |
| weeks | $6.048 \times 10^5$ | seconds |

Source: *Pump Handbook* by I. J. Karassik et al. Copyright 1976, McGraw-Hill, Inc.

| Nom. pipe size, in | Relative humidity, % | | | | | | | | | | | | | | | |
|---|---|---|---|---|---|---|---|---|---|---|---|---|---|---|---|---|
| | 20 | | | 50 | | | 70 | | | 80 | | | 90 | | | |
| | THK* | HG† | ST‡ | THK | HG | ST | THK | HG | ST | THK | HG | ST | THK | HG | ST | |
| 0.50 | | | | 0.5 | 4 | 64 | 0.5 | 4 | 64 | 0.5 | 4 | 64 | 1.5 | 2 | 68 | |
| 0.75 | | | | 0.5 | 4 | 64 | 0.5 | 4 | 64 | 0.5 | 4 | 64 | 1.5 | 3 | 67 | |
| 1.00 | | | | 0.5 | 6 | 63 | 0.5 | 6 | 63 | 1.0 | 4 | 66 | 1.5 | 3 | 67 | |
| 1.25 | | | | 0.5 | 6 | 63 | 0.5 | 6 | 63 | 1.0 | 5 | 65 | 1.5 | 3 | 67 | |
| 1.50 | | | | 0.5 | 8 | 62 | 0.5 | 8 | 62 | 1.0 | 5 | 66 | 1.5 | 4 | 67 | |
| 2.00 | | | | 0.5 | 8 | 63 | 0.5 | 8 | 63 | 1.0 | 6 | 66 | 1.5 | 4 | 67 | |
| 2.50 | | | | 0.5 | 10 | 63 | 0.5 | 10 | 63 | 1.0 | 6 | 66 | 1.5 | 5 | 67 | |
| 3.00 | | | | 0.5 | 12 | 62 | 0.5 | 12 | 62 | 1.0 | 8 | 65 | 1.5 | 6 | 67 | |
| 3.50 | Condensation | | | 0.5 | 14 | 61 | 0.5 | 14 | 61 | 1.0 | 7 | 66 | 1.5 | 6 | 67 | |
| 4.00 | control not | | | 0.5 | 15 | 62 | 0.5 | 15 | 62 | 1.0 | 9 | 65 | 1.5 | 7 | 67 | |
| 5.00 | required for this | | | 0.5 | 16 | 63 | 0.5 | 16 | 63 | 1.0 | 11 | 65 | 2.0 | 7 | 67 | |
| 6.00 | condition | | | 0.5 | 22 | 61 | 0.5 | 22 | 61 | 1.0 | 13 | 65 | 2.0 | 8 | 67 | |
| 8.00 | | | | 1.0 | 16 | 65 | 1.0 | 16 | 65 | 1.0 | 16 | 65 | 2.0 | 10 | 67 | |
| 10.00 | | | | 1.0 | 20 | 65 | 1.0 | 20 | 65 | 1.0 | 20 | 65 | 2.0 | 11 | 67 | |
| 12.00 | | | | 1.0 | 22 | 65 | 1.0 | 22 | 65 | 1.0 | 22 | 65 | 2.0 | 13 | 67 | |

*THK—Insulation thickness, inches.
†HG—Heat gain/lineal foot (pipe) 28 ft (flat).
‡ST—Surface temperature.

**Figure A.47** Insulation thickness to prevent condensation, 34 degree F service temperature and 70 degree F. ambient temperature.

| Nom. pipe size, in | Relative humidity, % | | | | | | | | | | | | | | | |
|---|---|---|---|---|---|---|---|---|---|---|---|---|---|---|---|---|
| | 20 | | | 50 | | | 70 | | | 80 | | | 90 | | | |
| | THK* | HG† | ST‡ | THK | HG | ST | THK | HG | ST | THK | HG | ST | THK | HG | ST | |
| 0.50 | | | | 0.5 | 2 | 66 | 0.5 | 2 | 66 | 0.5 | 2 | 66 | 1.0 | 2 | 68 | |
| 0.75 | | | | 0.5 | 2 | 67 | 0.5 | 2 | 67 | 0.5 | 2 | 67 | 0.5 | 2 | 67 | |
| 1.00 | | | | 0.5 | 3 | 66 | 0.5 | 3 | 66 | 0.5 | 3 | 66 | 1.0 | 2 | 68 | |
| 1.25 | | | | 0.5 | 3 | 66 | 0.5 | 3 | 66 | 0.5 | 3 | 66 | 1.0 | 3 | 67 | |
| 1.50 | | | | 0.5 | 4 | 65 | 0.5 | 4 | 65 | 0.5 | 4 | 65 | 1.0 | 3 | 67 | |
| 2.00 | | | | 0.5 | 5 | 66 | 0.5 | 5 | 66 | 0.5 | 5 | 66 | 1.0 | 3 | 67 | |
| 2.50 | | | | 0.5 | 5 | 65 | 0.5 | 5 | 65 | 0.5 | 5 | 65 | 1.0 | 4 | 67 | |
| 3.00 | | | | 0.5 | 7 | 65 | 0.5 | 7 | 65 | 0.5 | 7 | 65 | 1.0 | 4 | 67 | |
| 3.50 | Condensation | | | 0.5 | 8 | 65 | 0.5 | 8 | 65 | 0.5 | 8 | 65 | 1.0 | 4 | 68 | |
| 4.00 | control not | | | 0.5 | 8 | 65 | 0.5 | 8 | 65 | 0.5 | 8 | 65 | 1.0 | 5 | 67 | |
| 5.00 | required for this | | | 0.5 | 10 | 65 | 0.5 | 10 | 65 | 0.5 | 10 | 65 | 1.0 | 6 | 67 | |
| 6.00 | condition | | | 0.5 | 12 | 65 | 0.5 | 12 | 65 | 0.5 | 12 | 65 | 1.0 | 7 | 67 | |
| 8.00 | | | | 1.0 | 9 | 67 | 1.0 | 9 | 67 | 1.0 | 9 | 67 | 1.0 | 9 | 67 | |
| 10.00 | | | | 1.0 | 11 | 67 | 1.0 | 11 | 67 | 1.0 | 11 | 67 | 1.0 | 11 | 67 | |
| 12.00 | | | | 1.0 | 12 | 67 | 1.0 | 12 | 67 | 1.0 | 12 | 67 | 1.0 | 12 | 67 | |

*THK—Insulation thickness, inches.
†HG—Heat gain/lineal foot (pipe) 28 ft (flat).
‡ST—Surface temperature.

**Figure A.48** Insulation thickness to prevent condensation, 50 degree F service temperature and 70 degree F. ambient temperature.

| Fixture | Pressure, psi |
|---|---|
| Basin faucet | 8 |
| Basin faucet, self-closing | 12 |
| Sink faucet, ⅜ in (0.95 cm) | 10 |
| Sink faucet, ½ in (1.3 cm) | 5 |
| Dishwasher | 15–25 |
| Bathtub faucet | 5 |
| Laundry tub cock, ¼ in (0.64 cm) | 5 |
| Shower | 12 |
| Water closet ball cock | 15 |
| Water closet flush valve | 15–20 |
| Urinal flush valve | 15 |
| Garden hose, 50 ft (15 m), and sill cock | 30 |
| Water closet, blowout type | 25 |
| Urinal, blowout type | 25 |
| Water closet, low-silhouette tank type | 30–40 |
| Water closet, pressure tank | 20–30 |

Figure A.49  Minimum acceptable operating pressures for various fixtures.

| Use | Minimum temperature, °F |
|---|---|
| Lavatory: | |
|   Hand washing | 105 |
|   Shaving | 115 |
| Showers and tubs | 110 |
| Commercial and institutional laundry | 180 |
| Residential dishwashing and laundry | 140 |
| Commercial spray-type dishwashing as required by National Sanitation Foundation: | |
|   Single or multiple tank hood or rack type: | |
|     Wash | 150 |
|     Final rinse | 180 to 195 |
|   Single-tank conveyor type: | |
|     Wash | 160 |
|     Final rinse | 180 to 195 |
|   Single-tank rack or door type: | |
|     Single-temperature wash and rinse | 165 |
| Chemical sanitizing glasswasher: | |
|   Wash | 140 |
|   Rinse | 75 |

Figure A.50  Minimum hot water temperature for plumbing fixtures and equipment.

| Pipe | Fittings Schedule | Fittings Sizes | Maximum working pressure |
|---|---|---|---|
| 160 psi (SDR 26) (1102.4 kPa) | 40 | ½″ thru 8″ incl. (12.7 mm–203.2 mm) | 160 psi–1102.4 kPa |
| | 80 | ½″ thru 8″ incl. (12.7 mm–203.2 mm) | 160 psi–1102.4 kPa |
| 200 psi (SDR 21) (1378 kPa) | 40 | ½″ thru 4″ incl. (12.7 mm–101.6 mm) | 200 psi–1378 kPa |
| | 80 | ½″ thru 8″ incl. (12.7 mm–203.2 mm) | 200 psi–1378 kPa |
| 250 psi (SDR 17) (1722.5 kPa) | 40 | ½″ thru 3″ incl. (12.7 mm–76.2 mm) | 250 psi–1722.5 kPa |
| | 80 | ½″ thru 8″ incl. (12.7 mm–101.6 mm) | 250 psi–1722.5 kPa |
| 315 psi (SDR 13.5) (2170.4 kPa) | 40 | ½″ thru 1½″ incl. (12.7 mm–38.1 mm) | 315 psi–2170.4 kPa |
| | 80 | ½″ thru 4″ incl. (12.7 mm–101.6 mm) | 315 psi–2170.4 kPa |
| Schedule 40 | 40 80 | ½″ thru 1½″ incl. (12.7 mm–38.1 mm) | 320 psi–2204.8 kPa |
| | 40 80 | 2″ thru 4″ incl. (50.8 mm–101.6 mm) | 220 psi–1515.8 kPa |
| | 40 | 5″ thru 8″ incl. | 160 psi–1102.4 kPa |
| Schedule 80 | 40 | ½″ thru 1½″ incl. (12.7 mm–38.1 mm) | 320 psi–2204.8 kPa |
| | 40 | 2″ thru 4″ incl. (50.8 mm–101.6 mm) | 220 psi–1515.8 kPa |
| | 40 | 5″ thru 8″ incl. | 160 psi–1102.4 kPa |
| | 80 | ½″ thru 4″ incl. (12.7 mm–101.6 mm) | 320 psi–2204.8 kPa |
| | 80 | 5″ thru 8″ incl. (127 mm–203.2 mm) | 250 psi–1722.5 kPa |

Figure A.51  Maximum working pressures.

# Index

Abbreviations, p.348, pp.408 – 411
   metric, p.431
Absorption area, p.71
Access in the well-drilling process, p.113
Acid neutralizers, p.184
Acidic water, p.183
Activated carbon filters, p.185
Additives, p.254
Administration of code requirements, pp.1 – 4
   applicability, pp.2 – 3
   approval, p.5
After a well is in, pp.138 – 139
Aggregate, p.11, p.39
Alluvial deposits, p.21
Alluvium, p.11
Appeal, code, pp.8 – 9

Backfill, p.39, p.51
Bacteria, p.184
Bacteria, beneficial, pp.252 – 254
Bad ground, p.227
Bedrock, p.11, p.21, pp.25 – 26
Beds, p.69
Biological factors, p.183
Black-water sumps, pp.241 – 242, p.243
Bored wells, p.133
Brick and block, p.218
Burying a septic tank, pp.192 – 193

Cast iron pipe, p.413
Centrifugal pumps, p.394
Cesspool, p.11, p.73
Chamber systems, pp.191 – 192, p.208, pp.229 – 231
Change in occupancy, p.3
Chemical restoration, p.53
Chemicals, pp.180 – 183
Chlorides, p.181
   Cleaning, p.254
Clear-water wastes, p.11

Clogged drain fields, p.261
Code officials, pp.4 – 5, p.11
Code requirements, administration of (see Administration)
Code regulations, p.15 – 16
   limitations, p.15
Colluvial deposits, p.21
Colluvium, p.11
Color patterns, p.20
Color, soil, p.11, p.20
Confusion about quality, pp.90 – 93, pp.95 – 96
Connections, p.31, p.33 – 34
   prohibited connections, p.34
Conserving water, p.415
Construction, p.38 – 39
Construction documents, p.11
Contaminants, p.167
Control, p.226-227
Copper, p.181
   plumbing tube, p.412
Cost overruns, p.275
Cover, p.71
Cycling, pp.178 – 179

Demand, estimating, pp.369 – 370, p.384
Depth, p.108
Designs, septic, p.199 – 204, pp.206 – 208
Diaphragm tanks, p.151
Discontinued disposal systems, p.2
Distribution system, p.71
Dosing, p.45
   chambers, p.52
Dowsing, pp.85 – 86
Drilled wells, pp.107 – 108, p.133 – 134, p.168
   basics, pp. 110 – 113
   problems, pp.126 – 127
Drilling, supervising the, p.117
Driven wells, pp.166 – 167, p.282
Driveways, p.194

Effluent, p.11
Erosion, p.194
Excavation, p.38 – 39
Excess water in septic system, p.254

Faulty systems, p.2
Fiberglass, p.218
Fill, p.71
Filled areas, p.26
  design requirements, p.27
Find water, reading signs to, p.83
First aid, p.333 – 334
  for back injuries, p.340
  for bleeding, pp.334 – 336
    tourniquets, p.336
  for blisters, pp.340 – 341
  for burns, pp.3443 – 344
  for cramps, p.344
  for exhaustion, pp.344 – 345
  for eye injuries, pp.338 – 339
  for facial injuries, p.339
  for hand injuries, p.341
  for heat-related problems, p.344
  for infection, p.336
  for legs and feet, p.340
  for nose bleeds, p.340
  for open wounds, p.334
  for scalp injuries, p.339
  for shock, pp.341 – 343
  for splinters, p.337 – 338
Flood fringe, p.12
Flood plain, p.16
Floodway, p.12
Flow, p.398, p.415
Fluorides, p.181
Formulas, p.349, p.351 – 352, p.397,
  p.399 – 405
Foundation, inside the, pp.124 – 125
Friction loss, p.372, pp.374 – 379, p.382,
  pp.385 – 392

Garbage disposals, p.252, p.262
General regulations, p.1
Grade, p.190
Grass, p.249
Gravel-and-pipe septic systems, p.211 – 213,
  pp.215 – 216
Gray-water sumps, pp.240 – 241
Grease interceptors, p.393
Ground water, p.19 – 20

Hard water, p.183
High ground water, p.11

High-tech stuff, p.86
High-water level, p.12
Historic buildings, pp.3 – 4
Holding tank, p.12, pp.53 – 54,
  pp.234 – 235
Horizontal reference point, p.12

Inlets, pp.47 – 48
In-line tanks, pp.148 – 149
Inspection openings, p.49
Inspections, p.7, p.75
Installations
  do your own, p.288
Insulation, p.441
Iron, p.181, pp.184 – 185

Jet pumps, troubleshooting, pp.169 – 173
Jetted wells, p.134
Joints, p.31, p.33 – 34
  prohibited joints, p.34

Landscaping septic-system areas,
  pp.247 – 250
Large tanks, pp.148
Leach field, problems, p.260
Lead, p.181
Legal description, p.12
Load value, p.368
Location of water wells, p.81, p.83

Maintenance, p.3, p.53, p.72
  septic system, p.251
Manganese, p.181
Manhole, p.12, p.48
Manhole covers, p.49
Maps, p.83, p.85
Materials, p.29
Measurement conversion, p.421 – 424,
  pp.428 – 429, p.433 – 440
Metal, p.217
  melting points, p.416
  weights, p.416
Mobile unit, p.12
Mobile unit park, p.12
Modifications, p.3
Money, pp.225 – 226, pp.316 – 317
Monitoring wells, design of, p.23
Mound systems, pp.55 – 56, pp.233 – 234
  construction methods, pp.70 – 72
  design, p.56, p.69, p.208

Negotiations, p.288 – 289
New wells, p.99 – 100

# Index 447

Nitrates, p.182
Nuisance, p.12

Observation pipes, p.40
Odors
  inside, pp.262 – 263
  outside, pp.263 – 264
Offsets, pp.359 – 360
Older fields, p.261
Outlets, pp.47 – 48

Pan, p.12
Parking areas, p.194
Pathogens, p.252
Percolation test, p.12, p.21
Perk rate, p.26
Permeability, p.12
Permit expiration, pp.6 – 7
Permit suspension, p.7
Permits, p.5
Pesticides, p.182
Pesticides in potable water, pp.103 – 104
  health effects, p.104
Pipe, p.31, p.350
  capacities, p.418
  discharge, p.417
  rod size, p.421
  sizing, p.367, p.381
  thermal expansion, p.418 – 419
Piping requirements, p.39 – 40
Pitch, pp.261 – 262
Plans and specifications, p.5 – 6
Plants, p.85, pp.248 – 249
Plowing, p.70
Potable water, p.89
Precast concrete, p.217
Pressure distribution system, p.12, p.43, p.442, p.443
  design, pp.43 – 44
Pressure loss, p.373
Pressure tanks, pp.147 – 152, p.284
  installation procedures, pp.152 – 153
  keep it dry, p.153
  pump-stand models, pp.150 – 151
  stand models, pp.149 – 150
Prices, pp.284 – 286
Private sewage-disposal system, p.12
Privy, p.12
Problems, well, pp.166 – 168
Protecting private wells, pp.104 – 105
Public sewer connections, p.2
Puddles, p.264
Pump installers. pp.114 – 115

Pump stations, p.192
Pump system, installing a, pp.119 – 120, p.126, p.234
  selection for shallow well, p.139, pp.141 – 142
Pumping chambers, p.52
Pumps, pp.44 – 45, pp.282 – 284
  cheap, 280
  cycles too often, p.172
  doesn't produce water, p.178
  runs but gives no water, pp.171 – 172
  will not run, pp.169 – 171, pp.177 – 178
  won't develop pressure, 172 – 173
  won't start, pp.174 – 175

Quality of septic systems, pp.226 – 227
Quality of water, pp.115 – 117, p.179
Quantity of water, pp.115 – 117

Radiological factors, p.184
Records, written, p.315
Referenced standards, pp.77 – 78
Registered design professional, p.12
Relief valve, p.153
Relocated buildings, p.4
Riser joints, pp.51 – 52

Safety, pp.190 – 191, pp.319 – 322
  air-powered tools, pp.329 – 330
  co-worker, pp.331 – 332
  drill and bits, pp.327 – 328
  eye and ear protection, p.324
  hand-held drain cleaners, pp.328 – 329
  in clothing, p. 323
  in vehicles, p.322 – 323
  ladders, pp.330 331
  large sewer machines, p.329
  lead pots and ladles, p.327
  pads, p.324
  powder-actuated tools, p.330
  power pipe threaders, p.329
  power saws, p.328
  screwdrivers and chisels, p.331
tool, pp.325 – 326
torches, pp.326 – 327
with jewelry, p.324
Sandy soils, p.21 – 22, p.167
Seepage bed, p.13, p.38
Seepage pit, p.13, p.38
Seepage trench, p.13
  excavating, p.36
Septage, p.13
Septic tank, p.13, pp.29 – 30

capacities, p.414
designs, pp.199 – 204, pp.206 – 208
failure symptoms, p.253
installation, p.49, p.51, p.218 – 219,
    p.222 – 225
    troubleshooting, pp.255 – 256
low-cost, pp.272
types, pp.217 – 219, pp.222 – 223
set and travel relationships, p.358
Setbacks, p.194 – 195
Shallow wells, pp.129 – 131, p.135, p.137,
    pp.167 – 168, pp.281 – 282
    preliminary work, p.138
Single-fixture pumps, pp.238 – 240
Single-pipe jet pumps, p.139, p.141, p.284
    installing, p.143 – 145
Site evaluation, p.17 – 18, pp.137 – 138,
    p.273
    observation requirements, pp.23 – 24
Site limitations for septic systems,
    pp.187 – 190
Site plans, p.6
Site requirements, p.24 – 25
Site visits, pp.196 – 197
Sizing tanks, p.49, p.152
Slope, p.18
Small tanks, pp.147 – 148
Sodium, p.182
Soil, p.13, pp.21 – 22
Soil absorption system, conventional,
    p.11, pp.35 – 36
Soil boring, p.13
Soil mottles, p.13, p.19
Soil reports, p.6
Soil saturation, p.13
Soil studies, p.199
Soil testing verification, p.22 – 23
Soil types, pp.208 – 209
Spec builder, pp.273 – 275
Specs, pp.286 – 288
Spring, pp.278 – 279
Standard systems, p.191
Stop work order, pp.7 – 8
Submersible pumps, p.142, pp.282 – 283
    troubleshooting, pp.173 – 174
Supervision, pp.264 – 265
Suppliers, pp.312 – 314
Switch fails, p.173

Tank pressure, p.178
Tanks, p.47, pp.258 – 260
Taps and tees, p.371
Temperature, p.252, p.442
    conversion, p.353 – 354
Test boring, p.18 – 19
Testing recommendations, p.96, pp.98 – 99,
    pp.100 – 103
Toilet, overflowing, pp.256 – 257
Trap sizes, p.432
Treatment tanks, pp.52 – 53
Trees, p.192
Trenches, p.69, pp.231 – 232
    land area, pp.232 – 233
Trenching, p.118
True dug wells, p.131, p.133
Turbidity, p.185
Two-pipe jet pumps, pp.141 – 142, p.147,
    p.283

Underground tanks, p.151
Underground water, pp.193 – 194
Unsafe conditions, p.8
Unwanted objects in septic system,
    p.252

Vent cap, p.13
Vertical elevation reference point, p.13
Violations, code, p.7
Volume deals, p.289

Wastewater pump systems, p.237
Water
    chemical characteristics, pp.180 – 183
    physical characteristics, pp.179 – 183
Water-quality problems, p.184
Water service, pp.123 – 124
Water softeners, p.185
Water wells, site evaluation for, pp.79 – 81
Watercourse, p.13
Weather, p.40 – 41
Well casing, pp.121 – 123
Well-drilling equipment, p.108, p.110
Well systems
low-cost, p.277
selection, p.281
troubleshooting, pp.165 – 166
whole-house backups, pp.257 – 258
Whole-house pump systems, pp.244 – 245
Wood, p.218
Working on your well site, p.114

Zinc, p.182

## ABOUT THE AUTHOR

R. Dodge Woodson is a licensed master plumber, licensed master gasfitter, and licensed building contractor. He has over 25 years of experience in construction and real estate. As a builder of as many as 60 homes a year, Woodson knows what it takes to create a successful building project. His experience as a plumber and plumbing contractor gives him detailed experience with wells and septic systems. Much of the land that Woodson has developed over his career has required private water and sewage systems. His hands-on knowledge makes his work as an author invaluable.